Life
cycle
Funds

One Fund
approach

Lifecycle
20-20
IRA
;
Find research

704
(401)
TIAA
Cref

8
-988-

EMERGENCE IN MIND

MIND ASSOCIATION OCCASIONAL SERIES

This series consists of occasional volumes of original papers on predefined themes. The Mind Association nominates an editor or editors for each collection, and may cooperate with other bodies in promoting conferences or other scholarly activities in connection with the preparation of particular volumes.

Emergence in Mind

Edited by
CYNTHIA MACDONALD AND
GRAHAM MACDONALD

OXFORD
UNIVERSITY PRESS

OXFORD
UNIVERSITY PRESS

Great Clarendon Street, Oxford OX2 6DP

Oxford University Press is a department of the University of Oxford.
It furthers the University's objective of excellence in research, scholarship,
and education by publishing worldwide in

Oxford New York

Auckland Cape Town Dar es Salaam Hong Kong Karachi
Kuala Lumpur Madrid Melbourne Mexico City Nairobi
New Delhi Shanghai Taipei Toronto

With offices in

Argentina Austria Brazil Chile Czech Republic France Greece
Guatemala Hungary Italy Japan Poland Portugal Singapore
South Korea Switzerland Thailand Turkey Ukraine Vietnam

Oxford is a registered trade mark of Oxford University Press
in the UK and in certain other countries

Published in the United States
by Oxford University Press Inc., New York

First published 2010

British Library Cataloguing in Publication Data
Data available

Library of Congress Cataloging in Publication Data
Library of Congress Control Number 2009943760

Typeset by Laserwords Private Limited, Chennai, India
Printed in Great Britain
on acid-free paper by the
MPG Books Group, Bodmin and King's Lynn

ISBN 978–0–19–958362–1

1 3 5 7 9 10 8 6 4 2

Contents

Notes on Contributors

John Ross Churchill is lecturer in the Department of Philosophy at Indiana University. He is a specialist in philosophy of mind, philosophy of science, and metaphysics. He is the co-author of 'Non-Reductive Physicalism or Emergent Dualism? The Argument from Mental Causation', in G. Bealer and R. Koons, eds, *The Waning of Materialism* (forthcoming) and 'Reasons Explanation and Agent Control: In Search of an Integrated Account', *Philosophical Topics* 32 (2004).

Tim Crane is Knightbridge Professor of Philosophy at the University of Cambridge and a Fellow of Peterhouse. He was previously a professor of Philosophy at University College London, and the Founding Director of the Institute of Philosophy in the University of London. His recent publications include *Intentionalität als Merkmal des Geistigen* (2007) and 'Intentionalism', in A. Beckermann, B. McLaughlin, and S. Walter, eds, the *Oxford Handbook to the Philosophy of Mind* (2008).

Michele Di Francesco is Full Professor at the University Vita-Salute San Raffaele of Milan, where he teaches philosophy of mind and philosophy of cognitive science. He has published, edited and co-edited several books and many articles in Italian and English. His most recent co-edited books are *Neurophilosophy*, special issue of *Functional Neurology* 4 (2007), with Patricia Churchland, and *Il soggetto: L'io e la scienza della mente* (2009), with Massimo Marraffa. He is the president of the European Society for Analytical Philosophy and Dean of the Faculty of Philosophy of the University San Raffaele.

Michael Esfeld is Full Professor of Philosophy of Science at the University of Lausanne (Switzerland). He received the Cogito award for his work on the philosophy of physics in 2008. His recent publications include *Naturphilosophie als Metaphysik der Natur* (2008); 'Mental Causation and the Metaphysics of Causation', *Erkenntnis* 67 (2007); 'Moderate Structural Realism about Space-time' (with V. Lam), *Synthese* 160 (2008); 'The Modal Nature of Structures in Ontic Structural Realism', *International Studies in the Philosophy of Science* 23 (2009); and 'Psycho-neural reduction through functional sub-types' (with P. Soom and C. Sachse), *Journal of Consciousness Studies* 17 (forthcoming 2010).

Simone Gozzano is Professor of Philosophy of Science at the Università di L'Aquila-Italy. He has published four books in Italian and many articles. He is the co-editor (with Francesco Orilia) of *Tropes, Universals and the Philosophy of Mind*, Ontos (2008), and author of papers that appeared in *Acta Analytica, Axiomathes* and of 'Levels, Orders and the Causal Status of Mental Properties', *European Journal of Philosophy* 17.3 (2009).

Robin Findlay Hendry is Reader in Philosophy at Durham University. He studied Chemistry and Philosophy of Science at King's College London and received his PhD from the London School of Economics. He has also taught at the University of Edinburgh. His publications include *The Metaphysics of Chemistry* (forthcoming 2010), and *Philosophy of Chemistry* (forthcoming 2010).

Max Kistler is professor in the Department of Philosophy of Université Pierre Mendès France, Grenoble, and member of Institut Jean Nicod, Paris. He is author of *Causation and Laws of Nature* (2006); (with A. Barberousse and P. Ludwig) *La Philosophie des sciences au XXe siècle* (2000); and has edited *Dispositions and Causal Powers* (with B. Gnassounou) (2007); and special issues of *Philosophie* 89 on *Causation* (2006); *Synthese* 151 on *New Perspectives on Reduction and Emergence in Physics, Biology, and Psychology* (2006); and *Philosophical Psychology* 22 on *Cognition and Neurophysiology: Mechanism, Reduction, and Pluralism* (2009).

Stephan Leuenberger is lecturer in the Department of Philosophy at the University of Glasgow. He gained his PhD at Princeton (2006). Recent and forthcoming publications include 'Ceteris absentibus Physicalism' (Winner of the Oxford Studies in Metaphysics Younger Scholar Prize 2006), *Oxford Studies in Metaphysics* IV (2008); 'A New Problem of Descriptive Power', *Journal of Philosophy* 2006; 'What Is Global Supervenience?', *Synthese* (2009); and 'Humility and Constraints on O-Language', *Philosophical Studies* (forthcoming).

Christian List is Professor of Political Science and Philosophy at the London School of Economics. He works in social choice theory, political philosophy, formal epistemology, and the philosophy of social science. A graduate of the University of Oxford, he held research and visiting positions at Oxford, the Australian National University, Massachussets Institute of Technology, Harvard, Princeton, and the University of Konstanz. He was awarded a Nuffield Foundation New Career Development Fellowship, a Philip Leverhulme Prize in Philosophy, and the 5th Social Choice and Welfare Prize (jointly with Franz Dietrich). He is also an editor of *Economics and Philosophy*. His webpage is available at <http://personal.lse.ac.uk/list/>.

Cynthia Macdonald is Professor of Philosophy at Queen's University Belfast and Emeritus and Adjunct Professor of Philosophy at the University of Canterbury, New Zealand. Her research interests focus on the metaphysical foundations of mental causation and explanation and on authoritative self-knowledge. Related recent publications include *Varieties of Things: Foundations of Contemporary Metaphysics* (2005); 'Consciousness, Self-Consciousness, and Authoritative Self-Knowledge'; *Proceedings of the Aristotelian Society* 108 (2008); and 'Introspection', in A. Beckermann, B. McLaughlin, and S. Walter, eds, the *Oxford Handbook to the Philosophy of Mind* (2008). She is currently completing a monograph with Graham Macdonald, *Mental Causation and Explanation in the Special Sciences*, funded by the Royal Society of New Zealand Marsden Foundation and the Mind Association.

Graham Macdonald was educated in South Africa and England. He is presently Distinguished International Fellow in the Institute of Cognition and Culture, Queen's University Belfast, and Emeritus and Adjunct Professor of Philosophy at the University of Canterbury, New Zealand. Recent publications include 'The Metaphysics of Mental Causation' (with Cynthia Macdonald); *Journal of Philosophy* (2006); and 'The Two Natures: Another Dogma?', in C. Macdonald and G. Macdonald, eds, *McDowell and His Critics* (2007).

Ausonio Marras is Professor Emeritus at the University of Western Ontario, Canada. His recent research has been devoted to an articulation and defense of a supervenience-based version of Nonreductive Physicalism. His recent and forthcoming publications include: 'The Prospects for Nonreductive Physicalism' (forthcoming); 'The 'Supervenience Argument': Kim's Challenge to Nonreductive Physicalism' (with Juhani Yli-Vakkuri), in F. Orilia and S. Gozzano, eds, *Tropes, Universals, and the Philosophy of Mind* (2008); 'Kim's Supervenience Argument and Nonreductive Physicalism', *Erkenntnis* 66 (2007); and 'Consciousness and Reduction', *British Journal for the Philosophy of Science* 56 (2005).

Peter Menzies is Professor of Philosophy at Macquarie University, Sydney. He is a Fellow of the Australian Academy of the Humanities, and co-editor with Helen Beebee and Christopher Hitchcock of the (2009) *Oxford Handbook of Causation*. Recent publications include (with Huw Price) 'Is Semantics in the Plan?', in D. Braddon Mitchell and R. Nola, eds, *Naturalistic Analysis* (2009); 'The Folk Theory of Colours and the Causes of Colour Experience', in Ian Ravenscroft, ed., *Mind, Conditionals and Metaphysics: Essays in Honour of Frank Jackson* (2009); and 'Causal Exclusion, the Determination Relation, and Contrastive Causation', in Jesper Kallestrup and Jakob Hohwy, eds, *Being Reduced: New Essays on Reductive Explanation and Special Science Causation* (2008).

Paul Noordhof is Anniversary Professor of Philosophy at the University of York. Related publications include 'Causation by Content', *Mind and Language*, 14.3 (1999) and 'Not Old . . . but Not that New Either: Explicability, Emergence and the Characterisation of Materialism', in Sven Walter and Heinz-Dieter Heckman, eds, *Physicalism and Mental Causation: The Metaphysics of Mind and Action* (2003). His views on causation, developed over a series of papers, will be set out in detail in *A Variety of Causes* (forthcoming).

Timothy O'Connor is Professor and Chair of the Department of Philosophy at Indiana University and a member of its Cognitive Sciences Program. He is a specialist in metaphysics, philosophy of mind, and philosophy of religion. He is the author of *Persons and Causes: The Metaphysics of Free Will* (2000) and *Theism and Ultimate Explanation: The Necessary Shape of Contingency* (2008). He is also the editor and co-editor of several volumes in the philosophy of mind and action.

David Papineau was educated in Trinidad, England, and South Africa. Since 1990 he has been Professor of Philosophy of Science at King's College London. Previously he had lectured at Reading University, Macquarie University, Birkbeck College London, and Cambridge University. He has written *For Science in the Social Sciences* (1978); *Theory and Meaning* (1979); *Reality and Representation* (1987); *Philosophical Naturalism* (1993); *Introducing Consciousness* (2000); *Thinking about Consciousness* (2002); and *The Roots of Reason* (2003); and he has edited *Philosophy of Science* (1996) and (with Graham Macdonald) *Teleosemantics* (2006). He was President of the British Society for the Philosophy of Science from 1993 to 1995 and editor of the *British Journal for the Philosophy of Science* from 1993 to 1998. He will be President of the Mind Association for 2009–10.

Philip Pettit is L. S. Rockefeller University Professor of Politics and Human Values in Princeton University, where he teaches in philosophy and political theory. Among his recent books are *Rules, Reasons and Norms: Selected Papers* (2004), and *Made with Words: Hobbes on Language, Thought and Politics* (2008). *Common Minds: Themes from the Philosophy of Philip Pettit*, ed. G. Brennan *et al.* appeared in 2007.

Achim Stephan is Professor of Philosophy of Cognition at the Institute of Cognitive Science at the University of Osnabrück, Germany. He had visiting positions at the Vreije Universiteit (in English, the Free University of) Amsterdam and the University of Ulm, and has held research fellowships at both the Hanse Institute of Advanced Study (Delmenhorst) and the Konrad Lorenz Institute (Altenberg and Vienna). Recently, he was a member of two ZiF Research groups at the Center for Interdisciplinary Research (Bielefeld) on Emotions as Bio-Cultural Processes and Embodied Communication in Humans and Machines, respectively. He has written extensively on the topic of emergence, including 'The dual role of "emergence" in the philosophy of mind and in cognitive science', *Synthese* 151 (2006), *Emergenz. Von der Unvorhersagbarkeit zur Selbstorganisation* (1999), and is currently working in the field of affective science and the philosophy of emotions (see <http://www.animal-emotionale.de/index_en.htm>).

Peter Wyss is an Associate Research Fellow at Birkbeck College London, where he submitted his PhD thesis on the coherence of emergentism. He also teaches philosophy for The Open University, Oxford University Continuing Education, and a sixth-form college. He is currently working on the relation between emergence and individuation.

Juhani Yli-Vakkuri is a graduate student at the University of Oxford. His current research focuses on context-sensitivity in natural language. His recent and forthcoming publications include: (with James McGilvray) 'Reference and Extension', in P. C. Hogan, ed., *The Cambridge Encyclopedia of the Language Sciences* (2010) and (with Ausonio Marras) 'The Supervenience Argument: Kim's Challenge to Non-Reductive Physicalism', in S. Gozzano and F. Orilia, eds, *Tropes: Universals, and the Philosophy of Mind* (2008).

1

Introduction*

Cynthia Macdonald and Graham Macdonald

1. SOME HISTORICAL BACKGROUND

The issue of whether there are emergent properties, and if so, what their relation to other properties is, has been much debated over the past two centuries. One critical focus of these debates is on the very intelligibility of such 'emergence': critics of the notion are suspicious that emergent properties inhabit a kind of halfway house that is, by its nature, unstable, given that such properties are characterized by two features that are in tension with one another. On the one hand, an emergent property is said to be distinct from the properties from which it emerges; on the other hand, it is said to be dependent on those very properties. This combination of features creates the instability that threatens either to collapse emergent properties into their 'base' ones (the reductionist option) or to make them so different from their base properties that their relation to those properties is left mysterious or non-existent (the dualist option).

As a consequence of this tension, there are discernible fluctuations in the fortunes of emergence over different periods in the history of science. Emergence-style arguments go back at least as far as the time when psychophysical dualism was being promoted, but it was in the nineteenth and early twentieth centuries that doctrines bearing a resemblance to what we now call emergentism were most hotly disputed in various, mainly physiological and biological-medical, contexts. During this time it appeared to many that physics was not, and would not be able to explain chemical bonding, vital activity, or mental processes. Various 'additions' were thought necessary if the gap left by physical and chemical explanations was to be filled. One was the vitalist option mooted initially by Xavier Bichat, who thought that in order to explain life it was necessary to introduce 'sensibility' and 'contractility' as properties of tissues,

* We are indebted to an anonymous reader at Oxford University Press for comments and advice. Work on this volume and on the conference from which this volume originated was generously supported by a Mind Association Major Conference Grant, a Marsden Grant awarded by the Royal Society of New Zealand to Cynthia Macdonald and Graham Macdonald, and funding from the School of Politics, International Studies, and Philosophy, Queen's University Belfast.

these properties themselves not resulting from the combination of only physical or chemical properties. The processes of metabolism and fermentation were thought to require the activity of these 'vital' properties (for a brief history, see Bechtel and Richardson 1998). Another option was to invoke non-physical entities to do the explanatory work, such as Driesch's 'entelechies' which were introduced in order to explain the difference between organic and inorganic matter. It was questionable, however, whether the non-physical elements were genuinely explanatory, and what the nature of their relation to the physical was. Given their 'categorial' difference from the physical (Driesch conceived of 'entelechy' as essentially immaterial), it was difficult to see how they could be suitable candidates for interacting causally with the physical.

Another proposal did away with special entities altogether in favour of new forces: it was thought that at a certain level of complexity of physical organization, novel forces 'emerged'. The 'newness' of the forces was due to their being more than mere sums of the forces of the physical elements making up the complex structure. At this level of complexity, so it was claimed, some force that was not just the resultant of the antecedently given forces, the lower level forces, was discernible. As a consequence, not all of the forces operating in the world were physical; from the complexity of physical structures vital, mental, and social forces emerged, these novel forces enabling the explanation of the non-physical features of the world. The general doctrine was known as emergentism. With it there developed a hierarchical conception of the relation between the sciences, with physics being ontologically fundamental and the rest of the sciences stacking up in layers above it, each having its own laws describing the new forces arising at the relevant of complexity.[1]

This tendency to postulate supra-physical (and chemical) properties, entities, or forces did not go unopposed. In addition to generating some philosophical unease about the idea of independent causal forces operating 'downwards', threatening the pattern of purely physical causation of physical changes, the emergentist programme suffered blows from some major empirical discoveries, especially in the fast-developing field of biology. In the second half of the nineteenth century Darwinian accounts of design and speciation prompted many biologists to seek a unification of their theories with those of experimental physiologists. Particularly influential was the group of 'medical materialists' formed in Berlin in the 1840s under the leadership of Hermann von Helmholtz, who pleaded for the introduction of physico-chemical methods into biology.[2] The approach of the medical materialists was explicitly reductionist: they attempted to study the organism by studying isolated parts of it, using the experimental methods of physics and chemistry to reveal the underlying mechanisms of biological processes. Amongst those who were influenced by their work were Wilhelm Roux (via an

[1] Emergentism in Britain is illuminatingly described in Brian McLaughlin 1992.
[2] The relevant history here is derived from Garland Allen 1975.

embryologist Wilhelm Preyer, who had studied in Berlin), Jaques Loeb, and, later, Ivan Pavlov (who was introduced to experimental work on reflexes by a student of the Berlin school, Ivan Sechenov). Roux was an influential embryological experimentalist and also a persuasive propagandist for a programme he called *Entwicklungsmechanik*, roughly translated as 'developmental mechanics' (see Allen 1975: 33). According to this, the study of developing embryos is directed toward explaining how cell differentiation is caused by the internal physical and chemical constitution of the embryo.

Loeb, having imbibed much of the medical materialist philosophy at the University of Strasbourg, moved to the University of Würzburg where he was struck by the work of Julius Sachs on plant tropisms, automatic responses of organisms to specific circumstances, such as light. Loeb was convinced that the physico-chemical explanation applied to tropistic phenomenon could be applied more widely to life processes in general. He experimented on unfertilized sea urchin eggs, producing developmental changes in them by altering their chemical environment. His experiments on artificial parthenogenesis opened up the possibility of laboratory-created life, reinforcing Loeb's outlook that *all* living processes (including mental and social) could be explained as resulting from 'chemical mechanisms'. This determinist and mechanistic outlook was further strengthened by Pavlov's work on conditioned reflexes, which promised to show how learned behaviour could be explained as the result of prior conditioning, which in turn could be accounted for in neurophysiological terms.

It was not so much that any crucial experiments disproved the postulation of immaterial entities or emergent forces, but that a climate of reductionist optimism was fortified. The new developments earned respect because the theories produced were testable, at times mathematically expressible, and gave rise to detailed experimental work both on cells and supra-cellular tissues. Philosophers and scientists responded to these developments by retaining the hierarchical conception of science, with reduction replacing emergentism as the favoured relation between the levels. Arguably, this picture has been further supported by twentieth-century developments in molecular biology, with the discovery of the nature of genes, how they are transmitted, and how they are expressed in the course of the development of an individual. The use of chemical theory in all these developments has been crucial, suggesting that biology was reducible to chemistry and thereby to physics, given that the reducibility of chemistry to physics was thought to have been demonstrated by the physical explanation of chemical bonding.

The major trend in all of this scientific work was to explain processes at the macro-level by discovering more of the detail of microprocesses. Reductionism looked to be an eminently suitable research strategy. The appearance of emergent elements was to be explained away, and it was the reductionist's prediction that our increasing knowledge of physical processes would obviate the need for any special entities or forces. The difficulty of downward causation was avoided, since reduction would place all such causation at the level of the physical. The problem of the causal powers of the higher-level properties was solved: the reductionist's picture would endorse their causal efficacy, but would do so at a cost, robbing them of any causal autonomy. For, according to the reductionist, any higher-level property that has causal powers has them because it is, really, a physical property. Physics is fundamental, where 'fundamental' means that the physical is causally, ontologically, and explanatorily all-encompassing.

However, other developments within the favoured science, physics, supported a somewhat different view. Quantum mechanics deals with the ultimate microparticles, but not everyone is convinced that all explanations in quantum mechanics fit the reductionist mould: not all quantum events, it appears, can be explained by citing properties of subatomic particles acting independently of one another. It has been suggested that entangled states in quantum mechanical systems are ontologically emergent, as 'the individual states before interaction do not determine the joint state after the interaction, whereas the joint state does determine the individual states. This feature makes compositional accounts of the joint system implausible.' (Humphreys 2008: 586). The example is important in that it shows that debates about emergence are not necessarily confined to interdisciplinary contexts (biology in relation to physics and chemistry, psychology in relation to biochemistry, and so on); issues to do with emergence can arise within a single domain, provided that the domain is partitioned into a 'substrate' and a level that putatively emerges from the substrate. As Humphreys puts it, emergence is essentially a relational phenomenon: a property is emergent only in relation to another set of properties from which it can be said to have emerged, and this relation can exist in both intra- and interdisciplinary contexts.

Apart from the controversies within quantum mechanics, there were other developments that supported an emergentist view, particularly ones connected with considerations about complex phenomena. In an important article, P. W. Anderson questioned what many assumed to be a corollary of the reductionist hypothesis that our minds and all animate and inaminate matter are controlled by the same set of fundamental laws. The putative corollary was that 'the only scientists who are studying anything really fundamental are those who are working on those laws' (Anderson 1972: 393). Against this, Anderson argued that understanding more complex phenomena may require new laws and concepts, so that psychology need not be mere applied biology, nor biology mere applied chemistry. And again, even in the physical domain, processes resulting in phase transitions, such as a solid becoming a liquid, require new properties to emerge, and so need to be understood at their own level. Anderson's message is conveyed by the title of his article: 'More is Different'. In some cases, he says, we can see 'how the whole becomes not only more than but very different from the sum of its parts' (Anderson 1972: 395).

But caution is needed: the type of emergence exemplified by phase transitions, or, to take another example of a supposedly 'emergent' phenomenon, by termite organization (see Johnson 2001), does not seem to raise any significant philosophical problems. These are cases that have been 'tamed' by science itself, in its ability to explain how the transitions are effected, or how termites can follow an individual path *thereby* contributing to the organization of the larger colony. We are disinclined to enter into any argument about whether these cases are ones of 'real' emergence, or just apparent emergence; such an argument would presume that there is a widely understood, univocal sense of 'emergence', and that all that is in dispute is where and when it applies. This would be an oversimplification of complex debates. It is better to characterize the different types of relations that are involved in putative cases of emergence, noting where they differ and whether significant problems arise, given our present state of scientific knowledge. In what follows we shall set out in fairly broad terms different aspects of what could be called an emergence-relation, situating our discussion within recent developments in the philosophy of mind.

2. THE PHILOSOPHICAL CONTEXT

Anderson's 1972 paper was sandwiched between two seminal philosophical articles, those by Donald Davidson (Davidson 1970) and by Jerry Fodor (Fodor 1974), that set the agenda for debates concerning the reducibility of the mental and, more generally, the autonomy of the special sciences.[3] The context was one in which there was increasing dissatisfaction with what were thought to be only three possible views about the nature of mind: dispositionalism (e.g., forms of logical behaviourism), mind–body dualism, and mind–body type identity. Against the background of mounting pressure to 'naturalize' the mind, the identity option was favoured. However, its commitment to psychophysical type-type identities looked implausible; mental types, it was claimed, are 'multiply realized' by physical types, a single psychological property being realizable by different (sets of) physical properties, this prohibiting the type-type identities required for any reduction. This set the agenda: any solution to the mind–body problem must respect multiple realizability while satisfying the naturalist constraint that the mind be shown to be part of the natural (or more specifically, the physical) world. In different ways Davidson and Fodor proposed to satisfy the naturalist constraint by identifying the domain of the mental with that of the physical at the level of individual, or token, events while accommodating the possibility of multiple realization (or, more generally, non-reduction) at the level of mental properties or types. Fodor's argument was the more empirical argument. In the model he presented of the relation between higher-level special sciences

[3] Putnam's paper 'Psychological Predicates' (Putnam 1967) proved to be influential as well, but it did not have the immediate impact of those of Davidson and Fodor, partly due to being published in a less accessible volume.

and lower-level physics, the taxonomy imposed on physical events by physical theory was not required to align neatly with the taxonomies used by higher-level sciences; the same events could be subject to various taxonomies. The identity of the higher-level events with the lower-level (physical) events satisfied the physicalist requirement that all individual, or token, events be physical events, while the variation in taxonomies ensured non-reducibility. One result of this, as Fodor saw it, was that the higher-level explanations would require appeal to *ceteris paribus* laws, laws that allow for exceptions. The argument was 'empirical' because there was no way of knowing in advance of investigation whether a particular upper-level science would or would not be reducible. It could just happen that, given the results of such investigations, the most plausible conclusion to draw would be that no reduction was forthcoming: there could be just simple taxonomic divergence, this divergence corresponding to the differing interests of investigators. The proposed disunity of science was just a 'working hypothesis'.

Fodor's argument is clearly consistent with Anderson's view that different levels of reality may require new laws and concepts, and it was intended to apply quite generally to non-physical sciences (the 'special sciences'). One example used by Fodor is illustrative of higher-level multiple realizablity, that of the implausibility of supposing that descriptions of transactions involving money could ever be reduced to physical laws, given the diversity of (physical) objects that could serve as money and the variety of actions that could realize economic transactions. Davidson's argument for token identity was restricted to the psychophysical case, and so could make use of features of the mental not present in the subject matter of other special sciences. This restriction enabled Davidson to put forward an a priori argument, one making essential use of the fact of causal interaction between mental and physical events, the nomological character of causality (events causally related must be covered by a strict law), and the anomalous character of the mental (mental predicates being unfit to serve in strict laws). The argument for the anomalous nature of the mental also established non-reducibility, this being claimed to follow from the constraints that rationality imposes on the appropriateness of describing, for example, actions using mental descriptions, these constraints being inapplicable to the appropriateness of describing those same events in physical terms.

Despite these differences between the arguments of Fodor and Davidson, the resulting metaphysical picture was broadly the same: a non-reductive monism which embraced token identities between higher-level and lower-level (physical) events but which permitted re-descriptions of those events, and so explanatory autonomy, to higher-level disciplines. Explanatory autonomy could be achieved either by using essentially *ceteris paribus* laws (Fodor) or by employing a conceptual framework inimical to any psychophysical (or psychological) laws (Davidson). The autonomy was not absolute, however; naturalistic scruples would not permit free-floating higher-level descriptions. There had to be some connection to the physical level, however loose, and this was provided by construing the higher-level/lower-level domains as standing in a relation of supervenience: there could be no change in a higher-level phenomenon without a change at the lower (physical) level. Supervenience allowed for multiple realizability while satisfying the naturalist thought that the physical was 'basic and general'. It looked as though non-reductive monism could allow the naturalistically minded philosopher to have his or her cake and eat it, to have explanatory freedom with ontological respectability.

3. THE DEBATES

Harmony, such as it was, did not last long. In what follows we provide an overview of some of the problems raised, with an eye to the issues that connect the papers in this volume.

A. Structured Events and Causation

The metaphysical picture offered by Davidson was one that is suspicious of properties; a picture of 'structureless' events bearing or satisfying different descriptions, physical and mental. These event-descriptions qualified as physical or mental if they participated in the vocabularies distinctive of their subject matters. So, in brief, Davidson claimed that to be a mental description is to participate in a vocabulary whose application conditions are answerable to norms of rationality; to be a physical description is to participate in a vocabulary governed by what may be termed the norms of nomologicality. Physical events are those events truly described using the physical vocabulary; mental events are those truly described using the mental vocabulary. Given that physical events unproblematically cause their effects, so, too, do those selfsame events cause their effects even if they are mentally described (and so are mental events). This much is guaranteed by the extensionality of the causal relation: if *A* causes *B* then no matter how *A* is described, whether the vocabulary is mental or physical, *A* 'under that description' still causes *B*.

Many were unhappy with what they saw as too quick a fix; it seemed to make mental events causally efficacious only '*qua* physical' (i.e., only insofar as they have a physical description), a suspicion reinforced by Davidson's insistence that only physical descriptions could be used in the formulation of causal laws. Foremost amongst those unhappy with this 'solution' was Jaegwon Kim, who provided a metaphysical 'structure' to the events: an event was taken to be the exemplification of a property in an object at a time. In the simple (monadic) case, Kim represented it schematically by an expression of the form '[x,P,t]', this being construed as a singular term referring to an event, where x is the object in which the property is exemplified, P the property whose exemplification in x is an event, and t the time of exemplification. Talk of relations between mental and physical descriptions could now be replaced by talk of relations between mental and physical properties, and non-reducibility was expressed by the claim that mental properties are not identical to physical properties.[4] This non-identity claim can be stated with varying force, allowing for different types of dependence on physical properties, as we shall see.

Given this added structure, and given non-reductive monism, a mental event will have two properties, a mental one and a physical one. This being so, a question naturally arises: in virtue of which of its properties is it causally responsible for bringing about the effects it does? The question is complicated by a number of assumptions thought to be part of the physicalist's ideological baggage. The physicalist is thought to be committed to the 'basic' or fundamental character of the physical, and an expression of this is contained in the assumption that the physical domain is causally closed: any event that has a cause has a complete (sufficient) physical cause. The thought that a mental event (or a mental property) could cause an effect without relying on, or working through, physical events (or properties) was rightly deemed inimical to physicalism. The tension with the claim that mental properties do ineliminable causal work is palpable, especially since systematic overdetermination of effects by both mental and physical properties is highly implausible.

Many of the lead chapters in this volume grapple with finding a satisfactory answer to the questions arising from this tension. A number of alternative answers are available: (a) the hard-core physicalist one: only the physical property is causally efficacious, the mental property being either eliminable or epiphenomenal (Papineau); (b) both mental and physical properties are causally efficacious, this being ensured by the identity of their exemplifyings or instancings (Macdonald and Macdonald); (c) both mental and physical properties are causally efficacious, this requiring a departure either from physicalism and the causal closure of the physical (Crane, O'Connor and Churchill, Hendry in the case of chemistry), or from strict extensionality of the causal relation (Menzies and List). The latter option suggests that when a mental property brings about a physical effect it could be the case that no physical property is doing the same work, so again causal closure of the physical is rejected.

[4] An event is mental just in case P is a mental property, and physical just in case P is a physical property. Since the identity conditions on events require identity of their constitutive objects, properties, and times, and since Kim maintains both that each event has only one constitutive property and that mental properties are constitutive properties of mental events, his version of the property exemplification account has the consequence that psychophysical event identity entails the identity of mental and physical properties constitutive of events, effectively ruling out non-reductive monism. There are other versions of the property exemplification account, however, that do not have this consequence, and these are compatible with non-reductive monism. For more on this see Macdonald and Macdonald (2006).

B. The Distinction Between Causation and Causal Explanation

The extensionality of the causal relation (more precisely, the extensionality of sentences describing a causal relation) is predicated on causal realism, the assumption being that causes do their work 'in the world', and so the events implicated in the causal relation, however complicated they are, can be variously described without any change in the truth of the causal claim. This requires that causes are not description-dependent, and if they are not there may be many true causal statements which are irrelevant to an *explanation* of a given effect. Some think that certain recent theories of causality might cast doubt on this assumption. In particular, some of the work done in the formulation of the 'manipulation' account of causation (Woodward 2003; see also Hitchcock 2001 and other work) may lead to the idea that what is essential in causation is that interventions changing the value of a variable in the cause (e.g., increasing the pressure of a fixed volume of gas) be 'matched' by an appropriate change in a variable in the effect (an increase in the temperature of the gas). Applied to our topic, the thought is that if an intention to drink some beer causes an appropriate action (e.g., drinking some beer), then the *relevant* 'change in the value of the variable' will be a change in the *intention* leading to a different effect, and not a change in any underlying physical property or properties (see Menzies and List, this volume). This will make the intention the relevant cause of the action, rather than some underlying physical state or event.

The result is that the truth of causal statements is much more dependent on explanation than had previously been thought. A satisfactory resolution of these issues requires answers to a number of questions that are connected to the metaphysics of events. Chief among these are whether the causes here are events or properties, and what the relation is of mental causes to physical causes. One possible way of keeping the causal relation extensional is by resisting the identification of issues concerning explanation with ones concerning causation, insisting on the difference between property-instance-causation (causal efficacy) and property causation (relevant to explanation) (Macdonald and Macdonald 2006).

C. Multiple Realizability and Non-Reducibility

One way of making the mental properties do causal work without violating the causal closure of the physical is by identifying them with physical properties. Non-reductive physicalists are not attracted to this solution for obvious reasons: such identities are usually thought to be a consequence of the reduction they reject. At least part of what motivated this rejection was the thought that mental properties are 'multiply realized' by physical properties, one psychological property being 'realizable' by different (sets of) physical properties, this meaning that there could not be the one-to-one relation required for any identity between the two. But given the tension described above between the physicalist assumptions and the belief in the causal efficacy of the mental, reduction has been thought to be the best way out. Kim eventually opted for this route, avoiding the multiple realizability claim by 'slimming down' the mental properties to suitable size. Initially (Kim 1993), instead of taking properties such as pain to be the mental properties available for reduction, he took species-specific properties, such as human pain, or bat pain, to be the target ones, maintaining that 'within species' psychophysical identities escape the force of any argument from multiple realizability. More recently (1998, 2005), Kim has 'slimmed down' the mental properties still further, to individual types of mental events or states as experienced at specific times by particular subjects, such as Jones' pain at 3 p.m. on Monday, 29 June 2009, claiming that such 'individual types' are capable of being functionally reduced to particular physical types of states.

It is a moot point whether this ploy is successful or not. There is an important issue to be raised about what is required for different tokens to be of the same type (see Heil 2003 for discussion), and so about whether instances of, say, bat pain are *sufficiently similar to* instances of human pain to warrant their being co-typed as instances of pain, but Kim's strategy appears more radical than one proposing caution with respect to co-typing. The suggestion seems to be that if there is a sufficient difference at the physical, *realizing* level then we should count the realized (mental) tokens as type-different. One objection is that Kim's refined position is only terminologically distinct from a non-reductionist token-identity proposal: tokens of higher-level (including mental) properties are to be collected into similar types if, and only if, they are of the same realizing (physical) type; otherwise they are assigned to different types. But further, the ad hoc nature of this manoeuvre is not really congenial to non-reductionists, as it makes the typing of the special science properties work from bottom-up, the typing being dependent on subvenient similarities, rather than being controlled by the requirements of the special science in question. It would, for example, rule out what could be interesting economic generalizations just because monetary exchanges are effected by different 'materials' (electronic versus chapter exchanges, for instance). For this reason the position will be uncongenial not only to the Fodorian wing of the non-reductionist camp but also to the Davidsonian one, since Davidsonians are inclined to stress the 'categorial' difference between the supervening and subvening properties, with the consequence that control from below will be seen to conflict with what is essential to the supervening set of properties.

Despite the apparent artificiality of Kim's proposal, his insistence on the causal troubles facing the non-reductionist has been bracing. There has also been much discussion of what makes for multiple realization, a notion central to the non-reductionist argument, with different accounts being given. Aizawa and Gillett (2009, citing Endicott 2005) mark a distinction between a computational or mathematical account, where X is said to realize Y if the elements of Y map onto (are isomorphic with) the elements of X, and a *causal-mechanist* account where, very roughly, a higher-level property has causal powers which are determined by the causal powers of lower-level properties. According to the latter,

> A property G is multiply realized if and only if (i) under condition \$, an individual s has an instance of property G in virtue of the powers contributed by instances of properties/relations F1 -Fn to s, or s's constituents, but not vice versa; (ii) under condition \$* (which may or may not be identical to \$), an individual s* (which may or may not be identical to s) has an instance of a property G in virtue of the powers contributed by instances of properties/relations F*1 -F*m to s* or s*'s constituents, but not vice versa; (iii) F1 -Fn ≠ F*1 -F*m and (iv), under conditions \$ and \$*, F1 -Fn and F*1 -F*m are at the same scientific level of properties. (Aizawa and Gillett 2009: 188)

Several features of this account are worth noting. First, just as it is important to note the relational character of emergence, it is also important that multiple realization is a relation between distinct levels of properties. How one distinguishes the relevant levels is crucial. It may turn out that *every* higher-level property is multiply realized if at the quantum-mechanical level there are different properties and relations realizing the same higher-level property and relation. But this possibility would not have much relevance for a debate about the possibility of reducing psychological properties to, say, neurophysiological properties. That reduction is blocked only if the psychological properties are multiply realized *with respect to* the neurophysiological level.

Second, those wishing to defend the non-reducibility of, say, psychology via a claim of multiple realization have an obligation to specify what makes the different realizations relevantly multiple: not just *any* differences between different instantiations of a property will do. That a psychological property's instantiations are realized by neurophysiological processes that sometimes occur in the right, sometimes in the left, hemisphere of the brain may not be evidence supporting a claim of multiple realization; location just may not be an important difference from the perspective of the neurophysiologist (see Shapiro 2000 and 2008 for discussion of this issue). The differences between different realizations must be ones that are relevant to the science, or theory, imposing the taxonomy at that level.

Relatedly, there is a question as to what would constitute evidence for a claim of multiple realizability. Some have thought it sufficient to imagine creatures composed of different 'stuff' from humans but displaying, for example, the same pain behaviour, the idea being that we would not desist from attributing pain to such an alien merely on the grounds that the alien's pain was thus differently realized. Here it is the *possibility* of different realizations that is thought to suffice for the truth of multiple realizability. Others insist on actual multiple realizations, the evidence being provided by the relevant sciences, with some being more sceptical than others. (Sceptics include Bechtel and Mundale (1999);

Polger (2009); Shapiro (2008). Optimistic responses include Aizawa and Gillett (2009) and Aizawa (2009). For a nice example of multiple realization from animal behaviour, see Keeley 2000.)

Third, the emphasis on causal powers leads to a problem discussed in more detail below (sub-section 5): the same causal power (of property G) is comprised by the causal powers of different sets of lower-level properties (F1-Fn and F*1-F*m), the difference between these sets of lower-level properties being (at least partly) constituted by a difference in their causal powers.[5] How can different lower-level causal powers constitute or 'comprise' the same higher-level causal power? The emphasis on realization makes the question about the causal power of higher-level properties more urgent: if the higher-level property's causal power is constituted by the contribution from the lower-level realizing properties, then it is difficult to see how its causal power could fail to be exhausted by that contribution—it seems that it will contribute nothing of its own to the effects it is said to cause.

These worries reflect the previously noted inherent tension in an emergentist perspective. Bedau specifies two characteristics of emergent phenomena:

(1) Emergent phenomena are somehow constituted by, and generated from, underlying processes.
(2) Emergent phenomena are somehow autonomous from underlying processes. (Bedau 1997: 376)

The complications arising from reconciling (1) with (2) are grist to the mill for some of those who, being inclined to reject the reductionist perspective, opt for a more radical position, rejecting the monism entailed by 'orthodox' non-reductive physicalism (Crane, O'Connor, and Churchill, Hendry, Menzies and List, all this volume). In doing so they embrace emergentism, and for them the question becomes one of how their position differs from orthodox non-reduction, given that this is already a property-dualist view.

D. Property Dualism and Emergence?

Once structured events appeared on the scene it became apparent that non-reductive physicalism was committed to a dualism of properties (at least). One suspicion was that this was just old-fashioned dualism in a different guise, and that property dualism was just as 'mysterious'. Defenders took the view that non-reductive physicalism is really very different from substance dualism, since a respectable naturalism can still be defended by making the mental properties *dependent* on physical properties. Non-reductionists took this dependence to require supervenience, the general idea being that one set of properties supervenes

[5] Aizawa and Gillett (2009) assume that, for scientific properties, a difference of property is sufficient for a difference in the causal powers of those properties.

on another 'base' set of properties if there can be no change in an object with regard to a supervening property without a change in the base set of properties. A great deal of work has been done to make this intuitive thought more precise (see, for example, Kim 1993 and papers therein; Horgan 1993; McLaughlin 1995, 1997a, 1997b).

The relevance for our purposes is that some emergentists wish to remain naturalists, thus rejecting the complete autonomy of non-physical properties (Crane, Noordhof, O'Connor and Churchill, Stephan, all this volume). For many, even emergent properties must be suitably related to physical ones if the mysteries of substantial dualism are to be avoided and naturalism is to prevail. Given that both types of non-reductive naturalism (non-reductive monism and emergence) require distinctness of properties (a requirement for non-reduction) and dependence (a requirement of naturalism), two ways of marking the distinction between non-reductive monism and emergence recommend themselves. The first would be to vary the type of *distinctness* of properties that generates the difference between the base and emergent properties, one suggestion being that non-reductive monism insists only on numerical distinctness of the properties, with emergence requiring mereological distinctness (see Stoljar 2008: 276). The second would be to vary the strength of the dependence between the base and supervening properties, one suggestion being that emergent properties are only nomologically dependent on physical properties, whilst non-reductive physicalism is committed to a stronger, metaphysical, dependence (Noordhof, this volume). This second way of drawing the distinction seems more promising to us, relating as it does to (sets of) properties, whereas the first seems to relate more properly to property instances.

Matters are in fact more complicated than this, since there is in the literature a third way of marking a relevant distinction, by distinguishing between 'weak' emergence and 'strong' emergence, this distinction being drawn in terms of causation. Strong emergence requires 'direct' downward causation, a causal power irreducible to the causal powers of the base set of micro-properties (O'Connor 1994 done); weak emergence then requires that the higher-level property be a structural property, one 'constituted wholly out of its microstates' (Bedau 1997: 378). Consequently, there is no 'mystery' of irreducible downward causation, and weak emergence becomes ubiquitous. For the 'emergence' of weakly emergent properties Bedau requires (roughly) that the higher-level (macro)state be derivable from the environmental input and the dynamic governing the microstates, but only by simulation; this guarantees a certain autonomy for the higher-level states without any 'metaphysical illegitimacy' (Bedau 1997: 396). The question remains as to whether strong emergence, so characterized, can maintain metaphysical legitimacy. (For further discussion and characterizations of weak and strong emergence see Gillett 2002; Chalmers 2008; and Macdonald and Macdonald, Noordhof, and Stephan this volume.)

E. The Coherence of Laws

As long as there is some form of dependence on physical properties, the non-reductivist/emergentist is burdened with giving some account of the connection between base and supervening properties. This connection has two facets, a synchronic dependence already remarked upon (some form of supervenience), and a diachronic 'harmony'. The former dependence has been much discussed, the latter less so. Diachronic harmony is required by the structure of supervenience when this structure is 'put in motion', when one considers the structure over time. Suppose, for example, that an instantiation of a mental property, in a suitable context of other properties, causes an action. This effect will itself have both mental and physical properties, and the process of the causing of action will need to respect supervenience. That is, the resultant state of affairs, the effect, will have to accord with the general picture of the relation between mental and physical properties. In the case envisaged, the physical causes of bodily movements will have to ensure that the right action is performed, 'right' here meaning an action that intelligibly flows from that intention. The puzzle is this: how do the different levels 'march in step', so that upper-level causes mirror lower-level causes in bringing about effects that supervene in the right way? One way of ensuring harmony between levels is to rely on natural design (Papineau, this volume, Chapter 12). If the levels are arranged so that they 'fit', the right causal profile of supervening and subvening properties is assured. The question then becomes one of ascertaining how such a fit is obtained. Darwinian natural selection is one answer; it operates on the effects of certain causes, such causes being grouped together, and so co-typed, *just because* they produce the desired (fitness-enhancing) effect. Such biofunctional properties can be multiply realized, the realizing properties having one thing in common: they cause the 'right' effect.

It is an open question as to how much of the causal harmony between the putatively non-reducible levels is explicable by the mechanism of a selection process. Hendry (this volume, Chapter 14) argues for strong non-reducibility for chemistry without relying on any selection process; Pettit (this volume, Chapter 17) outlines conditions under which group-rationality may be said to emerge, where a group may be said to exhibit rational agency, also apparently without the operation of selection, though he does include a 'disciplining' condition which could have the same effect as natural selection. This general topic of the compatibility of differing causal processes is relevant to issues as diverse as whether the kinds delivered by the taxonomies of the special sciences are natural kinds (Papineau), or whether there can be room for freedom given the causal closure of the physical (Stephan, this volume, Chapter 15).

4. CONTRIBUTIONS

We conclude with a brief summary of the lead contributions to this volume.

Tim Crane defends the claim that any genuinely physicalist position must distinguish itself from (what traditionally has been known as) emergentism; it cannot afford to postulate inexplicable or 'brute' correlations or identities. As a result, he argues, physicalism is necessarily reductive in character—it must either give a reductive account of apparently non-physical entities, or a reductive explanation of why there are non-physical entities. Crane claims that many recent 'non-reductive' physicalists do not do this, and because of this they cannot adequately distinguish their view from emergentism. This, he argues, is the real challenge posed by Joseph Levine's 'explanatory gap' argument: if physicalists cannot close the explanatory gap in Levine's preferred way, they must find some other way to do it. The price of failing to close the explanatory gap is to give up on non-reductive physicalism, since the resulting position will be indistinguishable from emergentism. Emergentists can embrace the generality of physics, with emergent properties being the supervenient properties of a thing not identical to any properties of its parts and supervenience being inexplicable in physical terms.

The attempt to reconcile non-reductive physicalism with a causal powers metaphysics is the target of the chapter by **Timothy O'Connor** and **John Ross Churchill**. The authors first outline Kim's attack on the non-reductionist's thesis, noting its dependence on a 'causal exclusion' principle. They argue that this assumes what they call a 'causal-powers metaphysic', a metaphysics requiring the exercise of ontologically primitive causal powers or capacities of particulars, these powers providing the 'oomph' in causation. Kim's argument is then reformulated to make this dependence explicit (hence the 'power-exclusion argument'). O'Connor and Churchill claim that if (as they think) this causal-power metaphysics is correct, then the non-reductivist position is incoherent, short of accepting systematic overdetermination of the mentally caused effects. Their argument is strengthened by examining a notable attempt to defend non-reductionism assuming the causal-power metaphysic, that of Shoemaker. They find this defence unpersuasive, concluding that non-reductive physicalism and the causal-power metaphysic are incompatible. The non-reductionist is forced, as a result, to give up on non-reduction and opt for either reductionism or eliminativism. Their preferred alternative is to sacrifice some of the premises leading to the unwelcome conclusion, specifically premises asserting the realization of the mental by the physical and the causal completeness of the physical. This inclines them towards the acceptance of an ontological variety of emergence, one respecting 'the distinctive character and efficacy of the mental'.

The focus of **Paul Noordhof**'s chapter is on the conditions that need to be satisfied if we are justifiably to say that we have emergent property causation. A necessary condition for property causation is that an instance of the property cited is a cause. Property causation also involves a certain kind of generality, say a causal law linking cause and effect. Noordhof identifies a set of narrowly physical property causes, a subclass of the class of properties identified by current physics or a future physics sufficiently resembling our own, containing just those properties which are property causes in this way. These are not exhaustive of the physical properties: there are broadly physical properties that supervene on the narrow physical properties. There is an important distinction to be made between the broadly physical properties and emergent properties. Both supervene on the narrow physical properties, but the former do so with metaphysical necessity, the latter with nomological necessity. Noordhof uses this distinction to motivate a difference between non-reductive physicalism and emergent dualism: both suppose that the instantiation of narrowly physical properties determines the instantiation of the other target properties, the non-reductive or emergent ones. They differ over whether the instantiation of these other properties involves something genuinely new. Non-reductive physicalists deny this, whereas emergent dualists assert it. The problem is to make sense of when there is something new introduced.

Building on work done elsewhere, Noordhof develops a counterfactual theory of property causation, focusing on the possibility of emergent causation, and in particular on the question of whether all emergent causation involves emergent property causation. His negative conclusion leads him to identify a second kind of emergence: emergent non-reductive physicalism. He goes on to apply the conclusions of this discussion to two candidates for emergence: phenomenal consciousness and free will.

Causation is clearly a central concern for non-reductionists of all persuasions. **Peter Menzies** and **Christian List** contend that recent interventionist accounts of causation, and in particular that developed by James Woodward, help to shed light on the debates (introduced by Kim) surrounding causal closure of the physical. Menzies and List use Woodward's interventionist account to identify necessary and sufficient conditions for the causal autonomy of a higher-level property and to show that these conditions are satisfied when causal claims about higher-level properties have a special feature, that of realization-insensitivity. This feature consists in the fact that relevant causal claims are true regardless of the way the higher-level properties they describe are physically realized. If higher-level properties are realization-insensitive, then when such a property, say a mental property Ma causes an action Ba, the realizing physical property does *not* cause Ba. This 'Downwards Exclusion Result' ensures the causal autonomy of the realization-insensitive properties. Menzies and List go on to show that these findings are consistent with those of other philosophers (e.g., Alan Garfinkel), who have noted the realization-insensitivity of higher-level causal relations as a distinctive feature of the special sciences, and who have suggested that this feature ensures their independence from lower-level causal relations.

In their contribution, **Cynthia Macdonald** and **Graham Macdonald** support a form of strong emergence, one where emergent properties are not just complex properties derivable from the properties of more simple parts and their relations. The reason for the non-derivability of such emergent properties may differ from case to case, so there may be a different explanation of the non-derivability of biological properties than there is for the non-derivability of mental ones. The Macdonalds proceed to defend their version of strong emergence against critics, such as Kim, who think that it must lead to downward causation, which is claimed to be incompatible with the causal closure of the physical. The Macdonalds provide a metaphysics of events to show why this claim is false, arguing that this metaphysics allows two properties, say a mental and a physical property, to be co-instanced in a single event, and so allows the cause-event to be a single exemplifying of both a mental and physical property. This ensures the causal efficacy of the mental property *instance* while opening up space for distinctive causal (and explanatory) work to be done by the mental *property*. The chapter concludes by arguing against recent objections to this approach voiced by Philip Pettit (1996) and the alternative position developed, in different ways, by him and by Carl Gillett (2006a, 2006b).

Much of the discussion surrounding non-reductive physicalism has focused on its ability to allow supervenient causation. As **David Papineau** notes, there has been far less discussion of whether non-reductive physicalism can accommodate non-physical *laws*; Fodor assumed there were laws, even though these may be 'loose', or *ceteris paribus*, ones. The problem facing the Fodorian that Papineau points to concerns the compatibility of there being both supervening and subvening laws with multiple realizability: the *different* realizations of a supervening state, S_1, must all result in some physical state that determines the *same* effect, say S_2. The subvening heterogeneity, required for non-reduction, sits uncomfortably with the supervening homogeneity. The choice is either to give up on there being special science (supervening) laws, or to accept that the supposed underlying variability is an illusion, and that type-type reduction is available. Papineau queries whether this reduction seriously threatens the explanatory autonomy of the special sciences, given the often insuperable practical difficulties of providing reductions. The only way to avoid reducibility is to appeal to selection processes, processes in which the different physical states are 'designed' to lead to the same effect.

Although such 'selection-based patterns' may allow projectible correlations, Papineau worries that they result in a taxonomy of natural kinds that are too 'thin' to figure in a substantial science; for paradigm natural kinds, a multiplicity of their properties will figure in laws, whereas with selection-based kinds, only the property yielding the relevant effect is projectible. Pain, for example, leads to damage-avoidance, and so the concept *pain* will figure in a relevant law, but insofar as pain is variably realized, no further laws will be forthcoming; to the extent that it is not multiply realized (say, in humans), there will be further laws to be discovered. The special sciences may consist in some 'thick' natural kinds (where there is no multiple realization) and 'thin' selection-based kinds. As far as causation is concerned, Papineau argues (with Menzies and List, and Hendry, this volume) against the assumption that the subvenient, realizing, fact always usurps the causal power of the more general supervening fact. But even granting this, there is still the problem that with variably realized special facts there will be no uniform physical law linking cause to effect. And in this case selection-based laws don't help, since these are based on pre-existing causal powers and do not add to them. Papineau concludes that non-reduced special kinds are not causes. They range over cases with quite different causal structures.

Most of the argumentation concerning the causal autonomy of the properties of higher-level sciences has focused on either psychology or sociology. **Robin Hendry** turns to chemistry, arguing against what he sees as the orthodox view amongst physicalists, who hold (i) that successful quantum-mechanical explanations of chemical bonding render unlikely the existence of downward causation from the chemical to the physical, and (ii) that the very idea of downward causation is murky. In his chapter he argues against both claims: against (i) he argues that it does not withstand investigation of either the early history or the mathematical structure of quantum chemistry, and against (ii) he proposes a counternomic criterion for downward causation, one which attempts to capture the emergentist views of C. D. Broad. The emergentism Hendry argues for permits, he claims, the supervenience of chemical properties on physical properties, since the supervenience relation is consistent with downward causation. With regard to the specific case of chemistry Hendry makes a distinction between resultant and configurational Hamiltonians (which describe the evolution of the quantum-mechanical complex system). The reductionist claims that the quantum-mechanical explanations of chemical structure and bonding will involve only resultant Hamiltonians; the emergentist expects that they will involve configurational Hamiltonians. Only evidence supplied by science can decide the issue, and Hendry argues that the evidence supports the emergentist. He concludes by replacing the physicalist's 'completeness of physics' with the 'ubiquity of physics': physical principles constrain the motions of particular systems without necessarily fully determining them.

Within the debates about mental causation and emergence there has been little discussion of free will, and it is this gap that **Achim Stephan**'s chapter aims to fill. Stephan situates his discussion in the context of free-will debates between two German philosophers and a neuroscientist, these usefully lining up as a libertarian, a compatibilist, and a hard determinist. The central question for each is how to envisage the relation between 'person-level' psychological properties and processes, and the underlying subpersonal neurophysiological processes. Stephan discerns a surprising commonality amongst the disputants: all accept a synchronic dependency claim, that there can be no psychological change without a neurophysiological change. All are also committed to a diachronic 'principle of alternative possibilities' in characterizing what free will requires: the agent must have been able to have done otherwise. As Stephan sees it, the hard determinist must see the personal/subpersonal relation as one of reducibility: the person-level properties are explained (away) by the neurophysiological properties. The libertarian sees this relation as a case of emergent properties, where a property is said to be emergent if it is had by a system but not by that system's parts. Here the synchronic dependency claim is protected by postulating that non-deterministic psychological processes must rest on non-deterministic neurophysiological processes. Stephan argues that the compatibilist must embrace reductionism and rely on a notion of mental causation, where the subsequent actions and neurophysiological processes are caused by the agent's psychological properties, for example beliefs and desires. The nub of the dispute, as he sees it, relies on the plausibility of psychological reduction; given the imperfect state of knowledge concerning brain processes and their relation to psychological processes, no convincing conclusion is presently available.

The free-will advocate connects our supposed freedom with the exercise of our psychological, specifically rational, capacities. For individuals, **Philip Pettit** argues, a rational configuration of propositional attitudes is maintained by our ability to reason. The question he poses is whether there can be *group agents* and, if so, whether the rationality of the analogue of propositional attitudes can be maintained without reasoning. The possibility of group agents, and group rationality, also raises the question about the relation between the group and its members: must a group judgement supervene on the judgements of its members?

Reasoning, for Pettit, requires being able to monitor one's own propositional attitudes and actions, adjusting these where rationality so requires in response to evidence, to other attitudes, and/or to proposed actions. Pettit argues that under a range of plausible conditions, groups lacking feedback on their own decisions will not be able to reason, and so will not function satisfactorily as agents. This is supported by results on the aggregation of judgements that show, under certain conditions, the impossibility of ensuring that group judgements over connected issues will be complete and consistent.

Revising the way that group judgements are arrived at in any way that does not include systemic feedback will also fall short of ensuring group rationality, suggesting that any 'no-feedback constitution', one aiming to have rational but unreasoning group agents, is likely to fail. With feedback there is the possibility for group rationality to emerge: the group may be able to exercise the sort of control over its processes of judgement formation analogous to the personal control over judgements characteristic of individual reasoning. This control makes the group agent responsible for decisions reached, and elevates the group above that of a mere self-organizing collective to that of a self-governing collective.

REFERENCES

Aizawa, K. 2009. 'Neuroscience and Multiple Realization: A Reply to Bechtel and Mundale'. *Synthese* 167: 493–510.

—— and Gillett, C. 2009. 'The (Multiple) Realization of Psychological and other Properties in the Sciences'. *Mind & Language* 24: 181–208.

Allen, G. 1975. *Life Science in the Twentieth Century*. Cambridge, Cambridge University Press.

Anderson, P. W. 1972. 'More is Different'. *Science* 177, no. 4047: 393–6.

Bechtel, W. and Mundale, J. 1999. 'Multiple Realizability Revisited: Linking Cognitive and Neural States'. *Philosophy of Science* 66: 175–207.

—— and Richardson, R. 1998. 'Vitalism'. In E. Craig (ed.), *Routledge Encyclopedia of Philosophy*. London: Routledge and Kegan Paul: 9.

Bedau, M. 1997. 'Weak Emergence'. *Philosophical Perspectives* 11: 375–99.

Chalmers, D. 2006. 'Strong and Weak Emergence'. In P. Clayton and P. Davies (eds.), *The Re-Emergence of Emergence*. Oxford: Oxford University Press, 244–54.

Davidson, D. 1970. 'Mental Events'. In L. Foster and J. Swanson (eds.) *Experience and Theory*. London: Duckworth, 79–101.

Endicott, R. 2005. 'Multiple Realizability'. In D. Borchert (ed.), *The Encyclopedia of Philosophy*, Supplement, 2nd edn. New York: Macmillan, 427–32.

Fodor, J. 1974. 'Special Sciences: or the Disunity of Science as a Working Hypothesis'. *Synthese* 28: 77–115.

Gillett, C. 2002. 'The Varieties of Emergence'. *Grazer Philosopische Studien* 65: 95–121.

—— 2006a. 'Samuel Alexander's Emergentism: or, Higher Causation for Physicalists'. *Synthese* 153: 261–96.

—— 2006b. 'The Hidden Battles Over Emergence'. In P. Clayton and Z. Simpson (eds.), *Oxford Handbook of Religion and Science*. Oxford: Oxford University Press, 261–96.

Heil, J. 2003. *From an Ontological Point of View*. Oxford: Clarendon Press.

Hitchcock, C. 2001. 'A Tale of Two Effects'. *Philosophical Review* 110: 361–96.

Horgan, T. 1993. 'From Supervenience to Superdupervenience: Meeting the Demands of a Material World'. *Mind* 102: 555–86.

Humphreys, P. 2008. 'Computational and Conceptual Emergence'. *Philosophy of Science* 75: 584–94.

Johnson, S. 2001. *Emergence: The Connected Lives of Ants, Brains, Cities, and Software*. New York: Scribner.

Keeley, B. 2000. 'Shocking Lessons from Electric Fish: the Theory and Practice of Multiple Realization'. *Philosophy of Science* 67: 444–65.

Kim, J. 1993. *Supervenience and Mind: Selected Philosophical Essays*. Cambridge: Cambridge University Press.

—— 1998. *Mind in a Physical World*. Cambridge, MA: MIT Press.

—— 2005. *Physicalism, or Something Near Enough*. Princeton, NJ: Princeton University Press.

Macdonald, C. and Macdonald, G. 2006. 'The Metaphysics of Mental Causation'. *Journal of Philosophy* 103: 539–76.

McLaughlin, B.P. 1992. 'The Rise and Fall of British Emergentism'. In A. Beckermann, H. Flohr, and J. Kim (eds.), *Emergence or Reduction?*, Berlin: de Gruyter, 1992, 49–93.

—— 1995. 'Varieties of Supervenience'. In E. Savellos and U. Yalcin (eds.), *Supervenience: New Essays*. Cambridge: Cambridge University Press, 16–59.
—— 1997a. 'Supervenience, Vagueness, and Determination'. *Philosophical Perspectives* 11: 209–30.
—— 1997b. 'Emergence and Supervenience' *Intellectica* 25: 25–43.
O'Connor, T. 1994. 'Emergent Properties'. *American Philosophical Quarterly* 31: 91–104.
Pettit, P. 1996. *The Common Mind*. Oxford: Oxford University Press.
Polger, T. 2007. 'Realization and the Metaphysics of Mind'. *Australasian Journal of Philosophy* 85: 233–59.
—— 2009. 'Multiple Realization and Evidence'. *Synthese* 167: 457–72.
Putnam, H. 1967. 'Psychological Predicates'. In W. H. Capitan and D. D. Merrill (eds.), *Art, Mind and Religion*, 37–48. Reprinted. as 'The Nature of Mental States', in H. Putnam, *Mind, Language and Reality* (1975), 429–40.
Shapiro, L. 2000. 'Multiple Realizations'. *Journal of Philosophy* 97: 635–54.
—— 2008. 'How to Test for Multiple Realization'. *Philosophy of Science* 75: 514–25.
Stoljar, D. 2008. 'Distinctions in Distinction'. In J. Kallestrup and J. Hohwy (eds.), *Being Reduced: New Essays on Causation and Explanation in the Special Sciences*. Oxford: Oxford University Press, 263–79.
Woodward, J. 2003. *Making Things Happen: A Theory of Causal Explanation*. Oxford: Oxford University Press.

2

Cosmic Hermeneutics vs. Emergence: The Challenge of the Explanatory Gap*

Tim Crane

1. THE EXPLANATORY GAP

Joseph Levine is generally credited with the invention of the term 'explanatory gap' to describe our ignorance about the relationship between consciousness and the physical structures which sustain it.[1] Levine's account of the problem of the explanatory gap in his book *Purple Haze* (2001) may be summarized in terms of three theses, which I will describe and name as follows:

(SP) *Supervenience physicalism*: every minimal physical duplicate of the actual world is a duplicate in every respect.

(DE) *Deductive explanation*: the explanation of consciousness must consist in a deduction of the truths about consciousness from the physical truths.

(EG) *Explanatory gap*: we lack an adequate deductive explanation of all the truths about consciousness in physical terms.

SP, DE and EG are not inconsistent, since consciousness could supervene on the physical without its being explicable. Consciousness might be wholly determined by the physical but nonetheless be inexplicable. This is what 'mysterians' believe. Nonetheless, Levine thinks (and many agree with him) that there is a challenge here which physicalism has to meet. If physicalism is to be an adequate account of the world, it must not postulate too many 'brute' or inexplicable correlations and identities. Accepting a mere brute correlation is accepting a mystery. In this he is echoing Thomas Nagel, who said famously that someone who asserts that consciousness is a process in the brain would be in the same epistemological

* An earlier version of this chapter was read at the 'Explanatory Gap' conference in memory of Nikola Grahek, held in Belgrade in April 2005. I am grateful to Katalin Farkas, Mike Martin, Massimiliano Vignolo and Tim Williamson for discussion. I dedicate the chapter to the memory of Nikola Grahek.

[1] See Levine 1983, 1997, 2001. For a discussion of some interesting historical precursors, see Tennant 2007.

position as an ancient Greek who asserted that matter is energy: they would have said something true, but they would not have understood how it could be true (Nagel 1974).

That there is an explanatory gap (in the sense of EG) is surely undeniable. The question is what its significance is. Some physicalists think that it is of little significance, because it is only a matter of time before we arrive at an explanation. Others think it is of little significance because physicalism does not require such an explanation of consciousness. However, others think that with some additional assumptions, the explanatory gap shows that physicalism is false.

In this chapter I will argue that the real significance of the explanatory gap lies elsewhere. In my view, the explanatory gap creates a challenge for the proper formulation of *non-reductive* versions of physicalism. Non-reductive physicalists have attempted to distinguish their theories from reductive versions of physicalism on the one hand, and 'emergentism' on the other. But their standard response to the explanatory gap fails to distinguish their non-reductive physicalism from emergentism. Since whatever emergentism is, it is not physicalism, this is a problem for the proper formulation of physicalism. What the explanatory gap shows is what is properly required from an adequate non-reductive physicalism.

In section 2 I will sketch some physicalist responses to the challenge of the explanatory gap; in section 3 I will give an account of emergent properties; in section 4 I will argue that physicalism is necessarily reductive in character; and in section 5 I will show how the explanatory gap challenges so-called 'non-reductive' versions of physicalism to distinguish themselves from emergentism.

2. PHYSICALISM AND COSMIC HERMENEUTICS

Like Nagel before him, Levine did not argue that the explanatory gap shows that physicalism is false. As just noted, EG, DE and SP are mutually consistent. Nagel's view was that physicalism is true, but that we cannot fully understand it. Levine, similarly, thinks that physicalism is explanatorily inadequate until EG is shown to be false. The explanatory gap is precisely that: an *explanatory* gap. There is no *metaphysical* gap, on the Levine–Nagel view. But it is only if there is a metaphysical gap that physicalism is false.

Frank Jackson (1998) rejects this account of the situation. Jackson thinks that a serious version of physicalism needs not just SP, but an explanation of SP itself. SP makes a claim about the necessary connection between the physical truths and all the other truths. Jackson argues first that this necessity cannot be 'brute', and second that it must be explicable as conceptual necessity, that is, as the necessity deriving from relationships among concepts. The now well-known picture that Jackson arrives at is that all truths stated in non-physical vocabulary must in principle be derivable from the physical truths plus conceptual analyses of the non-physical concepts.

Jackson's thesis therefore implies what Terence Horgan (1984) calls 'cosmic hermeneutics': that full knowledge of the microphysical supervenience base would allow us in principle to 'read off' all the truths about the world from the truths about the supervenience base (see also Byrne 1999). Understood in Jackson's way, cosmic hermeneutics is the following doctrine:

(CH) *Cosmic hermeneutics:* there can be an a priori deduction of all the truths about the world from the microphysical truths about it plus the conceptual truths about non-physical concepts.

So stated, CH is a claim about possibility: it says merely that there *can* be such an a priori deduction of all the truths, not that anyone has actually done it. Jackson argues that if physicalism is true, then CH is true. Physicalism implies cosmic hermeneutics.

An oversimplified example can illustrate. Suppose that we achieve a complete analysis of the concept of water, in terms of concepts like *wet*, *transparent*, *odourless*, etc. And let's suppose that we have an analysis of these concepts in terms of the causal roles of the things to which they apply: we have a causal analysis of the concept of wetness, for example. Let's abbreviate our analysis of the concept of water to *the watery stuff*. Equipped then with a complete physical knowledge of the world—knowledge of all the particular matters of physical fact, and the physical laws—we will recognize the truth of the proposition:

(a) H_2O covers most of the earth

Given our grasp of the concept of *the watery stuff*, and our knowledge of physics and chemistry, we can also recognize the truth of the proposition:

(b) H_2O = the watery stuff

And now given our conceptual analysis of *water*, we can then deduce, without any need for any further empirical investigation:

(c) Water covers most of the earth

Given a conceptual analysis of mental concepts—even concepts of consciousness—we could extend this kind of a priori 'reading off' to the relationship between the mental and the physical.[2]

Since CH implies the a priori deduction of all truths from the physical truths and the conceptual analyses about the other truths, it implies DE. For it will imply a deduction of all the truths about consciousness from all the physical truths, including the laws of nature, and such a deduction would be of the form that a

[2] The origin of these ideas in David Lewis's analytical functionalism should be obvious: see Lewis 1972. For similar ideas, see Lewis 1995 and Pettit 1993.

Deductive Nomological explanation of Hempel's form requires (Hempel 1966). But DE does not imply CH, since it does not say that the deduction of all the truths about consciousness must be an a priori deduction from the supervenience base: that is, a deduction which requires no more empirical information than the information about the supervenience base.

So CH is stronger than the conjunction of SP and DE. It is for this reason that Jackson's view (that physicalism implies CH) makes him more vulnerable to attack from non-physicalists. For if Jackson is right about what physicalism involves, a successful attack on CH would undermine physicalism. For example, a non-physicalist could construct this familiar metaphysical extension of the explanatory gap argument:

(i) Physicalism implies that cosmic hermeneutics is possible;
(ii) Cosmic hermeneutics is impossible;
(iii) Therefore physicalism is false.

The defence of premise (ii) could come from an argument that there is no a priori incoherence in supposing the supervenience base of consciousness to be present in the absence of consciousness—no matter how good our conceptual analysis of consciousness. Given this, then there can be no a priori deduction of the sort envisaged by CH, and since physicalism implies CH, physicalism is false.

Support for (ii) could be bolstered by a zombie argument (Chalmers 1996): if zombies are conceivable, then they are possible, and if zombies are possible, then physicalism is false. However, if a non-physicalist took this route, then they would not need the detour via CH. It is a nice question whether CH or the move from conceivability to possibility is a more controversial one; but not a question we have to settle here.

Levine, in contrast to Chalmers, accepts that the zombie argument shows that zombies are conceivable, but denies that this implies their possibility. What the zombie argument shows is the presence of an explanatory gap, not a metaphysical gap:

> No matter how rich the information-processing or the neurophysiological story gets, it still seems quite coherent to imagine that all that should be going on without there being anything that it's like to undergo the states in question. Yet if the physical or functional story really explained the qualitative character, it would not be so clearly imaginable that the qualia should be missing. (Levine 1997: 549)

It is clear here that Levine thinks that it is perfectly conceivable that all the physical might exist without the qualia; but he will not go as far as Chalmers in saying that this is a real possibility. So he remains a physicalist, despite the explanatory gap.

Some physicalists, however, disagree with Jackson *and* with Levine. All these physicalists will deny that physicalism requires CH (see Byrne 1999) but some also deny that it requires DE (see, e.g., Block and Stalnaker 1999). That is, they deny that physicalism needs to explain consciousness in terms of a link to its physical basis which must be articulated by a deductive argument. I call these philosophers 'non-reductive physicalists'. They accept SP, but they do not claim that SP must have an explanation. The necessity postulated in SP is a brute necessity, it is what Chalmers calls a 'strong metaphysical necessity'.

I think these physicalists are right to reject CH and DE. But the problem now is what makes their view a genuinely physicalist one, in any interesting sense. Physicalism cannot simply be the denial of Cartesian dualism—it cannot simply be the assertion that there are no purely mental particulars or souls—since then there would be little debate about the truth of physicalism (Chalmers would be a physicalist on this definition). At the heart of physicalism, it seems to me, is a commitment to either the ontological or the explanatory priority of the physical ('physical' in the sense of 'physics' or 'physical science'). Non-reductive physicalism can certainly assert the ontological priority of the physical in the sense of SP. But, as we will now see, the doctrine known as *emergentism* can do the same. Yet ever since the advent of physicalism as a serious metaphysical view, physicalists have sought to distinguish their view from emergentism. If non-reductive physicalism cannot be distinguished from emergentism then it barely deserves the name of physicalism at all. In what follows I shall develop this line of thought.

3. EMERGENCE

In order to see the problem for non-reductive physicalism, we first need to distinguish emergence from reduction. The term 'emergence' could be used for a number of different kinds of phenomena. The philosophically interesting use of the term is to express the view that certain properties of things are fundamentally different from others: certain properties are 'emergent' properties and others are not. To distinguish this idea from the idea that the mind developed or emerged over time, I shall call this 'synchronic' emergence. Synchronic emergence is the view that it is true of a system or entity *at a time* that some of its properties are emergent and others not, regardless of how it evolved or whatever its history (i.e. the distinction between emergents and non-emergents would remain even if the world were created in an instant, with no evolution or historical development of mind).

Synchronic emergence in this sense is centrally a feature of *properties*. This is how the term was introduced by some philosophers of the nineteenth century to describe certain features of macroscopic objects.[3] The rough idea is that these features of objects are genuinely 'novel' in the sense that they are not purely

[3] For the historical background, and a brilliant discussion of the merits of the view, see McLaughlin 1992. For a sympathetic discussion of the views McLaughlin describes, see Crane 2001b.

'consequences' microscopic parts, and yet they are not 'added from outside' in the way that is claimed by (for example) a Cartesian conception of mental properties, or a vitalist conception of biological properties. (Vitalism is the view that the explanation of biological life cannot be given in chemical and physical terms alone, but requires the postulation of vital forces.) So emergentism—the idea that some properties are emergent—is intended to steer a middle path between reductionism and forms of dualism like Cartesianism and vitalism.

The roughness of this description of emergent properties is indicated by the metaphors and scare quotes. How can we make these ideas more precise? Let us start with the idea of novelty. Certainly there are properties of macroscopic wholes which are not identical with properties of their parts, yet which are in an obvious sense 'nothing over and above' the properties of their parts. An object's weight is an example. An object may weigh ten kilos, and yet in an obvious sense this weight is nothing over and above the weights of (say) its ten parts, each of which weighs a kilo. The weight of the object is a simple function of the weights of its parts; the object's weight is only 'novel' from what Ernest Nagel (1963) once called 'the additive point of view'. Properties like weight are what the nineteenth- and early twentieth-century British emergentists called 'resultants'. Emergents were contrasted with resultants, in that they were not a priori deducible from the properties of a thing's parts, by 'adding' the properties of the parts.

Yet although this is one of the traditional ways of introducing the term 'emergent', this criterion will not distinguish for us the precise sense in which emergent properties are novel. For there are many properties of macroscopic things which are plausibly reducible to properties of the parts without being reducible by this simple 'additive' method: the temperature of some types of substance is an example. And this is true even if the temperature of something is a numerically distinct property from any of its microproperties. Distinctness is not sufficient for novelty.

We have more success if we turn to the idea of 'adding something from the outside'. Though vague, the idea that emergent properties are properties which are not 'added from the outside' suggests the following, somewhat more precise thought: emergent properties of macroscopic objects are *dependent* on the properties of their parts, in such a way that there is no variation in the object's macroproperties without variation in its parts' microproperties. In other words, the macroproperties *supervene* on the microproperties, in one of the many senses of that term employed in recent philosophy. Indeed, Jaegwon Kim has claimed that this is all that emergence really means:

> According to emergentism, higher-level properties, notably consciousness and other mental properties, emerge when, and only when, an appropriate set of lower-level 'basal conditions' are present and this means that the occurrence of the higher properties is determined by, and dependent on, the instantiation of appropriate lower-level properties and relations. In spite of this, emergent properties were held to be 'genuinely novel'

characteristics irreducible to the lower-level processes from which they emerge. Clearly, then, the concept of emergence combines the three components of supervenience, namely, property co-variance, dependence and non-reducibility. In fact, emergentism can be regarded as the first systematic formulation of non-reductive physicalism. (Kim 1995: 576–7)

Kim here claims that emergence is supervenience, and that emergentism is a type of physicalism. I think Kim is right to connect the ideas of dependence and covariance with emergence, but wrong to think that emergence simply *is* supervenience. And he is also wrong to say that emergentism is a form of physicalism.

Supervenience is not sufficient for emergence because the supervenience of A on B is compatible with the reduction of A to B. Indeed, the idea of supervenience in general is compatible with almost any position in the metaphysics of the mind, so it can hardly be used to express such positions.[4] Certainly, if mental properties *are* emergent properties, then it ought to *follow* that they are supervenient—since this is what is meant by 'not being added from the outside'. But this only means supervenience is *necessary* for emergence, not that it is sufficient. What more is needed to pick out the idea of an emergent property?

Sometimes it is said that emergent properties are those properties of a thing whose instantiation cannot be predicted from knowledge of the thing's parts (Broad 1929: 67–8). We might be concerned here that prediction is an epistemic notion, relating to our ways of knowing things, and it is unwise to mark a distinction in kinds of properties (emergent vs. reducible) in epistemic terms. And there are other difficulties with the idea of predictability too, since whether we can actually predict something depends on having a vocabulary in which to describe it, and the existence of such a vocabulary cannot determine whether a property is emergent or not (see Crane 2001: §2).

Nonetheless, properly understood, the idea of predictability contains the key to emergence. For prediction is linked to the idea of explanation. When one has an explanation of X in terms of Y, then often there is some prospect of predicting X from the presence of Y. If we have no further explanation of Y, then we can say that Y is *explanatorily basic*.

Now, physicalists are those who think that physical science has some kind of priority in our account of the world. One way to understand this priority is merely in terms of the idea that everything has physical properties, or that everything is subject to the laws of physics. As noted above, this anodyne (though plausible) view is surely not strong enough to count as physicalism. We need to add to this the idea that the physical sciences are *explanatorily* more basic as well. Physicalism must also contain the idea that explanations of our world must come to an end with physical principles and the appeal to purely physical entities. Explanations of natural phenomena (of whatever form they take) must bottom

[4] Here I agree with Kim's later thoughts (1998: ch. 1).

out in terms of explanations in the physical sciences. And this is *why* they are also *metaphysically* basic: because, according to physicalism, physics provides the basis of the true metaphysics.

In a particularly insightful discussion of this matter, Terence Horgan says:

> A physicalist position should surely assert, contrary to emergentism . . . that any metaphysically basic facts or laws—any unexplained explainers, so to speak—are facts or laws within physics itself. (Horgan 1993: 560)

This, plus the supervenience thesis, is the key to emergentism. A doctrine that holds that mental and other higher-level phenomena supervene on the physical, but that the supervenience in question has no explanation from within physics, is not physicalism by Horgan's lights. It is, rather, a doctrine that fully deserves the name *emergentism*. This interpretation of the situation makes it intelligible as to why physicalists traditionally have wanted to distinguish themselves from emergentists.

In the next section, we will see where this leaves non-reductive physicalists.

4. PHYSICALISM AND REDUCTIONISM

Let me repeat again that if physicalism is to be an interesting doctrine, something worth asserting and also worth denying, then it must be more than the claim that all objects have physical properties. The denial of this is the claim that not all objects have physical properties; or, in other words, there are objects with no physical properties. Ignoring numbers for the time being, such objects are the traditional mental substances of Cartesian metaphysics. Such mental substances receive few supporters these days and I will pass them over in silence here.

The claim that all objects have physical properties (the denial of Cartesian dualism) I shall call *the generality of physics*, and I shall take it to be partly constitutive of our (contemporary) idea of the physical that physics has this generality (see Crane 2001a: ch. 2). We believe that the laws of physics apply unrestrictedly across the universe; there are no regions where these laws fail or break down, and there are no kinds of entities that are immune from the effects of gravity and so on. But for the laws to have this generality, then all the objects to which they apply must have the kinds of properties which these laws concern: physical properties. Everything in space-time has (or has parts that have) these properties: for example, mass, temperature, electric charge, and so on.

The generality of physics is consistent with there being many kinds of objects: some objects, for example, are mental, some biological or organic, and some are social. These objects pose no problem for the generality of physics, so long as these objects or their parts have uncontroversially physical properties. A mental object is just an object with a mental property, a social object one with a social property, and so on. So, in particular, the generality of physics is compatible with the existence of emergent properties, defined above as: the supervenient properties of a thing not identical to any properties of its parts, and where the supervenience has no explanation in physical terms.

It is for this reason that I insist that anything worth calling the name 'physicalism' must be a reductive or reductionist view about the mental, *either* in the sense that it requires that the mental be metaphysically 'grounded in', 'realized by' or identified with physical phenomena, *or* in the sense that it requires that the mental be *explained* in physical terms.

This is the distinction which in previous work (Crane 2001a: ch. 2) I described as the distinction between 'ontological reduction' and 'explanatory reduction'. The distinction can be summarized as follows:

(OR) *Ontological reduction*: All entities (objects, properties, relations, facts, etc.) belong to a subclass of the class of physical entities.

(ER) *Explanatory reduction*: All truths (particular truths, or general theoretical truths or laws) can be explained in principle in terms of broadly physical truths.

A classic example of OR in the philosophy of mind is Davidson's (1970) anomalous monism, which identifies the class of mental events as a subclass of physical events. Another classic example is D. M. Armstrong's (1968) type identity theory, which identifies the class of mental properties as a subclass of the physical properties (by means of an identification of both as the states apt for bringing about a kind of behaviour).

A classic example of ER is the Armstrong/Lewis reduction of mental concepts to functional role concepts. On this view, mental concepts imply certain generalizations about the typical causes and effects of the properties or states to which those concepts refer. If this is right, then the fact that someone is in a certain mental state can be explained in terms of their being in a state which is apt for bringing about certain states of affairs, given certain other input from other mental states or from the environment (see Lewis 1972). Since the proposal is that these causal truths can be characterized in non-mental terms, we have a reduction of the mental to the non-mental.

It is clear from these examples that the two kinds of reduction are independent of one another. Armstrong and Lewis hold both. But some physicalists hold OR without holding ER: Davidson (1970) is an obvious example. And some hold ER without holding OR: a recent example is Melnyck (2004), but see also Smith (1992). Those who hold ER without holding OR typically say that they have no concern to identify mental properties with physical properties—perhaps because they are persuaded by Hilary Putnam's 'multiple realization' argument. But nonetheless, they think that the relationship between the mental and the physical needs to be explained: it cannot be an inexplicable mystery.

5. THE CHALLENGE OF THE EXPLANATORY GAP

The upshot of the previous section is that any doctrine worthy of the name of physicalism holds either OR, or ER, or both. If this is right, then the problem becomes apparent for those non-reductive physicalists mentioned in section 2 above, who respond to the challenge of the explanatory gap by saying that CH or DE is false (Byrne 1999; Block and Stalnaker 1999). These philosophers reject Jackson's requirement of cosmic hermeneutics, and they also reject Levine's demand for a deductive-nomological explanation of consciousness. Yet unless they give some other explanatory account of the relationship between the physical and the mental, then this form of physicalism collapses into emergentism.

To put the matter another way, both non-reductive physicalism and many traditional emergentists (e.g. Broad 1929) hold the following two doctrines:

(SP) Any minimal physical duplicate of the actual world is a duplicate in every respect.[5]

(Not-OR) It is not the case that all entities (objects, properties, relations, facts, etc.) belong to a subclass of the class of physical entities.

Not-OR implies that there are some non-physical entities: for example, non-physical properties. By 'non-physical properties', I simply mean properties that do not figure in physical theories. Non-reductive physicalists hold not-OR because of their denial of the identity theory. But I argued above that physicalists must either hold OR, or ER, or both. So the difference between physicalism and emergentism must come down to their attitude to ER:

(ER) All truths (particular truths, or general theoretical truths or laws) can be explained in principle in terms of broadly physical truths.

The emergentist, of course, must deny ER. But I hope it is obvious now why a non-*ontologically* reductionist physicalist must accept it. For to deny ER is to hold that there is no explanatory reduction of the mental in physical terms. This amounts to accepting that the connection between the physical and the mental is a brute, inexplicable necessity. But this was supposed to be the characteristic thesis of emergentism. So if non-ontologically reductionist physicalists do not accept ER, then their physicalism collapses into emergentism.

Jackson's position can now be seen as a challenge to other forms of physicalism: either give your own solution to the explanatory gap, or become an emergentist. Recall that Jackson argued that (a) the necessity involved in SP cannot be 'brute'; and (b) that it must be explicable as conceptual necessity, in terms of an a priori

[5] I am assuming here that emergentism holds the supervenience in question to be necessary; this is clear in Broad 1929 and the argument can be found in McLaughlin 1992 and Horgan 1993.

analysis of mental concepts. Emergentism, as we have seen, denies (a) and (b). Physicalists may reject (b), along with their rejection of cosmic hermeneutics (CH), but they cannot reject ER, on pain of abandoning what is essential to physicalism.

What physicalists need to do, then, is to give another kind of explanatory account of the relationship between consciousness and the physical world. The existence of statistical models of explanation or teleological forms of explanation, for example, shows that explanation need not fit the deductive-nomological form. Not all explanation is deductive-nomological. I think this is what Brian Loar is getting at when he writes that 'it is a mistake to think that, if physicalism is true, consciousness as we conceive it at first hand needs explaining in the way that liquidity as we ordinarily conceive it gets explained' (Loar 1997: 609). Note that an identity theory itself might be such an explanation, so long as asserting an identity between A and B is a way of explaining why A is B. This is, however, somewhat controversial.

In an insightful passage at the end of his book on pain, Nikola Grahek opposes Levine's claim that there is an explanatory gap:

> The fact that we can conceive such states of affairs [as pain without injury] will not signal—contrary to Levine—the presence of an explanatory gap; it will not reveal the unintelligibility of the connection between pain and injury . . . That we can conceive of injury without pain as well as pain without injury is rather to be explained by the simple fact that they are, to use traditional terminology, distinct existences. . . . But this does not mean at all that pain and injury are arbitrarily stacked together; that their relationship is mysterious or unintelligible because we do not see at all why they should be tightly connected. It only speaks against the identification of pain with injury and merely shows that pain cannot be *a priori* analysed in terms of its causal role . . . The general lesson to be learned from these considerations is that the *a priori* analysis of phenomenal concepts in functional terms is not a prerequisite for adequate or intelligible psychophysical explanations. (Grahek 2001: 150)

It seems to me that what Grahek says about the relationship between pain and injury can be applied also to the case of pain and its neurophysiological basis. Non-reductive physicalists will want to join Grahek in rejecting the need for an a priori analysis of phenomenal concepts in functional terms. But they cannot rest here if they want to avoid emergentism: they should also join Grahek in looking for empirical explanations of why pain and its neural basis are not 'arbitrarily stacked together'.

So the explanatory gap does present a real challenge to physicalism; but not exactly in the way that Levine thinks. Levine thinks that the explanatory gap shows physicalism to be incomplete. Without a solution to the explanatory gap, physicalism might be true, but it would be unintelligible. The argument of this chapter has been, rather, that without closing the explanatory gap, there would be no way that a non-ontologically reductive physicalist could remain a physicalist. Although neither deductive explanation (DE) nor cosmic hermeneutics (CH) should be required for a physicalist account of the mental, physicalists cannot simply deny the need to give *some* kind of explanatory reduction of the mental. Rather, they must give some other explanatory reduction if physicalism is not to collapse into emergentism. They need to give an account that is consistent with and explains the necessity of (SP). So far, no such account has been forthcoming. But the tenability of non-reductive physicalism ultimately depends on giving such an account, and with it a solution to the problem of the explanatory gap.

REFERENCES

Armstrong, D. M. 1968. *A Materialist Theory of the Mind*. London: Routledge and Kegan Paul.

Block, N. and Stalnaker, R. 1999. 'Conceptual Analysis, Dualism, and the Explanatory Gap'. *Philosophical Review* 108: 1–46.

Broad, C. D. 1929. *The Mind and its Place in Nature*. London: Routledge and Kegan Paul.

Byrne, A. 1999. 'Cosmic Hermeneutics'. *Philosophical Perspectives* 13: 347–83.

Chalmers, D. 1996. *The Conscious Mind: In Search of a Fundamental Theory*. Oxford: Oxford University Press.

Crane, T. 2001a. *Elements of Mind*. Oxford: Oxford University Press.

Crane, Tim 2001b. 'The Significance of Emergence'. In C. Gillett and B. Loewer (eds), *Physicalism and its Discontents*. Cambridge: Cambridge University Press, 207–24.

Davidson, D. 1970. 'Mental Events'. In L. Foster and J. Swanson (eds), *Experience and Theory*. London: Duckworth, 79–101.

Grahek, N. 2001. *Feeling Pain and Being in Pain*. Oldenburg: Hanse Institute for Advanced Study.

Hempel, C. G. 1966. *Philosophy of Natural Science*. Englewood Cliffs, NJ: Prentice Hall.

Horgan, T. 1984. 'Supervenience and Cosmic Hermeneutics'. *Southern Journal of Philosophy* 22 (Spindel Conference Supplement on Supervenience): 19–38.

—— 'From Supervenience to Superdupervenience: Meeting the Demands of a Material World'. *Mind* 102 (1993): 555–86.

Jackson, F. 1998. *From Metaphysics to Ethics*. Oxford: Oxford University Press.

Kim, J. 1995. 'Supervenience'. In S. Guttenplan (ed.) *A Companion to the Philosophy of Mind*. Oxford: Blackwell, 575–87.

—— 1998. *Mind in a Physical World*. Cambridge, MA: MIT Press.

Levine, J. 1983. 'Materialism and Qualia: the Explanatory Gap'. *Pacific Philosophical Quarterly* 64: 354–61.

—— 1997. 'On Leaving out What it's Like'. In N. Block, O. Flanagan, and G. Güzeldere (eds), *The Nature of Consciousness*. Cambridge, MA: MIT Press, 543–56.

—— 2001. *Purple Haze*. Oxford and New York: Oxford University Press.

Lewis, D. 1972. 'Psychophysical and Theoretical Identifications'. *Australasian Journal of Philosophy*, 50: 249–58.

—— 1995. 'Reduction of Mind'. In S. Guttenplan (ed.), *A Companion to the Philosophy of Mind*. Oxford: Blackwell, 412–31.

Loar, B. 1997. 'Phenomenal States'. In N. Block, O. Flanagan, and G. Güzeldere (eds), *The Nature of Consciousness*. Cambridge, MA: MIT Press, 597–616.
McLaughlin, B. 1992. 'The Rise and Fall of British Emergentism'. In A. Beckermann *et al*. (eds), *Emergence or Reduction?* Berlin: De Gruyter, 49–93.
Melnyk, A. 2004. *A Physicalist Manifesto*. Cambridge: Cambridge University Press.
Nagel, E. 1963. 'Wholes, Sums and Organic Unities'. In D. Lerner (ed.), *Parts and Wholes*. New York: Free Press, 135–55.
Nagel, T. 1974. 'What is it Like to be a Bat?'. *Philosophical Review* 83: 435–50.
Pettit, P. 1993. 'A Definition of Physicalism'. *Analysis* 53: 213–23.
Smith, P. 1992. 'Modest Reductions and the Unity of Science'. In D. Charles and K. Lennon (eds), *Reduction, Explanation and Realism*. Oxford: Oxford University Press, 19–44.
Tennant, N. 2007. 'Mind, Mathematics and the *Ignorabimusstreit*'. *British Journal for the History of Philosophy* 15: 745–73.

3

Explanation, Emergence, and Causality: Comments on Crane

Michele Di Francesco

INTRODUCTION

Tim Crane's chapter on emergence and the explanatory gap (this volume, Chapter 2) is a valuable contribution to our understanding of the relation between physicalism, explanation, and emergentism, and offers a new reading of the significance of the 'explanatory gap' (Levine 1983)—a reading which amounts to a strong attack on non-reductive physicalism. Crane's main thesis is that non-reductive physicalism either can close the explanatory gap, addressing the challenge posed by Levine's argument, or it becomes identical to emergentism. But since no way of closing the gap is available, the result is that there cannot be an interesting philosophical position intermediate between physicalism and emergentism.

In this chapter I argue that if we look at the relation between physicalism and reductionism from the vantage point of reduction Crane's analysis is rather persuasive. However, if we switch from reduction to causality, its conclusions appear to be more doubtful. I shall proceed in the following way: first I focus on a few points in Crane's overall strategy, and in particular on (1) the thesis that the supervenience of the mental cannot be accepted as a 'brute fact' by a physicalist; (2) his reading of the idea that emergent properties are 'not added from outside'. Then I explore the possibility of the existence of an interesting form of non-reductive physicalism. To do this, I shift from 'novelty and explanation' to 'novelty and causality' as key features of emergent properties. I introduce a distinction between two kinds of emergentism (moderate and radical) based on their attitude towards the causal inheritance principle. By accepting the causal inheritance principle, (some versions of) moderate emergentism, I claim, may be considered a form of non-reductive physicalism.

1. CRANE'S OVERALL STRATEGY

According to Crane, the real significance of the explanatory gap lies in the challenge it creates for 'a proper formulation of non-reductive versions of physicalism':

> Non-reductive physicalists have attempted to distinguish their theories from reductive versions of physicalism on the one hand, and 'emergentism' on the other. But their standard response to the explanatory gap fails to distinguish their non-reductive physicalism from emergentism. Since whatever emergentism is, it is not physicalism, this is a problem for the proper formulation of physicalism. (Crane this volume: 23)

I shall not discuss Crane's overall strategy; rather I shall concentrate on a couple of points: one about physicalism, the other about emergence. The first one is the thesis that any position worth the name of physicalism cannot be content to accept as a brute fact the supervenience thesis (SP: 'every minimal duplicate of the actual world is a duplicate in every respect'—Crane this volume: 22). As Crane says elsewhere (2001a: 66), the thesis that there is no way to explain the relation between the emergent properties and their 'bases' characterizes emergentism rather than physicalism. I think that Crane is formulating here a crucial point for anybody interested in the possibility of a meaningful notion of non-reductive physicalism. In the discussion of this issue, however, we should give due attention to a second element, stressed by Crane himself: the fact that emergent properties are not 'added from outside'. What is at stake, I believe, is the dialectic between two elements that are constitutive both of non-reductive physicalism and emergentism: dependence and autonomy.[1] If a difference between (a non-empty notion of) non-reductive physicalism and emergentism can be found, it probably would depend on the balance between these two notions.

In order to look more closely at the question, we should ask how a physicalist could give sense to the claim that SP cannot be taken as a brute fact. The answer is that SP should be explained without taking it as 'metaphysically fundamental'(Horgan 1984: 21): it cannot be that, besides the basic entities and principles postulated by physics, we should admit other non-physical realities. If this were the case, physicalism would be false. Following Horgan (1984) and Jackson (1998), Crane introduces the idea that physicalism is committed to cosmic hermeneutics (CH): 'there can be an a priori deduction of all the truths about the world from the microphysical truths about it plus the conceptual truths about non-physical concepts' (Crane this volume: 24). Of course, CH is a rather radical thesis; to give the most obvious example, the recent debate

[1] Crane (2001b) refers to 'dependence' and 'distinctness' as characters of mental properties according to non-reductive physicalism. I prefer the term 'autonomy' just to stress the contrasts between these two competing requests.

about phenomenal consciousness strongly suggests there are aspects of mental phenomena that simply defeat any attempt to deduce them from physical truths.

In fact, this is why non-reductive physicalism is welcome. Non-reductive physicalists repudiate ontological reduction: mental properties cannot be reduced to physical properties, even if they supervene on the physical. But supervenience as a brute fact is not enough. This is the point of the difference between physicalism and emergentism. Emergentism may content itself with the generality of physics thesis (GP: all objects have physical properties—the laws of physics apply without limitation to everything), together with the assumption as a brute fact of the supervenience thesis. But to be physicalist, we have to explain supervenience. Putting aside ontological reductions, the remaining option is explanatory reduction (ER): the thesis that 'all truths (particular truths, or general theoretical truths or laws) can be explained in principle in terms of broadly physical truths' (Crane this volume: 30). But now the explanatory gap becomes relevant, since in order to adopt ER we need to close it. Typically, non-reductive physicalists deny ER (they believe that the gap cannot be closed by a reductive strategy), so they owe us some other way to close the gap if they want to avoid the collapse into emergentism. But since no other way is in fact available, the collapse threat is incumbent.

2. EMERGENT PROPERTIES ARE NOT 'ADDED FROM OUTSIDE'

To give an appropriate evaluation of the threat, we start noticing that in the case of emergentism, too, we find a tension between dependence and autonomy. Emergent properties, in fact, are both novel and not added from outside (in the way postulated, i.e. by Cartesian substance dualists). Putting aside novelty, let's focus on the idea that emergent properties are not added from outside. This idea implies SP (as Crane acknowledges), but SP is not enough to qualify a position as emergentism. What we need, as in the case of non-reductive physicalism, is an explanation of SP. Here, however, lies an important difference between non-reductive physicalism and emergentism. According to the former, SP should be explained 'within physics'; according to the latter it can be explained by non-physical correlation laws—whose existence has to be taken as a brute fact. This analysis explains both similarities and differences between (a) the idea that supervenience has to be explained within physics and (b) the idea that the new emerging properties are not added from the outside. Both appear as expressions of the request of dependence, but the former requires a stronger commitment to dependence, which may be associated with the idea that physics is explanatorily more basic than higher-level sciences. To quote Horgan again, 'any metaphysically basic facts or laws—any unexplained explainers, so to speak—are facts or laws within physics itself' (Horgan 1993: 560). We may call this view physical fundamentalism; physical sciences are explanatorily fundamental because physics provides the basis of true metaphysics. The next question we should address is whether non-reductive physicalism is committed to physical fundamentalism.

3. FROM REDUCTION AND EXPLANATION TO CAUSALITY

So far we have referred to emergentism as a unitary notion, and we have compared it with physicalism in terms of reduction and explanation, but a different approach is possible. In particular, in the next pages, I would like to try to show that, if we first shift from reduction and explanation to causality, and then we distinguish between moderate and radical emergentism, there may be a way to characterize a form of non-reductive physicalism, intermediate between physical fundamentalism and full-blooded emergentism—a form of non-reductive physicalism that takes physics as ontologically but not causally fundamental.

Speaking of causality is justified. Brian McLaughlin (1992: 50) describes emergentism as 'a view about the causal structure of reality'. It says that reality is structured on different levels and, when a given level reaches a certain degree of complexity new causal powers emerge. According to Jaegwon Kim, there are two groups of ideas associated with emergentism: the first group refers to the irreducibility of the emergent properties; the second focuses on the fact that emergent phenomena 'bring into the world new causal powers of their own, and, in particular, that they have powers to influence and control the direction of the lower processes' (Kim 1999: 5–6).

Now, it can be illuminating to apply to causation the idea we discussed with reference to reduction and explanation, namely, that (a) higher-level features of the world are not added from outside, but supervene on a physical basis, and that (b) this supervenience needs explanation. Physicalism and emergentism, I claim, differ to the extent of the required explanation. Or better, we may think of a continuum of 'explanations', from physical reductive explanation up to the mere reference to SP as a brute fact.

Let us then ask how emergentism can explain the existence of causally efficacious emergent properties. First, the emergence of the mind from physical phenomena is no miracle; no extra-ingredient need be added to the usual physically constituted stuff we meet in the natural sciences to produce (for example) a mind, a self, or a person. All the ingredients for the new emergent phenomena must be present at the 'basic level', but they manifest themselves only at the emergent level. They are visible and they can be detected only a posteriori, but—since emergentism is not dualism—they must be there from the very beginning. How can this be? A possible answer exploits the idea of causal inheritance. Kim (1999) proposes his 'causal inheritance principle' in terms of identity through realization.[2] I am not going into the details of Kim's proposal; what we need is an explanation of how emerging causal powers are grounded on basic levels. Identity through realization is of course a good explanation, but I am not committed to the exclusion of other forms of explanation. So I shall take causal inheritance in quite a loose way, as the claim that higher-order causal powers depend on 'basic'-level causal powers. What is important is that if we ask how dependence on basic causal interactions occurs—if we look for explanation of inheritance—we may have a plurality of answers. Moreover, we may take the causal inheritance principle as a sort of litmus test to distinguish two kinds of emergentism: moderate emergentism and radical emergentism:

> Moderate emergentism accepts the causal inheritance principle: the causal powers of the emerging properties are the product of the causal powers of the 'basic' properties; that is, the new causal relations among the elements of the emergent level originate out of properties, which belong to the basic level. Basic/physical level is causally fundamental.
>
> Radical emergentism denies causal inheritance: higher-level systems exhibit new kinds of causal organization. Such an organization should be understood as the result of the action of new kinds of properties, whose existence not only is not detectable at the basic level, but should be considered as new emergent features of the (causal organization of) the world.

Radical emergentism seems to imply ontological causal pluralism; moderate emergentism is better conceived as a combination of ontological causal monism plus epistemological causal pluralism—an attempt to bring together a physicalistic ontology and the plurality of explicative styles, conceptual tools, and languages we find in our current investigations of reality.[3]

If we deny the causal inheritance principle, we have a full-blooded form of emergentism. New causal powers are truly novel, and the involved downward causation is independent of causal intercourses at basic level. Radical emergence is a kind of emergence that undermines the (causal) unity of the world we associate with physicalism. And it is obviously at odds with physical fundamentalism.

Moderate emergentism is different; even if the new causal powers are not 'mechanistically reducible' in terms of their underlying properties (emergentism is not reductionism), the new causal powers are inherited from the basic level, which is the fundamental level, where the true causal intercourse is grounded. This means that—to a certain extent—we may ask how inheritance takes place

[2] Kim (1999: 16) writes: 'If a functional property E is instantiated on a given occasion in virtue of one of its realizers Q, being instantiated, then the causal powers of this instance of E are identical with the causal powers of this instance of Q.'

[3] Cf. Di Francesco (2005) for an analysis of the relation between emergentism and causal pluralism.

and why the conceptual resource of lower levels cannot be sufficient to explain the new emergent features.

For example, Sydney Shoemaker (2002) suggests that the elements of an emergent whole have latent properties that cannot be deduced from their manifest properties. While manifest properties are both present and detectable at the 'basic' level, latent properties are present but not detectable. They are grounded on the base level, but still not realized until the system in which they occur acquires the required complexity (Shoemaker 2002: 54). Obviously what makes latent properties interesting is that they explain why emerging systems have causal powers inherited but not predictable from manifest causal powers of their parts. Now, if we accept that latent properties are physical properties, there is a sense in which such a variety of moderate emergentism accepts, at least partially, physical fundamentalism, and may probably be considered as a form of non-reductive physicalism—a form of non-reductive physicalism that accepts an epistemic but not an ontological reading of causal pluralism (latent properties cannot be detected, but are still present at physical level).

However, that latent properties are physical can be disputed, since they are individuated only with reference to the emerging system in which they are detectable.[4] Were this the case, the commitment to physical fundamentalism would be minimal—and it would be imprudent to define such a position as physicalism—since it appears to be committed to ontological causal pluralism. So, if we look at the relation between causal inheritance and explanation, we have four positions:

(1) Causal inheritance depends on physical causal interactions, and in principle can be explained within physics. Causal pluralism is rejected.
(2) Causal inheritance depends on physical causal interactions; but to be explained it requires reference to latent properties. An epistemic reading of causal pluralism is tolerated.
(3) Latent properties are 'non-physical' properties already present at basic level; in this sense higher-order causal powers are inherited—but they do not originate from physical properties (alone).
(4) Causal inheritance does not occur. Higher-order causal powers are grounded on physical phenomena; but such grounding is a brute fact.

Positions (1) and (4) are neat ones: respectively reductive physicalism and radical emergentism. Positions (2) and (3), I do acknowledge, are vaguer and perhaps unstable, whose difference may be criticized. And the same holds for (3) and (4): it may be difficult to defend their difference.

Even if vague and in need of specification, the distinctions I propose can be of interest (I hope) to show that, when we investigate the psychophysical nexus,

[4] Shoemaker himself examines this question (2002: 60); cf. also Di Francesco (2005: 112–14).

there is an alternative to an 'all or nothing' strategy to close the explanatory gap.

In fact, if we look at the plurality of (philosophical and) scientific contexts in which we consider the general problem of the relation between different levels of reality, we find a good deal of variability: explanatory pluralism, coevolution of theories, and interfield integration are the norm.[5] Perhaps the kind of emergence involved in the hard problem of consciousness does not leave room for midway positions between full-blooded emergentism and reductive physicalism. In other fields of current scientific research, however, there are ways to tackle the interlevel connections which may suggest other ontological relations (from non-reductive physicalism to moderate emergentism). If we accept the existence of causal pluralism as a fact of our scientific practice, it is easy to find a plurality of explanatory gaps, so to speak, with different stories and geographies. Sometimes the gap will be accounted for by reference to reduction or elimination, sometimes other conceptual frameworks will be closer to the data: non-reductive physicalism, moderate emergentism, radical emergentism may be useful. To impose a single metaphysical orthodoxy may be good for our aesthetic sense and our love for ordered and clean distinctions, but perhaps we live in a messier world than we like to believe. If this is true, the best way to describe it may require a continuum from full-blooded physicalism to radical emergentism in which non-reductive physicalism represents a non-empty position.

REFERENCES

Bechtel, W. and Hamilton, A. 2007. 'Reductionism, Integration, and the Unity of the Sciences'. In T. Kuipers (ed.), *Philosophy of Science: Focal Issues* (vol. 1 of the *Handbook of the Philosophy of Science*). New York: Elsevier.

Crane, T. 2001a. *Elements of Mind*. Oxford, New York: Oxford University Press.

—— 2001b. 'The Significance of Emergence'. In C. Gillett and B. Lower (eds), *Physicalism and Its Discontents*. Cambridge: Cambridge University Press, 207–24.

Di Francesco, M. 2005. 'Filling the Gap, or Jumping Over it? Emergentism and Naturalism'. *Epistemologia* 28: 93–120.

—— 2007. 'Menti. Varietà dell'emergentismo'. In A. Bottani and R. Davies (eds), *Ontologie regionali*. Milan: Mimesis, 123–40.

——, Motterlini, M. and Colombo, M. 2007. 'In Search of the Neurobiological Basis of Decision-making. Explanation, Reduction and Emergence'. *Functional Neurology* 22: 197–204.

Horgan, T. 1984. 'Supervenience and Cosmic Hermeneutics' *Southern Journal of Philosophy* 22 (Supplement on Supervenience): 19–38.

[5] For explanatory pluralism in neuroscience see Bechtel and Hamilton (2007); McCauley (2007); McCauley and Bechtel (2001). For the connection between special science pluralism and emergentism see Di Francesco, Motterlini and Colombo (2007).

Horgan, T. 1993. 'From Supervenience to Superdupervenience: Meeting the Demands of a Material World'. *Mind* 102: 555–86.

Jackson, F. 1998. *From Metaphysics to Ethics*. Oxford: Oxford University Press.

Levine, J. 1983. 'Materialism and Qualia: the Explanatory Gap'. *Pacific Philosophical Quarterly* 64: 354–61.

Kim, J. 1999. 'Making Sense of Emergence'. *Philosophical Studies* 95: 3–36.

McCauley, R. N. 2007. 'Reduction: Models of Cross-Scientific Relations and Their Implications for the Psychology-Neuroscience Interface'. In P. Thagard (ed.), *Handbook of the Philosophy of Psychology and Cognitive Science*. Amsterdam: Elsevier, 105–58.

——and Bechtel, W. 2001. 'Explanatory Pluralism and Heuristic Identity Theory'. *Theory and Psychology* 11.6: 736–60.

McLaughin, B. 1992. 'The Rise and Fall of British Emergentism'. In A. Beckermann, H. Flohr, and L. Kim (eds), *Emergence or Reduction? Essays on the Prospects of Nonreductive Physicalism*. Berlin, New York: De Gruyter, 49–93.

Shoemaker, S. 2002. 'Kim on Emergence'. *Philosophical Studies* 108 (2002): 53–63.

4

Is Non-reductive Physicalism Viable within a Causal Powers Metaphysic?

*Timothy O'Connor and John Ross Churchill**

Throughout the 1990s, Jaegwon Kim developed a line of argument that what purport to be *non-reductive* forms of physicalism are ultimately untenable, since they cannot accommodate the causal efficacy of mental states. His argument has received a great deal of discussion, much of it critical. We believe that, while the argument needs some refinement, its basic thrust is sound. In what follows, we will lay out our preferred version of the argument and highlight its essential dependence on a causal powers metaphysics, a dependence that Kim does not acknowledge in his official presentations of the argument. We then discuss a recent physicalist strategy for preserving the causal efficacy of the mental in the face of this sort of challenge, a strategy that endorses a causal powers metaphysics of properties while offering a distinctive account of the physical realization of mental properties. We argue that the resulting picture cannot be satisfactorily worked out, and that seeing why it fails strongly suggests that non-reductive physicalism and a causal powers metaphysics are not compatible, as our original argument contends.

1. A CAUSAL POWERS ONTOLOGY

Let us first explain what we mean by the term 'causal powers'. One way of using this term involves no definite commitments on the metaphysics of causation. A person using the term this way might say, for example, that a defoliant has the causal power to kill plants, where this claim is neutral as to whether (a) the

* Versions of this paper were read at Queen's University Belfast, St Louis University, the Catholic University of Milan, and the University of Durham. We thank the audiences on these occasions for helpful criticism and advice, especially Tim Crane, Carl Gillett, John Heil, Stephan Leuenberger, E. J. Lowe, Cynthia Macdonald, Graham Macdonald, Peter Menzies, Paul Noordhof, David Papineau, and Jessica Wilson.

presence of the defoliant would or might stand in an ontologically basic relation of producing or bringing about the death of a plant, (b) there is a law of nature that relates properties of the defoliant and plant death, (c) plants regularly die after being sprayed with the defoliant, (d) there are subjunctive conditionals relating the properties of the defoliant and the death of plants in a certain way, (e) citing the presence of the defoliant satisfactorily explains the occurrence of plant deaths in certain contexts, or some (f) fitting a still different analysis of causation.

We do not use the term in this neutral manner. Our usage corresponds to the first of these: a power to produce or to bring about some event, where this is assumed to be a real relation irreducible to more basic features of the world. Our favoured technical term for this is 'causal oomph'. So understood, causation is not amenable to analysis in non-causal terms, but instead involves the exercise of ontologically primitive causal *powers* or *capacities* of particulars. Powers are either identical to, or figure into the identity conditions of, certain of the object's properties, which are immanent to those things as non-mereological parts. (Whether one thinks of these as immanent universals or tropes is not crucial in this context.)

It bears emphasis that this view is *not* committed to assuming that all causation must amount to something like pushing/pulling or the exertion of a force. What is assumed, rather, is solely this: when an instance of a property—the event of the particular's having the property—is a cause, the world unfolds in a certain way after the instance of that property, and that property instance is one of the factors that jointly *make* the world unfold this way. This is just another way of saying what's come before, that the property instance and others jointly produce or bring about certain effects; they jointly oomph the world into going on in *this* way rather than *that*. Because of this, there are certain counterfactuals true of the world ('were the property not to have been instanced, such-and-such effects would not have occurred'). But these counterfactuals are derivative from, and not to be equated with, or seen as the basis of, the causal facts themselves: it is because the property instance was among the factors that jointly produced the relevant happenings that certain corresponding counterfactuals are true. Causally efficacious properties have the power to make the world unfold in ways that otherwise it would not, and this is a fundamental feature about these properties upon which all else (counterfactuals true of them, regularities and patterns that encompass them, explanations that cite them) is derivative.

There is much debate, and not a little confusion, over how to delineate the finer points of this general picture. While we cannot delve deeply into these matters, we make the following two remarks to forestall confusion that might infect understanding of our subsequent argument.

First, there is a pervasive manner of speaking that appears on the surface to say that *objects* have and exercise causal powers. (Witness our example above with respect to defoliants.) In our view, such talk should be construed by the causal powers metaphysician as a shorthand way of expressing the claims that:

(1) the object's having the property *is* its having the causal power;
(2) the event of the property's being had by the object over a particular interval of time and in appropriate circumstances will causally contribute to the effect, where it occurs; and
(3) the exercise of the causal power just is this causal contribution.[1]

Second, a single property may contribute to a very wide array of effects, depending on the context in which it is instanced. A particle's being negatively charged may contribute to its accelerating at varying rates away from a similarly charged nearby particle, accelerating towards an oppositely charged nearby particle, even accelerating towards a similarly charged particle (though at a slower rate than would occur were the particle not to have been so charged), and countless other manifestations, all depending on the context of its occurrence. But in ordinary speech, again, there is a tendency to talk of a corresponding array of causal powers being exercised, 'each' of which is identified through the effect actually manifested. This sort of speech has encouraged some metaphysicians to posit a multiplicity of properties, or worse, to posit a distinct type of entity (a causal power), any number of which are 'conferred by' a single property. We should resist such moves on grounds of parsimony, and here science is a much better guide to property/power identifications. The key is to understand a basic power or disposition not in terms of this or that salient manifestation, but rather in terms of a unitary causal influence, something that is constant across circumstances while its manifestations will vary.[2]

In considering the prospects for a non-reductive physicalist view of the mental, we are assuming, rather than arguing for, this causal powers metaphysics. We are investigating its implications for the question at hand. Can the (by our lights) right-thinking metaphysician who has seen his way clear to this view of causation make out a non-reductive physicalist view on which mental states are causally efficacious in this sense? We will try to persuade you that the prospects are bleak.

[1] One *might* hold to a philosophical view leading one to insist that in certain cases, it is indeed the object that exercises the power, and not the event of the object's having the property/causal power. Such is the claim of the agent causationist, e.g., with respect to the forming of a free decision. But this is a substantive and controversial thesis, not a spelling out for one sort of case what is common to every case of causation. (For a discussion of the relationship of agent causation to the more usual 'event causation' within a causal powers metaphysics, see O'Connor [2008].)

[2] We are influenced here by Corry (2002, 2008) and Heil (2003).

2. CAUSAL POWERS AND THE DILEMMA OF REDUCTION OR CAUSAL EXCLUSION OF THE MENTAL

We will now present our preferred version of a Kim-style argument.[3] We begin with three related premises concerning causation and properties:

(1) Causation is a real relation irreducible to more basic features of the world (*causal non-reductionism*).
(2) Causation involves the exercise of ontologically basic causal powers or capacities of particulars (*production account of causation*).
(3) Properties are individuated in terms of causal powers, such that there are no distinct properties that confer exactly the same causal profile (*causal theory of properties*).

The next four premises flow from the distinctive commitments of non-reductive physicalists:

(4) No mental property is identical to any physical property (*distinctness thesis*).
(5) Mental properties supervene on physical properties (*supervenience thesis*).

The hoary slogan, of course, is 'no mental difference without a physical difference', intended to capture an appropriate dependence relation. What exact form the supervenience relation should take in this context, however, is a difficult and controverted issue. We will follow Kim in supposing that complication arising, e.g., from mental content externalism can be safely ignored. If this is correct, we may assume for the sake of argument that mental properties 'strongly supervene' on the physical properties of the individual (or on the physical properties and relations of the individual's parts). Next we have:

(6) Mental properties are realized by physical properties: a particular event M of a person S's having mental property **M** is either 'constituted by' (a kind of ontological posteriority) or is identical to various physical particulars—possibly including portions of the person's environment—having certain physical properties and standing in certain physical relations (*realization thesis*).

We will be non-committal on whether the realization of mental properties by physical properties involves constitution or identity of the corresponding events, since non-reductive physicalists' pronouncements on this matter are varied and often obscure.[4] Finally, physicalists typically wish to assert:

[3] See Kim (1993, 1997, 1998, 1999, 2003, and 2005).

[4] See for example the variation among Fodor (1974); Pereboom and Kornblith (1991); Pereboom (2002); Shoemaker (2001, 2007); and Gillett (2002).

(7) For every *physical* event, its objective chance of occurring is fully fixed by physical events (*causal completeness of physics*).

According to (7), nothing non-physical is *required* in order to causally account for the occurrence of any physical event.

We now contend that (1)–(7) are inconsistent with supposing:

(8) There is a causally efficacious mental event, M, that is the instancing of a particular mental property, **M**. The causal activity of M is distinct from the activity of the physical event, P, that is the instancing of **M**'s realizer property (or properties), and this activity in one way or another impinges the realm of physical events[5] (*assumption for reductio*).

Premise (8) is an instance of the general claim that the causal efficacy of mental properties does not reduce in every case to the causal efficacy of some physical properties. The singular causal action of the mental event of **M**'s being instanced does not reduce to the singular causal action of some physical event or events, such as the instancing of the physical property **P** that realizes **M** in the circumstances.

The argument that (8) is inconsistent with (1)–(7) proceeds as follows:

(9) The instance of **M** either
(a) directly produces a subsequent mental event, M*, or
(b) it directly produces a wholly physical event, P*.

The realization thesis (6) and non-reductive-productive account of causation (1–2) together strongly suggest that option (a) is a non-starter. On this view, mental events are ontologically dependent on their subvening realizers, wholly constituted by (if not identical to) them, and this is no less true of mental *effects* as of mental *causes*. Bringing about such a mental event *eo ipso* involves causally affecting the physical event which realizes it. So

(10) Not (9a).

But the thesis of causal completeness (7) implies that:

(11) If 9b, then the physical event P* is overdetermined by M and some other physical event.

Now, if we accept the non-reductive-productive account of causation, it will seem passing strange to suppose that, in regular fashion, there are physical events that are systematically 'overoomphed' by distinct events, even if—indeed, *especially* if—these causes might stand in a supervenience relation. If, say, a physical event P, the realizer of the mental event M, produces or oomphs P*, what causal work

[5] We will, for the sake of convenience, continue to refer only to **P**, the single realizer of **M**, though it should be understood that on some accounts of realization **M** may be realized by multiple properties ('the **P**'s', say) each time it is instanced. Gillett (2002) is one such account.

is left over for M? Note that on reductive accounts of causation, on which causal facts are not something additional to the totality of non-causal facts, the situation looks very different. Suppose, for example, that our effect P* is counterfactually dependent on both P and M. If we accept something like the counterfactual analysis of causation, there is nothing strange or objectionable about deeming M, as well as P, to be a cause of P*. For in doing so we are not making a commitment to anything additional—M's status as a cause of P* falls out of the facts that we already accept, along with our analysis. It comes for free. By contrast, on the non-reductive-productive account, we would be positing an additional fundamental relation between M and P*, when doing so is entirely unnecessary for accounting causally for P*.[6] Thus, we should conclude that:

(12) There is not systematic mental-physical overdetermination, as the consequent of (11) implies.

But this is the end of the road. We are forced to conclude, therefore, that:

(13) M does not make a distinctive contribution to occurrences in the physical world, whether wholly physical or supervening mental occurrences (*completing reductio of (8)*).

Finally, the causal theory of properties (premise 3) both rules out an epiphenomenalist retreat and suggests the proper ultimate conclusion: we ought either to reductively *identify* **M** with **P** or *deny* that **M** is a bona fide property—one that earns its causal keep—in the first place.

Here is our argument laid out in compact form:

(1) Causation is a real relation irreducible to more basic features of the world (*causal non-reductionism*).
(2) Causation involves the exercise of ontologically basic causal powers or capacities of particulars (*production account of causation*).
(3) Properties are individuated in terms of causal powers, such that there are no distinct properties that confer exactly the same causal profile (*causal theory of properties*).
(4) No mental property is identical to any physical property (*distinctness thesis).*
(5) Mental properties supervene on physical properties (*supervenience thesis*).
(6) Mental properties are realized by physical properties: a particular event M of a person S's having mental property **M** is either 'constituted by' (a kind of ontological posteriority) or is identical to various physical particulars—possibly including portions of the person's environment—having certain physical properties and standing in certain physical relations (*realization thesis*).
(7) For every *physical* event, its objective chance of occurring is fully fixed by physical events (*causal completeness of physics*).

[6] These points are made clearly and effectively by Loewer (2001), a review of Kim (1998).

(8) There is a causally efficacious mental event, M, that is the instancing of a particular mental property, **M**. The causal activity of M is distinct from the activity of the physical event, P, that is the instancing of **M**'s realizer property (or properties), and this activity in one way or another impinges the realm of physical events (*assumption for reductio*).

(9) The instance of **M** either

(a) directly produces a subsequent mental event, M*, or
(b) it directly produces a wholly physical event, P*.

(10) Not (9a) (*from 1, 2, 6*).

(11) If 9b, then the physical event P* is overdetermined by M and some other physical event (*from 7*).

(12) There is not systematic mental–physical overdetermination, as the consequent of (11) implies (*from 1,2*).

(13) M does not make a distinctive contribution to occurrences in the physical world, whether wholly physical or supervening mental occurrences (*completing reductio of (8)*).

The argument just presented, like earlier relatives, seeks a reductionist or eliminativist conclusion by way of arguing for the *exclusion* of irreducibly mental causation. Yet it does this by explicitly invoking the thesis of causal powers realistically construed. So let us refer to it hereafter as the *power exclusion argument*.

The commitments that drive the power exclusion argument are tenets of the causal powers metaphysics, on the one hand, and non-reductive physicalism, on the other. If we wish to preserve a realist and non-reductive view of the mind and its causal influence, we must reject one or another tenet of these two packages. Or so we believe. Sydney Shoemaker, however, disagrees, and he has recently attempted to provide a way out for the non-reductive physicalist who is a realist with respect to causal powers. Since Shoemaker has bona fides as both a causal powers metaphysician and as a physicalist and has attempted to work out the metaphysics of realization and causation with much greater care than is usual in these discussions, it is fitting that we investigate his approach in detail.

3. SHOEMAKER ON NON-REDUCTIVE MENTAL CAUSATION

Shoemaker thinks that the key to vindicating the causal efficacy of mental properties without reduction lies in a distinctive account of the *realization* of mental properties by physical properties. In broad strokes, his proposal is that mental and other realized properties—those that do real causal/explanatory work—belong to a special class of *disjunctive* properties, with their disjuncts as their realizers: the relation of realizer to realized is simply the relation of disjunct to disjunction.[7] On this view, realized properties have a proper subset of each of their realizers' *forward-looking causal features*—what instances of the properties can causally suffice for—while having a superset of their realizer properties' *backward-looking causal features*—what can causally suffice for instances of the properties. Shoemaker then exploits the conclusion that realized properties have a subset of their realizers' powers to argue that mental causation is not reducible to causation by the physical realizers, owing to a certain proportionality thesis explained below concerning what counts as a cause of what. The following schema captures Shoemaker's picture.

B1 —> C1	B1 —> (C1 v C2)	C1 —> E1	C1 —> (E1 v E2)
B2 —> C2	B2 —> (C1 v C2)	C2 —> E2	C2 —> (E1 v E2)
	(B1 v B2) —> (C1 v C2)		(C1 v C2) —> (E1 v E2)

Figure 4.1. Bold font (**C1**) indicates a property, while regular font (C1) indicates a property instance, or the event of an object's having the property at a particular time. The '—>' denotes causal sufficiency. C1 and C2 represent instances of different realizing physical properties of (C1 v C2), the multiply realized property instance. B-type events are possible causal determinants of C-type events, while E-type events are possibly determined by C-type events, according to the patterns indicated. A realized event such as (C1 v C2) has more possible determinants than any of its realizers while it suffices for fewer effects.

How is accepting this picture of realization supposed to make things easier for non-reductive physicalism? Says Shoemaker: begin by observing that if the realized property has a subset of the forward-looking causal features of the realizer, then the realizer property instance is causally sufficient for everything the realized property instance is causally sufficient for, *plus more*. So, for example, C1 is causally sufficient for (E1 v E2), just as (C1 v C2) is, but unlike the latter it is also sufficient for an instance of E1. Now, if C1 and (C1 v C2) overlap in this way in what they causally suffice for, and if causal considerations ought to drive our conclusions about the identity of properties, a natural conclusion is that (C1 v C2) is a proper *part* of C1 (in the sense of entailment). More generally: the instances of realized properties are parts of the instances of the corresponding realizers and so are not identical to them.[8]

From here, Shoemaker invokes a version of Stephen Yablo's 'proportionality' constraint[9] on what we ought to count as the cause in a causal interaction: while it is true that C1 is causally sufficient for (E1 v E2), (C1 v C2) is, Yablo and Shoemaker say, a better candidate for being the cause. For (C1 v C2) is also causally sufficient for the specified effect, but only just so—it causally suffices for the effect *and nothing more besides*. The only features of C1 that contribute

[7] See Shoemaker 2007: section II of ch. 2, especially 17–18, 55–6, and section V of ch. 4, especially 79 and 82.

[8] See Shoemaker (2007: 13, 53).

[9] Shoemaker (2001: 81). See Yablo (1992).

to the bringing about of (E1 v E2) are features had by (C1 v C2), a part of C1. Compare, says Shoemaker, a more familiar sort of case: Jones fires a single shot as part of the volley of a firing squad, but his shot arrives just ahead of the others, killing the condemned. In the realization case as Shoemaker describes it, just as with the firing squad analogy, we are invited to conclude that while the whole (C1; the firing squad's firing) was causally sufficient for the effect ((E1 v E2); the death of the condemned), proportionality constraints argue in favour of counting a particular part of the event ((C1 v C2); Jones's firing) rather than the whole as *the cause*.[10] This is how realized events in general qualify as causes in certain scenarios. They do not overdetermine their effects alongside their realizers. Instead, they always cause effects which their realizers suffice for but do not cause.

We now have before us Shoemaker's account of realization and the way it provides for non-reductive mental causation. But how exactly, we may wonder, does the account underwrite a response to the power exclusion argument? Notice, first, that there is no rejection of the distinctness, supervenience, or realization theses (premises 4–6). What of (7), the causal completeness principle? Shoemaker insists on the importance of a controversial distinction between causal determination or sufficiency and causation proper. But this position is consistent with the way that we have formulated (7), and it seems clear that Shoemaker accepts it.

In our view, the *way* that Shoemaker applies the (mere) sufficiency/causation distinction to cases of realization is untenable. We argue for a disjunctive conclusion: either Shoemaker is soft-pedaling the irreducible efficacy of the mental (as embodied in premise 8) or his commitment to the causal powers metaphysics (as expressed in premises 1–3) is less than it appears.

To bring the problem into focus, consider first that, for all his distinctive claims, Shoemaker clearly gives ontological priority to the physical realizer event. He tells us that P realizes M just in case P is metaphysically sufficient for (but not identical to) M and 'constitutively makes it real' (2007: 4, 10). He goes so far as to say that realized states are 'nothing over and above' their realizers (2007: 2). If all this is so, then how is a case of M's causing an effect, E, not also a case whereby P, M's constituting realizer, is likewise causing E? Indeed, how is this not a case where P is causally prior to M, so that, by the power exclusion argument, we should conclude that P is the sole true cause?[11] We see two alternatives:

(A) What he terms causal 'determination' or 'sufficiency' is, at least in the context of realizer/realized events, metaphysically primary. His talk of

[10] See Shoemaker (2001: 81, 2007: 13–14, 52–3).

[11] A bolstering consideration comes from certain indeterministic scenarios. We take it to be evident that, *assuming the causal completeness of physics*, the chance of E given M cannot be greater than the chance of E given a total physical cause (here, our P). But there seems to be no reason to think that it cannot be less. Now consider a case where Pr (E/M) is significantly less than Pr(E/P). Surely, in such a case, where E in fact occurs, it is highly implausible to insist that nevertheless M, not P, is the cause of E. While this is a special case, if our conclusion about it is accepted, it seems to indicate that there is something wrong about Shoemaker's method for assigning causes.

'causation' in these contexts should be interpreted as something like 'causal relevance'—i.e., an *explanatory* relation. But if this is what he intends, then mental causation is not *metaphysically* irreducible, only explanatorily ineliminable for certain purposes. This sort of view has its defenders, and we shall say something about it below, when considering alternatives to non-reductive physicalism proper.

(B) If Shoemaker insists that mental causation as he describes it is no less metaphysical than basic physical causation, then we suspect some sort of retreat from a causal powers metaphysics, for only in this way could you say that P is somehow 'merely' causally sufficient whereas M is the proper cause. If P is ontologically prior to M, *able* to bring about E, and in the circumstances necessary to do so, how can it get out-oomphed by M?

We might try out a modified picture that gives up some of what Shoemaker claims in exchange for better prospects for a distinctive causal efficacy of the mental. For example, we might ignore Shoemaker's talk of P's being ontologically prior to and constitutively making real M and focus instead on his notion that M is a part of P, owing to the subset-of-powers thesis and the causal theory of properties. (In this reconstruction, we would emphasize Shoemaker's invocation of the firing squad analogy (2007: 53) and also such statements as 'It is only because the c-fiber [stimulation] instance realizer contains the pain instance realizer that it has the relevant effects' (2007: 48).) In the resulting picture, we would have what amounts to a radical inversion of the reductionist's vision, such that it is the *physical* properties that resolve into an assemblage of mental properties plus some non-mental causal features. That would allow for mental causation that is irreducible in one sense, at least: it does not reduce to causation by the corresponding *macro*physical events.

We think it would be very hard to make the picture a plausible one, though we set that worry aside. It is enough to note that making the view out will probably require us to analyse the associated macrophysical property as a structural property, the instancing of which just consists in, or is constituted by, the instancing of properties of the object's parts and relations between them. The mental property then comes out, on the view being considered, as an overlapping structural property, perhaps somehow abstracted from the full physical structural property. But if we do so, *both* mental properties and the larger structural physical properties in which they are embedded turn out to be derivative structures, entities that are constructions out of *micro*physical properties and relations. The spectre of reductionism menaces again.

Now, Shoemaker allows that macro-level properties are in a sense structural, but he resists their reductive *identification* with microphysical states of affairs. He suggests that instances of macro-level properties are microphysically realized, where this latter realization relation, like that of same-level realization, involves only constitution, not identity. However, his case for this rests on two claims about property identity that should be unacceptable to a causal powers theorist (2007: 48–9).

First, he lays down that, in general, a property instance has just one constituent object and one constituent property, so a mental property instance cannot be identical to a state of affairs involving many distinct properties and objects. He seems to put this forward as a definitional truth or platitude. But a causal powers theorist does not take quasi-grammatical considerations to be final arbiters concerning the structure of reality. One might just as well take Shoemaker's supposed platitude together with facts (assumed for now) about microphysical constitution and draw the conclusion that there are not, strictly speaking, mental properties at all—not in the sense of entities that contribute directly to how the world unfolds.

Shoemaker's second claim is that the *modal* properties of macro-level property instances and their microphysical realizers will generally differ. (Consider familiar claims made in discussions of the statue of Goliath.) This claim rests on intuitive judgements about possible variation in the material constitution of composite objects. But the status of composite objects, no less than that of 'their' properties, is very much in question on the powers metaphysics. One cannot simply assume that there are robustly *objective* modal facts about them and use these to ward off what otherwise appears to be a powerful reductionist challenge. For the causal powers theorist, the question of which candidates for being true (non-conventional) macro-level objects will turn on whether they manifest properties that play an ineliminable causal role. Shoemaker is alive to this issue, drawing a distinction between 'genuine' and 'phoney' properties (2007 ch. 4, sections 5–6) on the basis of what is 'directly detectable' by us. (We have a genuine macro-level property when we are able to reliably and directly detect its instantiation independently of its realizer.) Shoemaker freely acknowledges, however, that he has no argument that direct detectability suffices for genuineness. Nor does he say anything to allay the immediate worry that this will make the concept of genuineness to be epistemic, not metaphysical. Once again, we think he is forced to retreat to a defence that does not sit well with a metaphysics that revolves around causal powers.

It is time for a recap of our argument thus far. We have defended a power exclusion argument for the untenability of non-reductive physicalism. It is a variant of Kim's argument that makes explicit an assumption of a causal theory of properties. We then tried to show that Sydney Shoemaker's recent attempt to harmonize the two positions fails. Note that the causal theory of properties pretty directly entails a sparse, rather than abundant ontology of properties, as most predicates do not correspond to anything that makes a causal difference in the world. Shoemaker's strategy, in effect, is to try to make room for a less-than-austere, though still limited, inventory of properties, one that allows for irreducible macrophysical and mental properties, while being consistent with the physicalist's vision that everything supervenes on the microphysical. We believe that this strategy is bound to fail, since the causal theory requires that irreducible properties earn their keep, and there is no room for this at the macrophysical level if physicalism is true.

4. ALTERNATIVES TO NON-REDUCTIVE PHYSICALISM

The sparse metaphysics of causal powers forces a choice limited to three stances concerning macroscopic structures: reduction or elimination, if strict physicalism is maintained, or a rejection of one or more of the characteristic claims of physicalism—that is, towards the acceptance of an ontological variety of emergence.

Non-reductive physicalists see an obstacle to the first option, reductionism, in the fact that, as functional properties, intentional properties are multiply realized. What counts as a belief that *Q* in humans may be quite distinct, at any physical level of description, from what counts as that same belief in, say, an intelligent extraterrestrial or a sophisticated artificial machine built out of steel and silicon. Kim and some other reductionists recommend that we seek local, species-specific reductive identities for intentional properties—*human* belief that such-and-such as identical with physical property so-and-so—and so preserve the status of these intentional properties as causal powers. That is, we characterize both **M** and **P** in terms of *highly* specific mental and physical types, respectively, and move to a type-type identity theory.

The second, eliminativist option is to interpret apparent reference to mental properties as properly denoting mental *concepts* only. There are far fewer properties possessed by an object than the vast number of concepts it falls under. Genuine *properties* are immanent to their instances and make a non-redundant difference to how the objects act in at least some circumstances. As critics of Kim have observed, causal exclusion arguments appear to generalize beyond mental properties to all properties posited in the special sciences (sciences other than basic physics).[12] And since, contra Kim, it is highly plausible that special science categories are not typically ontologically reducible (owing in part to their own multiple realizability),[13] the argument ultimately leads (they say) to an eliminativist conclusion. This is often taken as a *reductio ad absurdum*: surely the terms of well-established biological and chemical theory pick out genuinely efficacious properties!

Now one response to this proposed *reductio* of the exclusion argument is to note the availability of a third alternative. Rejecting premise (6) and (7), the realization and causal completeness theses, suffices to block the final conclusion

[12] For discussion, see Baker (1993); Burge (1993); van Gulick (1993); Kim (1996, 1997, 1999, 2003, 2005); Block (2003); Ross and Spurrett (2004).

[13] See Fodor (1974); Dupré (1993); and Rosenberg (1994).

of the power exclusion argument. We will discuss shortly the viability of this strategy in relation to mental properties. We will not discuss whether this is viable as a general strategy for special science properties, though we observe that recent philosophy of science has seen a significant challenge to the completeness thesis in particular.[14]

Suppose that one takes the case for the completeness of physics with respect to some or all of the special sciences to be convincing, setting aside sciences impinging on mentality. In that case, contra Fodor and many others, it would not be absurd to embrace eliminativism. For so-called high level theories can be enormously useful and illuminating, and even necessary to the progress of human knowledge of how the world works, without answering to ontological 'levels' or layers populated by distinctive properties and their objects.[15] And the further fact that such theories are not generally reducible to more fundamental theories is a highly interesting one about our world (and necessary for science to get off the ground, as in practice we inevitably work our way in, not out), but it cuts no ontological ice. An alternative to the levels picture of physical reality has already been hinted at above: there is a vast array of microphysical entities (for simplicity, 'the particles') bearing primitive, dynamical features and standing in primitive relations. Talk of composite objects and their properties, at least in the general case, is the imposition of a conceptual scheme that selectively picks out coarse-grained patterns running through the vast storm of particles. These concepts really are (*objectively*) satisfied by the world, but not in virtue of a one-one relation between general concepts and properties, or individual concepts and particulars.

This second, eliminativist response to the powers exclusion argument might be thought to entail, implausibly, the devaluation of the special sciences. Such a conclusion would be too hasty, however. For it is simply false that science is of value only as a source of representing the world's causal joints in more and more accurate ways. It is, in addition, a source of means for intervening in and manipulating the world so as to change it for the better, and much of its value is due to this rather than to its representational fruits. We value science—we fund it, prioritize it, give special social status to many of its practitioners, etc.—because of its role in improving the world, and not just because of its role in representing the world. (The development of methods for effectively preventing and treating myriad diseases serves as just one example of such improvement.) But qua sources of improvement, some of the special sciences are at least as valuable, and perhaps more so, than fundamental physics. For very often we are *better* able to intervene and manipulate in ways that improve the world by using the resources of the non-fundamental special sciences.

[14] See Cartwright (1999) and Dupré (1993, 2001). And for a powerful challenge to the case for completeness in the special scientific domain of chemistry in which it is widely thought to be most secure, see Hendry (2006).

[15] On this point, see Heil (2003, chs 2–7).

So much for the first two alternatives left open by the power exclusion argument. Though we have suggested that eliminativism, in particular, is a viable approach to the predicates that appear in many of the special sciences, we believe that, when it comes to mentality, both reduction and elimination are implausible.

To secure a robust efficacy for mental properties, we should reject not only causal completeness but also mental-physical realization. Such are the negative commitments of what we call *ontological emergence*. (Carl Gillett's so-called *strong* emergence rejects completeness only. We have argued elsewhere that this position is too weak to secure a causal role for putative emergent properties.)

The term 'emergence' is used to cover a multitude of sympathies (in some cases, sins). So we want to indicate in clear, albeit very abstract, terms what an emergentist picture would look like, in our way of thinking.

Properties are *ontologically emergent* just in case:

(i) They are ontologically basic properties (token-distinct from, and unrealized by, any structural properties of the system).

(ii) As basic properties, they constitute new powers in the systems that have them, powers that non-redundantly contribute to the system's collective causal power, which is otherwise determined by the aggregations of, and relations between, the properties of the system's microphysical parts. Such non-redundant causal power necessarily means a difference even at the microphysical level of the system's unfolding behaviour. (This is compatible with the thesis that the laws of particle physics are *applicable* to such systems. It requires only that such laws be supplemented to account for the interaction of large-scale properties with the properties of small-scale systems.)

In respects (i) and (ii), emergent properties are no less basic ontologically than unit negative charge is taken to be by current physics. However, emergent and microphysical properties differ in that

(iii) emergent properties appear in and only in organized complex systems of an empirically specifiable sort and persist if and only if the system maintains the requisite organized complexity. The sort of complexity at issue can be expected to be insensitive to continuous small-scale dynamical changes at the microphysical level.[16]

We are inclined to further suppose that

(iv) the appearance of emergent properties is *causally originated and sustained* by the joint efficacy of the qualities and relations of some of the system's

[16] Concepts of emergence have a long history—one need only consider Aristotle's notion of irreducible substantial forms. Their coherence is also a matter of controversy. For an attempt to sort out the different ideas that have carried this label, see O'Connor and Wong (2002). And for a detailed exposition and defence of the notion we rely on in the text, see O'Connor and Wong (2005).

fundamental parts. (This would involve fundamental properties having latent dispositions to contribute to effects, dispositions that are triggered only in organized complexes of the requisite sort.)

One cannot give uncontroversial examples of emergent properties, of course. Though there are ever so many macroscopic phenomena that seem to be governed by principles of organization highly insensitive to microphysical dynamics, it remains an open question whether such behaviour is nonetheless wholly determined, in the final analysis, by ordinary particle dynamics of microphysical structures in and around the system in question.[17] Given the intractable difficulties of trying to compute values for the extremely large number of particles in any medium-sized system (as well as the compounding error of innumerable applications of approximation techniques used even in measuring small-scale systems), it may well forever be impossible in practice to attempt to directly test for the presence or absence of a truly (ontologically) emergent feature in a macroscopic system. Furthermore, it is difficult to try to spell out in any detail the impact of such a property using a realistic (even if hypothetical) example, since plausible candidates (e.g., phase state transitions or superconductivity in solid state physics, protein functionality in biology, animal consciousness) would likely involve the simultaneous emergence of multiple, interacting properties. Suffice it to say that if, for example, a particular protein molecule were to have emergent properties, then the unfolding dynamics of that molecule *at a microscopic level* would diverge in specifiable ways from what an ideal particle physicist (lacking computational and precision limitations) would expect by extrapolating from a complete understanding of the dynamics of small-scale particle systems. The nature and degree of divergence would provide a basis for capturing the distinctive contribution of the emergent features of the molecule.

Now, many contemporary philosophers seem to think that such a view is too extreme to be plausible. When pressed, such critics often cite the alleged consequence that an emergentist view compromises *the unity of nature*. But unity does not require the reductionist vision of the world as merely a vast network binding together local microphysical facts, with a pervasive and uniform causal continuity underlying all complex systems. It is enough that at every juncture introducing some new kind of causally discontinuous behaviour, there is a causal *source* for that discontinuity in the network of dispositions that underlie it. In short: unity in the order of the unfolding natural world need not involve causal continuity of behaviour, only continuity of dispositional structure.[18] For

[17] For numerous examples of such phenomena, see Laughlin et al. (2000).

[18] This is not to concede that it is *ipso facto* a theoretical virtue for a metaphysics that it entails greater unity in nature, nor that it is *ipso facto* a theoretical vice if the converse is true. The issue of the unity of nature, and the related issue of unity in science, is deep and complex. Our point in the text is that there is *a* kind of unity in nature if the emergentist account I have proposed is correct. For more on the topics of unity in science and nature, see Cat (2007).

the emergentist, the *seeds* of every emergent property and the behaviour it manifests are found within the world's fundamental elements, in the form of latent dispositions awaiting only the right context for manifestation.

We make no assertion one way or the other as to whether anything is like this for any chemical or biological properties, though we note that present evidence allows for the possibility that some perfectly respectable biological and chemical features are ontologically emergent in this way.

We do, however, propose that the conscious intentional and phenomenal aspects of the mind strongly favour an emergentist account. A human person's experiences and other conscious mental states exhibit features quite unlike those of physical objects, whether as revealed in ordinary sense perception or as uncovered in the physical and biological sciences. And the maximally direct nature of our first-person awareness of the intentional and phenomenal features of our conscious states blocks the a posteriori ascription to them of underlying physical microstructure hidden to introspection. The upshot of this familiar reflection, if it stands, is that our experiences and other conscious mental states have fundamentally distinctive characteristics. But these very characteristics are also prima facie causally efficacious. (Indeed, on a causal powers metaphysics, to countenance them as properties is to accept them as efficacious.) Thus, certain mental properties appear to be (1) resistant to analysis in terms of physical structural properties and so plausibly ontologically basic, (2) causally efficacious, and (3) borne only by highly organized and complex systems. Though we cannot argue the matter at length here, we find extant materialist attempts to overcome this prima facie case to be implausible.[19] (It goes without saying that we take the grounds for an emergentist account of the mental to be defeasible.)

Some philosophers acknowledge that the sort of broadly 'Cartesian' picture sketched above captures how we naively think about conscious experience but contend that it is an illusion. For our part, we think that such philosophers underestimate the difficulties for a theory of empirical knowledge that maintains that we are subject to a radical and pervasive cognitive illusion at the very source of all of our empirical evidence. And if the central argument of this paper is correct, then for any of these philosophers likewise committed to a causal powers metaphysics, the seemingly *paradoxical* position of denying the causal efficacy of mental states must be added to those difficulties.

REFERENCES

Baker, L. R. 1993. 'Metaphysics and Mental Causation'. In J. Heil and A. Mele (eds), *Mental Causation*. Oxford: Clarendon Press, 75–95.

Block, N. 2003. 'Do Causal Powers Drain Away?'. *Philosophy and Phenomenological Research* 67: 133–50.

[19] For argument on this point, see O'Connor and Kimble (n.d.).

Burge, T. 1993. 'Mind-Body Causation and Explanatory Practice'. In J. Heil and A. Mele (eds), *Mental Causation*. Oxford: Clarendon Press, 97–120.
Cartwright, N. 1983. *How the Laws of Physics Lie*. Oxford: Clarendon Press.
—— 1989. *Nature's Capacities and Their Measurement*. Oxford: Clarendon Press.
—— 1999. *The Dappled World: A Study of the Boundaries of Science*. Cambridge: Cambridge University Press.
Clapp, L. 2001. 'Disjunctive Properties: Multiple Realizations'. *Journal of Philosophy*, 98: 111–36.
Corry, R. 2002. 'A Causal-Structural Theory of Empirical Knowledge'. PhD thesis, Indiana University.
—— 2008. 'Scientific Analysis and Causal Influence'. In Toby Handfield (ed.), *Dispositions and Causes*. Oxford: Oxford University Press.
Crisp, T. and Warfield, T. 2001. 'Kim's Master Argument: a Critical Notice of *Mind in a Physical World*'. *Noûs*: 304–16.
Dupré, J. 1993. *The Disorder of Things: Metaphysical Foundations of the Disunity of Science*. Cambridge, MA: Harvard University Press.
—— (2001). *Human Nature and the Limits of Science*. Oxford: Oxford University Press.
Fodor, J. 1974. 'Special Sciences'. *Synthese* 28: 97–115.
Gillett, C. 2002. 'The Dimensions of Realization: a Critique of the Standard View'. *Analysis* 62: 316–23.
Heil, J. 2003. *From an Ontological Point of View*. Oxford: Oxford University Press.
Hendry, R. 2006. 'Is there Downward Causation in Chemistry?'. In D. Baird, L. McIntyre, and E. Scerri (eds), *Philosophy of Chemistry: Synthesis of a New Discipline*. Boston Studies in the Philosophy of Science, vol. 242. Springer, 173–89.
Kim, J. 1993. *Supervenience and Mind*. Cambridge, MA: MIT Press.
—— 1997. 'Does the Problem of Mental Causation Generalize?'. *Proceedings of the Aristotelian Society* 97: 281–97.
—— 1998. *Mind in a Physical World*. Cambridge: Cambridge University Press.
—— 1999. 'Supervenient Properties and Micro-based Properties: a Reply to Noordhof'. *Proceedings of the Aristotelian Society* 99: 115–18.
—— 2003. 'Blocking Causal Drainage and Other Maintenance Chores with Mental Causation'. *Philosophy and Phenomenological Research* 67: 151–76.
—— 2005. *Physicalism, or Something Near Enough*. Princeton, NJ: Princeton University Press.
Laughlin, R. B., Pines, D., Schmalian, J., Stojkovic, B., and Wolynes, P. 2000. 'The Middle Way'. *Proceedings of the National Academy of Sciences* 97: 32–7.
Lewis, D. 1980. 'Mad Pain and Martian Pain'. In N. Block (ed.), *Readings in the Philosophy of Psychology*, vol. 1. Cambridge, MA: Harvard University Press, 216–22.
Loewer, B. 2001. 'Review of J. Kim, *Mind in a Physical World*'. *Journal of Philosophy* 98: 315–24.
O'Connor, T. 2008. 'Agent-Causal Power'. In Toby Handfield (ed.), *Dispositions and Causes*. Oxford: Oxford University Press.
—— and Kimble, K. n.d. 'The Argument from Consciousness Revisited.' Unpublished manuscript.
—— and Wong, H. Y. 2002. 'Emergent Properties'. *Stanford Online Encyclopedia of Philosophy*. http://plato.stanford.edu/entries/properties-emergent/.

O'Connor, T. and Wong, H. Y. 'The Metaphysics of Emergence'. *Noûs* 39 (2005): 659–79.
Pereboom, D. 2002. 'Robust Non-Reductive Materialism'. *Journal of Philosophy* 99: 499–531.
—— and Kornblith, H. 1991. 'The Metaphysics of Irreducibility'. *Philosophical Studies* 63: 125–45.
Rosenberg, A. 1994. *Instrumental Biology or the Disunity of Science*. Chicago, IL: University of Chicago Press.
Ross, D. and Spurrett, D. 2004. 'What to Say to a Skeptical Metaphysician: a Defense Manual for Cognitive and Behavioral Scientists'. *Behavioral and Brain Sciences* 27: 603–47.
Shoemaker, S. 1980. 'Causality and Properties'. In P. Van Inwagen (ed.), *Time and Cause*. Dordrecht: Reidel, 109–35.
—— 1998. 'Causal and Metaphysical Necessity'. *Pacific Philosophical Quarterly* 79: 59–77.
—— 2001. 'Realization and Mental Causation'. In C. Gillett and B. Loewer (eds), *Physicalism and its Discontents*. Cambridge: Cambridge University Press.
—— 2002. 'Kim on Emergence'. *Philosophical Studies* 108: 53–63.
—— 2007. *Physical Realization*. Oxford: Oxford University Press.
Tooley, M. 1987. *Causation: A Realist Approach*. Oxford: Oxford University Press.
—— 1993. 'Causation: Reductionism versus Realism'. In E. Sosa and M. Tooley (eds), *Causation*. Oxford: Oxford University Press, 172–92.
van Gulick, R. 1992. 'Non-Reductive Materialism and Inter-theoretic Constraint'. In A. Beckermann, H. Flohr, and J. Kim (eds), *Emergence or Reduction? Essays on the Prospects of Non-Reductive Physicalism*. Berlin: De Gruyter, 157–79.
—— 1993. 'Who's in Charge Here? And Who's Doing All the Work?'. In J. Heil and A. Mele (eds), *Mental Causation*. Oxford: Clarendon Press.
Woodward, J. 2003. *Making Things Happen: A Theory of Causal Explanation*. Oxford: Oxford University Press.
Yablo, S. 1992. 'Mental Causation'. *The Philosophical Review* 101: 245–80.

5

Exclusion and Physicalism:* Comments on O'Connor and Churchill

Stephan Leuenberger

1. A SCHEMA FOR EXCLUSION ARGUMENTS

Exclusion arguments are a versatile tool to squeeze out reductionist or eliminativist consequences from certain metaphysical theses. They come in different varieties. There are, for example, causal exclusion arguments against non-physical causes, explanatory exclusion arguments against composite objects, spatiotemporal exclusion arguments against non-material objects, and metaphysical exclusion arguments against non-Humean truthmakers.

These different arguments share a common structure. In this section, I present a general schema of which they are instances. Recognizing parallels between arguments in different debates may help us become clearer about the dialectical options. The next two sections then focus on causal exclusion arguments, in particular the one presented in Timothy O'Connor and John Ross Churchill's 'Is Nonreductive Physicalism Viable within a Causal Powers Metaphysic' (this volume).

Schematically, the two premises of an exclusion argument can be stated as follows (Φ is mnemonic for 'physical', C for 'cause', and I for 'identity' or 'intimate metaphysical relation'):

Completeness For all z, there are xx such that Φxx and $Cxxz$.
Exclusion For all xx, yy, z, if $Cxxz$ and $Cyyz$, then $Ixxyy$.

According to Completeness, Φ is complete relative to C in the sense that one can always find Φ-things that are C-related to a given z. According to Exclusion,

* Many thanks to the organizers and to the participants at the emergence conference in Belfast, where an earlier version of this response was presented. Research for this work was partially funded by the Swiss National Research Foundation project 'Properties and Relations' (100011 – 113688).

the relation C is exclusive in its first argument place: if some things are C-related to z, no other things are, unless they are I-related to the first ones.[1]

Together, Completeness and Exclusion entail a thesis that I call 'Monopoly':

Monopoly For all *yy*, z, if *Cyyz* then *Iyyxx* for some *xx* with Φ*xx*.

According to Monopoly, Φ monopolizes C in the sense that no things can stand in the C-relation to anything unless they are suitably related (i.e. I-related) to Φ-things. I will discuss later to what extent Monopoly leads to reductionist or eliminativist consequences.

To obtain an instance of this schematic argument, we need to specify what Φ, C, and I stand for. Suitable choices yield the four examples of exclusion arguments mentioned in the first paragraph.

A simple causal exclusion argument results from the following interpretation of the schematic letters: Φ*xx* is true if each one of the *xx* is a physical event; *Cxxz* if the events *xx* jointly cause physical event z, and *Ixxyy* if the *xx* are (plurally) identical to the *yy* (that is, each one of the *xx* is also one of the *yy*, and each one of the *yy* is also one of the *xx*). Monopoly is then the claim that whenever some events *yy* jointly cause something, then all of them are physical events.

For a simple explanatory exclusion argument, we take Φ*xx* to be true if all the *xx* are mereologically simple; *Cxxz* if the individuals *xx* all essentially figure in an explanation of a fact z, and *Ixxyy* if the *xx* are (plurally) identical to the *yy*. Exclusion, thus understood, is in the spirit of Ockham's razor, denying that there are redundant explainers, while Completeness claims that everything can be explained in terms of mereological simples. Monopoly is then the claim that whenever the *yy* all essentially figure in an explanation of z, then they are all mereologically simple.[2]

For a simple spatiotemporal exclusion argument, we take Φ*xx* to be true if all the *xx* are material objects; *Cxxz* if the individuals *xx* exactly occupy the whole of spatiotemporal region z; and *Ixxyy* if each one of the *xx* overlaps one of the *yy*, and each one of the *yy* overlaps one of the *xx*. Exclusion then expresses a version of the Lockean thesis that no two things can be in the same place at the same time, while Completeness denies that there are any things that are not in space-time. Monopoly, in this context, is the claim that whenever an individual occupies a spatiotemporal region, then it overlaps a material object.

For a simple metaphysical exclusion argument, we take Φ*xx* to be true if all the *xx* are Humean entities; *Cxxz* if the *xx* jointly are truthmakers for true non-disjunctive proposition z; and *Ixxyy* if the *xx* are (plurally) identical to the

[1] For greater generality, I state these premises in a semi-regimented language that has plural (*xx*, *yy*, *zz*) as well as singular (x, y, z) variables and non-distributive plural predicates. For example, the exclusion argument against explanatorily relevant composite objects is not easily captured using singular quantification only. For an introduction to plural quantification, see Linnebo (2008).

[2] Cousins of that argument are offered by Dorr (2001) and Merricks (2001).

yy.[3] Exclusion is again in the spirit of Ockham's razor, while Completeness claims that a Humean ontology can provide truthmakers for all propositions. Monopoly, thus interpreted, is the claim that whenever some things are truthmakers for a proposition, then each of them is a Humean entity.[4]

In argumentative practice, simple versions as those presented will rarely do. However, the above schema should be general enough to be able to accommodate versions that are more complex and realistic. In the next section, I will discuss this in the case of the causal exclusion argument.

The claim I called 'Monopoly' does not entail the reductionist or eliminativist claim that everything (or at least everything in the domain of the pertinent quantifiers) is Φ. That sweeping claim follows from Monopoly only in conjunction with two auxiliary claims. First, that Φ is closed under I; that is, that whenever some things stand in I to Φ-things, then they are themselves Φ. In those instances of the schema where I is (plural) identity, that auxiliary claim follows staightforwardly from Leibniz's Law. In other instances, it may be more substantive. The second auxiliary claim is that everything stands in C to something, or more precisely, that everything is one of some things that stand in C to something. In the above instances, this boils down, respectively, to the claim that everything has physical effects (belongs to some things that jointly cause something physical), that everything is explanatorily relevant, that everything has a spatiotemporal location, and that everything serves as a truthmaker for something.

An exclusion argument need not use Completeness as a premise, but may instead be directed against it. Of course, Exclusion together with the negation of Monopoly yields an argument against Completeness. The negation of Monopoly is equivalent to an existential claim, which is most effectively supported by offering an instance, a claim of the following form:

Paradigm $Caab$ and $Iaaxx$ only for xx with $\neg\Phi xx$.

The schematic *aa* and *b* are to be interpreted by terms that denote things are paradigms for the relation C, and paradigmatically not I-related to things that are Φ.

A third form of exclusion arguments, exemplified in the chapter by O'Connor and Churchill (this volume), neither tries to establish Monopoly nor refute Completeness, but rather tries to establish the disjunction of Monopoly and Incompleteness, the negation of Completeness.

In section 2, I discuss the role that a causal powers metaphysic plays in a defence of the causal exclusion premise. I argue that O'Connor and Churchill's argument

[3] Existential propositions here count as disjunctive. If z could be any true proposition, then Exclusion would be implausible. According to standard truthmaker theory, any thing that satisfies Ψx is a truthmaker for the proposition that there exists something such that Ψx.

[4] The term 'metaphysical exclusion argument' is mine, and as far as I know, no argument of that sort has been explicitly given. Nonetheless, the argument is similar in spirit to the extended argument for Humeanism that David Lewis sketches in the Introduction to Lewis (1986) (even though Lewis does not appeal to truthmakers there).

lacks dialectical force because of its reliance on such a strong background view. In section 3, I argue that even if the exclusion premise is granted, exclusion arguments have less bearing on the debate about physicalism than is often assumed.

2. CAUSAL EXCLUSION AND CAUSAL FUNDAMENTALISM

The version of the causal exclusion argument sketched in the last section is too simplistic, and needs amendment in a couple of respects. First, partial causes need to be distinguished from complete causes. Both Suzy's pushing and Fred's pushing may be partial causes of the boulder's fall. Of course, such partial causes do not exclude each other. Exclusion applies at best to complete causes. Second, the universal quantification over events z should be restricted. Exclusion applies only to those that are not overdetermined. Intuitively, the death of the convict has as many complete causes as there are members of the firing squad. However, since such overdetermined events are no doubt exceptional, exclusion still has wide applicability.

With Φ and C spelled out more adequately, the exclusion premise thus reads:

Causal Exclusion For all *xx*, *yy*, *z*, if *z* is a non-overdetermined physical event and both the *xx* and the *yy* jointly constitute a complete cause of *z*, then the *xx* are identical to the *yy*.[5]

Why think that Causal Exclusion holds, that distinct complete causes exclude each other? On some conceptions of causation, that principle is highly implausible. If causation is counterfactual dependence, for example, then one event can have several distinct causes. O'Connor and Churchill take causal exclusion to follow from what they call a 'causal powers metaphysic'.

Indeed, the causal powers metaphysic plays a double role in the argument by O'Connor and Churchill. It vindicates the exclusion premise, as just noted, and it saddles non-reductive physicalism with causal commitments, as I discuss in the next section. It is thus important to become clearer on what a causal power metaphysic is supposed to be.

It is a prominent metaphysical view (Shoemaker, 1980) that properties are individuated by what causal powers they confer: properties X and Y are identical if and only if they confer the same causal powers to their bearers. However, that claim certainly does not imply Causal Exclusion, and a causal power metaphysic will thus have to go beyond it. O'Connor and Churchill do not give us a canonical formulation, but rather present it as a cluster of claims about

[5] Some versions of the causal exclusion argument also take I to be a weaker relation than plural identity. Presumably, I would be some relation of dependence. I will return to this point below.

causation, properties, and how properties figure in causation. They say that causation 'involves the exercise of . . . causal *powers* . . . of particulars', that causation is production rather than counterfactual dependence, and that properties are 'immanent to . . . things as non-mereological parts' (this volume: 44).[6] Some of these claims may suggest that causation obeys an exclusion principle, but they hardly entail it. Is there an argument for exclusion from the causal powers metaphysic, after all?

It seems to me that such an argument can be constructed from their assumption that causation is a 'real, irreducible' relation. Intuitively, exclusion applies to causes that are not just distinct, but independent of each other. Thus the above exclusion principle, where *I* is identity rather than dependence, seems too strong. However, the stronger and the weaker exclusion principle coincide if all causes are independent of each other. The irreducibility of causation, suitably understood, entails that causes are thus independent.

We can distinguish a collective and an individual independence or irreducibility of causation. Causal facts are collectively independent if the class of causal facts does not globally supervene on any class of non-causal facts. Causal facts are individually independent if no causal fact supervenes on any class of facts, possibly including other causal facts, to which it itself does not belong. Individual independence is a form of modal recombinability, which is often taken as the mark of fundamental, irreducible facts.

Given that causal facts are individually independent, causal exclusion is defensible. But why make that assumption? To me, even the weaker claim that causal facts are collectively independent seems highly implausible. It is perhaps fair to say that for the most part, the specialized literature on causation does not take causation to be fundamental, to belong to the metaphysical ground floor. In a survey article, Ned Hall (2006) criticizes the view that causation fails to supervene on 'spatiotemporal structure, a complete history of instantaneous physical states, and fundamental laws governing their evolution':

> [T]he arguments in favor of this position are tissue thin . . . And the arguments against are rock-solid. In particular, [it] generates a skepticism about our knowledge of causal facts for which the usual appeals to inference to the best explanation provide no cure (for the relevant explanatory work is already done by the fundamental laws), and makes an unacceptable mystery out of what should be a straightforward supervenience relation: namely, the supervenience of causal facts at the macroscopical on causal facts at the microscopical level. (Hall 2006: 2)

Among the cluster of theses that characterize a causal powers metaphysic, it is the thesis that causation is an irreducible, fundamental relation that supports the strong exclusion principle above, I claim. If that sort of causal fundamentalism is in trouble, so is that exclusion principle.

[6] According to them, both tropes and immanent universals satisfy this constraint; presumably, resemblance classes do not.

O'Connor and Churchill acknowledge that a causal powers metaphysic is among their assumptions, and is not argued for. The reliance on that assumption raises questions about the dialectical force of that argument. It is not clear to me which non-reductive physicalists subscribe to a causal powers metaphysic in the sense of O'Connor and Churchill. This worry gets reinforced in the second part of their chapter, when it turns out that not even Sydney Shoemaker is a causal-power metaphysician by their lights. If he is not in the target area for the argument, who is?

In the next section, I will grant the exclusion premise, though not the whole package of the causal powers metaphysic from which O'Connor and Churchill derive it. I will argue that the causal exclusion principle has only limited significance for the debate about physicalism.

3. CAUSAL EXCLUSION AND PHYSICALISM

Causal exclusion has been a double-edged sword in the discussion of physicalism: it has been used both in arguments for and in arguments against the view. Roughly, physicalists argue from Completeness and Exclusion to Monopoly, and anti-physicalists from Exclusion and Paradigm to Incompleteness. Thus considerations from causal exclusion have been widely taken to be pivotal in the discussion of physicalism. In my view, their significance has been overstated. For the arguments on either side rely on auxiliary assumptions that are highly controversial. Unsurprisingly, these auxiliary assumptions concern the place of causation in the world.

Does Monopoly imply physicalism? Typically, physicalism is committed to the global supervenience of all facts upon physical facts.[7] Physical facts are exemplifications of physical properties or relations, and the latter are those denoted by predicates in a final and complete theory of the world. It seems that Monopoly does not imply such a supervenience claim, and hence that the exclusion argument does not establish physicalism. What auxiliary assumption could bridge the gap? Say that a fact is causally active if it belongs to some complete cause of some physical fact.[8] The claim that everything globally supervenes on the causally active facts, together with Monopoly, entails physicalism.[9]

[7] This is, roughly, the definition of 'minimal materialism' given in Lewis (1983). To evaluate that global supervenience claim at a world *w*, not all possible worlds are considered, but only those in the 'inner sphere of possibility' around *w*.

My own preferred definition of physicalism (Leuenberger 2008) differs from Lewis, but only in ways that do not matter for present purposes.

[8] Causally active facts are to be distinguished from causal facts, i.e. facts consisting in the exemplification of the causal relation. Causal facts, as opposed to causally active facts, have themselves facts as constituents.

[9] In this context, there is no harm in going back and forth between taking the relata of causation to be events and taking them to be facts. I assume that they are exemplifications of properties or relations, to which both the terms 'facts' and 'events' are appropriately applied.

However, the assumption of global supervenience on causally active facts may deprive an argument for physicalism of some of its dialectical force. Many philosophers think that the most pressing worries for physicalism derive from facts about conscious experience. If sound, the Conceivability Argument (Chalmers 1996) and some versions of the Knowledge Argument (Jackson 1982) show that such phenomenal facts fail to supervene on the physical. Typically, proponents of those arguments acknowledge that such facts are epiphenomenal relative to the physical facts, and would thus resist an argument from Monopoly to physicalism.

An argument against physicalism, on the other hand, needs to show that the supervenience claim that characterizes physicalism entails Completeness (or perhaps that it entails it in conjunction with Exclusion, which is granted here). Causal fundamentalism aside, it is by no means clear why causal completeness should be among the commitments of physicalism. To be sure, the view entails that causal facts, like all facts, supervene on the physical ones. However, that causal facts thus supervene does not imply anything about whether the relata of the causal relations are themselves physical facts.

So far, I have spoken indiscriminately of 'physicalism', while O'Connor and Churchill specifically target so-called 'non-reductive' physicalism. Unfortunately, that term is used in different ways in the literature. Sometimes, it characterizes any physicalist view that allows that there are non-physical properties and facts. Thus construed, the claim that everything globally supervenes on physical facts is a version of non-reductive physicalism. Sometimes, the term characterizes a physicalist view that denies that every domain of facts is related to the physical by relatively simple, informative laws.

O'Connor and Churchill set up a dilemma. From Exclusion, they infer the disjunction of Incompleteness and Monopoly, and then try to show that non-reductive physicalism is not compatible with either disjunct. It is incompatible with Incompleteness in virtue of being physicalist, and with Monopoly in virtue of being non-reductive.

I have already suggested that physicalism may well be compatible with Incompleteness. Adding the qualification 'non-reductive' makes no difference in this respect. Likewise, I am unconvinced that non-reductive physicalism, even in the second, stronger sense, is incompatible with Monopoly. Perhaps there are phenomenal facts which supervene on, but do not cause, physical facts. Perhaps there are informative causal laws linking mental events, but no simple and informative bridge laws linking the mental to the physical.[10] Of course, one could stipulatively define 'non-reductive physicalism' in such a way as to make

[10] It is sometimes taken as obvious that the mind has physical effects, such as when my beliefs and desires cause me to raise a glass. However, as argued convincingly by Sturgeon (1998), these effects are macrophysical, and do not involve the fundamental physical properties and relations appealed to in the definition of supervenience.

it incompatible with Monopoly. But this would again undermine the dialectical force of the argument, by leaving it unclear whether any actual philosophers are in its target area.

REFERENCES

Chalmers, D. J. 1996. *The Conscious Mind: In Search of a Fundamental Theory*. Oxford: Oxford University Press.

Dorr, C. 2001. 'The Simplicity of Everything'. PhD thesis, Princeton University.

Hall, N. 2006.'Philosophy of Causation: Blind Alleys Exposed; Promising Directions Highlighted' *Philosophy Compass* 1: 84–94.

Jackson, F. 1982. 'Epiphenomenal Qualia'. *Philosophical Quarterly* 32:127–36.

Leuenberger, S. 2008. '*Ceteris Absentibus* Physicalism'. In D. Zimmerman (ed.), *Oxford Studies in Metaphysics* IV. Oxford: Oxford University Press, 145–70.

Lewis, D. 1983. 'New Work for a Theory of Universals'. *Australasian Journal of Philosophy*, 61: 343–77. (Reprinted in Lewis 1999, 8–55.)

—— 1986. *Philosophical Papers*, vol. 2. Oxford: Oxford University Press.

—— 1999. *Papers in Metaphysics and Epistemology*. Cambridge: Cambridge University Press.

Linnebo, Ø. 2008. 'Plural Quantification'. In *Stanford Encyclopedia of Philosophy* (Spring 2009 Edition), E. N. Zalta (ed.), <http://plato.stanford.edu/archives/spr2009/entries/plural-quant/>.

Merricks, T. 2001. *Objects and Persons*. Oxford: Oxford University Press.

Shoemaker, S. 1980. 'Causality and Properties'. In P. van Inwagen (ed.), *Time and Cause*. Dordrecht: Reidel, 109–35.

Sturgeon, S. 1998, 'Physicalism and Overdetermination'. *Mind* 107: 411–32.

6

Emergent Causation and Property Causation

Paul Noordhof

When we say that one thing caused another in virtue of certain of its properties, we attribute a case of property causation. A necessary condition for property causation is that an *instance* of the property cited is a cause. This is not a sufficient condition because property causation, as opposed to property instance causation, involves a certain kind of generality. A sufficient condition for property causation is that there should be a law that things with F (a putative property cause) cause things with G; in brief, that the Fs cause the Gs. To fix ideas—the ideas that introduce the problem upon which this chapter will focus—suppose that a subclass of the class of properties identified by current physics, or a future physics sufficiently resembling our own, contains just properties which are property causes in this way. Call these the *narrowly physical property causes*. They are narrowly physical because, presumably, our intuitive notion of a physical property is not exhausted by the properties identified by physics, and certainly not those which are property causes. Suppose, further, that there are properties that stand in a certain highly specific relation of supervenience to the narrowly physical properties, the exact character of which will be set out and defended in section 1. Call these the *broadly physical properties*. A very familiar thought is that if we get the characterization of the supervenience relation right, then the properties classified as either narrowly or broadly physical will capture our intuitive notion of a physical property. Properties that stand in this relationship of supervenience are not emergent properties *relative to the narrowly physical properties*. That does not mean that all emergent properties fail to stand in the relationship of supervenience. There may be some members of the class of narrowly physical property causes that are emergent with regard to other members of the class (for more discussion see, e.g., Hüttemann and Papineau 2005; Papineau forthcoming). Such properties would, of course, trivially supervene upon themselves.

The issue of property causation with which I am concerned is whether there may be any other property causes than the narrowly physical property causes, and the implications of my answer for the proper characterization of emergent causation (if such there be).

Positive answers to the question can take one of two forms. Either it can be argued that there are other property causes because these other properties stand in a certain relationship to the narrowly physical property causes, or an independent analysis can be given of the nature of property causation which serves to classify both the narrowly physical and other properties as property causes. The options, at this point, in part depend on the account of causation, in general, that is favoured. I have argued elsewhere that a counterfactual theory of causation ought to be adopted for particulars, amongst which I include property instances. My aim is to defend its application here and to derive from it an account of property causation.

Thus I shall argue that a univocal independent analysis of property causation can be provided. Appeal to the relationship between other broadly physical properties and those which are narrowly physical provides a justification of the verdicts arrived at by the independent account. It is no part of the correct theory of property causation. Nevertheless, one dimension of the relationship needs to be recognized in order to identify the right kind of generality indicative of property causation. Such an appeal is not required in the characterization of emergent property causation.

In the first section of this chapter, I will provide a characterization of the different relationship of broadly physical properties and emergent properties to narrowly physical properties, and defend it against objections, in particular, that the characterization I favour will not work if a causal theory of properties is true. I will also criticize the alternatives which have been offered as either not allowing for the existence of broadly physical properties or lacking appropriate independence from the key idea to which I appeal. In the second section, I will explain why, given Kim's exclusion argument, the verdicts that a counterfactual theory of property instance causation supplies for broadly physical properties require independent justification where those for emergent properties do not. In the third section, I outline how such a justification may be provided and develop an account of property causation which builds upon it. In the fourth section, I examine the implications of my analysis for the question of emergent causation, focusing in particular upon the question: does all emergent causation involve emergent property causation? My negative conclusion here will enable me to identify a second kind of emergence: emergent non-reductive physicalism. In the concluding fifth section, I briefly consider the consequences of my discussion for two candidates for emergence: phenomenal consciousness and free will.

1. THE PROPER CHARACTERIZATION OF EMERGENCE

Both non-reductive physicalism and emergent dualism involve the idea of property determination. In some sense, both suppose that the instantiation of narrowly physical properties determines the instantiation of the other target properties: the non-reductive or emergent ones. They differ over whether the instantiation of these other properties involves something genuinely new. Non-reductive physicalists deny this, whereas emergent dualists assert it. The problem is to make sense of when there is something new introduced.

According to the view I favour, a preliminary characterization can be provided in terms of different strengths of strong supervenience. As is familiar, Jaegwon Kim formulated it as follows.

> A *strongly supervenes* on B just in case, necessarily, for each x and each property F in A, if x has F, then there is a property G in B such that x has G, and *necessarily* if any y has G, it has F. i.e. □ (x)(F)(Fx & F *e* A □ (G) (G *e* B & Gx & □ (y)(Gy □ Fy))). (Kim 1993 [1984]: 65)
>
> (where A and B are families of properties, the supervening and supervenience-base (or subvenient) properties respectively, '*e*' is 'is a member of', and '□' is the necessity operator with a force to be specified).

Let the family of B properties be the narrowly physical property causes or spatiotemporal arrangements of such properties (I will supplement this shortly). Let the family of A properties be our target properties: mental properties, biological properties, or whatever. Whether one is a non-reductive physicalist or emergentist, it is plausible that the first modal operator should be understood in terms of nomic necessity. Neither position is to be distinguished by the fact that they require underlying supervenience-base properties in all possible worlds. The difference between non-reductive physicalism and emergent dualism lies in the interpretation of the second operator. Non-reductive physicalists should take it to be metaphysical necessity, emergent dualists should take it to be nomological necessity. The intuitive thought is that the reason why emergent dualists reject appeal to metaphysical necessity is that they suppose that some of the target properties determined by narrowly physical property causes are wholly distinct from them, whereas non-reductive physicalists are committed to thinking that they are not.

Of course, both non-reductive physicalists and emergent dualists deny property *identity*. Denying property identity (rather than property instance identity) is what makes non-reductive physicalists non-reductive. However, they add to this thought that even though the properties are not identical, instances of their target properties are, in some sense, nothing more than instances of narrowly physical properties. It is this that the appeal to metaphysical necessity is meant to articulate. The reason why emergent dualists should, at least, appeal to nomological necessity is that they claim that their target properties emerge rather than float free of the narrowly physical property causes. The fact that emergentist dualists need to appeal to fundamental laws to explicate this emergence provides a relatively ontologically robust sense in which the emergence is inexplicable. In brief, non-reductive physicalists believe in M-strong supervenience whereas emergent dualists believe in N-strong supervenience for their target properties (where M and N indicate the interpretation of the second modal operator: metaphysical or nomological necessity respectively). Others who take a similar line demarcating this position include James Van Cleve (1990: 222) and David Chalmers (2006).

Although the idea has been articulated with respect to a particular characterization of strong supervenience, the latter is not essential to it. A currently (and rightly) popular formulation of physicalism is:

> Any world which is a minimal physical duplicate of our world is a duplicate simpliciter of our world. (Jackson 1998: 12)

A minimal physical duplicate is a duplicate in terms of instantiations of narrowly physical properties and stopping right there. Such a world would be a duplicate in terms of instantiations of broadly physical properties as well. However, if emergent dualism were true, then the characterization of physicalism would be false. Instead, the emergent dualist holds that

> Any world which is a minimal physical duplicate of our world and has the same laws (where these might include fundamental physico-psychological laws for instance), is a duplicate simpliciter of our world.

The formulation of physicalism characterizes a relationship that should hold in all possible worlds—as we would expect from the appeal to metaphysical necessity in the strong supervenience claim—whereas the second limits it to worlds with the same laws.[1]

The favoured characterization of non-reductive physicalism does not rule out the possibility of the supervenience-base properties being relational so as to include features of the environment in which individuals are located, and their interactions with those features. It also does not rule out the possibility that some interactions between instances of mental properties and instances of narrowly physical property causes are on the same level, for instance the macro-level. It should not be assumed that all narrowly physical property causes are micro-properties. Nor does the characterization rule out the possibility that there is an interplay between a number of levels. The important point is that any property that is a potential threat to the truth of physicalism will not be identified by a physics resembling our own. For these properties, if there is a lower level supervenience-base for them, the particular characterization of supervenience I have given explains under what circumstances these properties will be classified as broadly physical as opposed to non-physical emergent properties. As a result, the kind of ontological emergence favoured by Michael Silberstein, who emphasizes the features I have just listed, is best classified as another version of non-reductive physicalism (Silberstein 2006). Silberstein rejects this classification on

[1] For the present purposes, weaker notions of emergence can be classified as types of non-reductive physicalism (for discussion see Chalmers 2006: 252–3; Bedau 1997: 377–9, 393–5; McGinn 1989).

the grounds that the kind of emergence he envisages involves emergent causation. I argue later that emergent causation is compatible with the properties it involves being broadly physical (and hence weakly emergent at best).

Apparent counter-examples to my characterization of emergence in terms of N-supervenience without M-supervenience are causal role properties or second order properties, those possessed in virtue of the fact that a certain causal role is occupied (hereafter, occupant attributing properties). On the assumption that the laws of nature are independent of the properties which are instantiated (though not necessarily the pattern of instantiations), the connection between narrowly physical property causes and causal role properties (say) must be a matter of nomological necessity alone (O'Connor 1994: 96). In different worlds, the very same narrowly physical property causes may have different causal roles in virtue of different laws that hold. Nevertheless, intuitively, causal role properties are not thought of as emergent properties. Functionalism, to take an obvious example, is one way in which non-reductive physicalism may be true (indeed, it is the most commonly cited example). Yet, whether functionalism takes mental properties to be causal role properties or occupant attributing properties, the connection between these properties and narrowly physical property causes is only one of nomological necessity.

It is at this point that I need to qualify my characterization of the supervenience-base of broadly physical properties. The supervenience-base should not just include narrowly physical property causes but also any laws concerning them alone. Once we include these, it will be metaphysically necessary that, given these laws and the narrowly physical property causes instantiated, their causal role will be instantiated. Of course, if we allowed brute physico-psychological laws to be part of the supervenience-base, emergent psychological laws would be necessitated too. The key difference is that there is no reason to suppose that brute physico-psychological laws are part of an intuitive characterization of the narrowly physical. Indeed, if there are emergent non-physical psychological properties, then clearly these laws are not part of the proper characterization of the narrowly physical. Hence, we have a sense in which physico-psychological laws would be explicable by narrowly physical properties and laws about them, if psychological properties were causal role properties or occupant attributing properties, that we would not have if they were emergent (cf. Sober 1999: 142–4). Other potential counter-examples—for instance, if mental properties are environment-dependent—can be accommodated by allowing their supervenience-base to include narrowly physical relational properties.

Another concern arises if a causal theory of properties is true so that the natures of all properties, or at least scientifically fundamental properties, are given by their causal role (e.g. Shoemaker 1980). Then appeal to the contrast between metaphysical and nomological necessity does not appear to be available. According to such a theory, it seems it is not possible for something to be the very same property and yet lack some aspect of its causal role. Suppose that the coinstantiation of a certain group of narrowly physical property causes yields an intuitively non-physical emergent property. Then, the thought runs, it won't be metaphysically possible for the group in question to be present with the emergent property absent. Yet, the property is supposed to be an emergent one and not a broadly physical property (O'Connor 1994: 97; Wilson 2005: 436–47).

The first point to make is that, if the nature of a property is given by its causal role, it doesn't follow that its causal properties are essential to it or, if you pack essence into nature then you should not assume that properties characterized by causal role alone have that causal role as their nature. Thus a property P may have causal role R_1 in world w_1, R_2 in w_2, R_3 in w_3 etc. There are a number of different ways in which this point can be developed in face of the obvious objection that these different world-relative roles imply that the property instantiated is different. Suppose, first, that properties are not world bound. They can be instantiated in different possible worlds. Then it is very plausible that, to be instantiated, they must have at least some of their causal role. Nevertheless, it is not clear why they should have all of it. If the laws of our world are holistic, as Jessica Wilson remarks, this is, at best, evidence that the properties of our world are not emergent and, hence, can provide no counter-examples to my way of demarcating emergence (Wilson 2005: 446).[2]

If the very same property is instantiated in different worlds, it might be wondered what could explain why its causal role is different. We could hardly appeal to different laws if the properties themselves are the basis of laws (see, e.g., O'Connor 2000a: 117–18). The obvious answer is that it, the property, is a bit different. These differences are not sufficient, though, to lead us to conclude that we are talking about a different property. We would say that if we considered two such properties, side by side, in our world (as it were). However, what we would say in our world does not imply that we should say the same when considering a property in our world and one in another possible world. Perhaps the background thought is that, if a property has no other nature than its causal role, then we could not point to something common to the property in each world, as we could if there were a common nature independent of causal role. This betrays a certain presumption as to what must count as common. A certain degree of similarity of position in the causal nexus might suffice. This would be common to each occurrence of the property.

Suppose now we consider properties to be world bound (Heller 1998). The question of whether a particular causal role property may be instantiated in other worlds becomes the question of whether counterparts of that property in other worlds may have a different causal role. There is no reason to reject this possibility. Those who claim that properties are to be understood entirely in terms of their causal roles don't deny that there are worlds with different laws. We may argue that, for any world-bound property, the laws hold in virtue of its

[2] Unfortunately, for reasons of space, I cannot discuss Wilson's challenge further.

nature. Nevertheless, in different worlds, the counterpart of that property may have a different nature and hence different laws hold. Laws may be intrinsic to properties without being essential to their instantiation in the same way that particulars may have accidentally intrinsic properties (for details on the latter, see Lewis 1986b: 198–209).

A second line of response, to the worry that the truth of a causal theory of properties would undermine the attempted demarcation of emergent property dualism from non-reductive physicalism, is that, if a causal theory of properties is true, there is no reason to be an emergent *property* dualist. Emergent property dualism has two features. First, it holds that properties emerge from narrowly physical property causes. Second, it claims that these emergent properties are not even broadly physical. Given a causal theory of properties, these two features are in tension.

If the nature of a property is determined by all the entities mentioned in the causal role, then the fact that putatively narrowly physical properties have non-physical consequences—instances of emergent properties—impugns their physical status. In which case, it would not be true that narrowly *physical* property causes stand in a metaphysically necessary relationship to emergent properties. If only a subsection of the causal role determines the nature of the narrowly physical properties—presumably their narrowly physical role—then this problem can be avoided. However, it seems to be avoided only at the expense of throwing into question whether the non-physical mental role should be taken to be essential to narrowly physical properties. If it should not, then the first line of response comes into play again.

In addition, it is questionable whether, if a causal theory of properties is true, the grounds on which we might claim that a property is non-physical remain. It is not as if, in these circumstances, they have, or are, phenomenal properties in a damaging non-physical sense. This is not to deny that there are phenomenal properties in the circumstances envisaged. If a functionalist theory of phenomenal properties is correct, then the causal theory of properties will allow that some properties are phenomenal properties without them being non-physical. The point is simply that the motivation for emergent dualism—drawing on a different view about the character of phenomenal properties—is undermined. In which case, there is no need to accept that emergent property dualism is compatible with metaphysically necessary relations between narrowly physical property causes and their emergent properties.

Two cautions are needed. The first is that the argument I have just run does not establish that mental properties are not emergent. The claim is rather that, if they are emergent, this is not because they are distinctively non-physical. As I have already noted, certain narrowly physical properties may be emergent with regard to other narrowly physical properties of that class. My claim is just that there would be no reason to suppose that mental emergent properties are any different. It is emergent property *dualism* which is threatened. In section 4, I will discuss the nature of emergent physicalism.

The second caution is that the argument is not intended to establish that, if a causal theory of properties is true, then physicalism is true (see, e.g., Shoemaker 1981: 274–8). For all I know, non-physical worlds of causally characterized properties are possible. Rather, my more limited aim is to establish that, if we have a world of pervasively instantiated narrowly physical property causes and the causal theory of properties is true, it is unclear how emergent property dualism is true.

Philosophers, convinced that the appeal to metaphysical necessity in the characterization of supervenience is mistaken, stick with nomological necessity and attempt to supplement it with the requirement that, in the case of non-reductive physicalism, we have an explanatory relationship between narrowly physical property causes and broadly physical properties whereas, in the case of emergent properties, the relationship is not explanatory. Matters turn on the favoured type of explanation. The two standard approaches take broadly physical properties to be functionalizable or macro-properties respectively. According to the former, favoured by Jaegwon Kim, broadly physical properties are those which are instantiated if another property occupies a certain causal role (see, e.g., Kim 1998: 100–1). According to the latter, defended by Timothy O'Conner and co-writer, Hong Yu Wong, broadly physical properties are structural properties in which the proper parts of particulars which possess them have properties, not identical with the structural property, jointly standing in relation R.

The problem with the first option, appealing to second order causal role properties, is that it is unclear whether this supplies us with an account of non-reductive physicalism in which mental properties exist. It seems more plausible to suppose that we have mental concepts, the concept of some property or other that occupies a certain causal role, which we apply to particulars partly in virtue of the property that occupies the role. Indeed, this is the conclusion that Kim draws in the development of his position (Kim 1998: 104; see also Wilson 2005: 453).

The problem with the second option, appealing to the idea of structural properties, concerns variable realization. O'Connor and Wong claim that if a particular possesses the relevant parts with properties standing in the appropriate relation R, then they possess the structural property. There is *nothing more* to the possession of the structural property than that (O'Connor 1994: 93; O'Connor and Wong 2005: 663). The question is how to understand this 'nothing more'. In certain cases, identity seems plausible. For instance, possession of the structural property of being methane is nothing more than instances of the properties of being carbon and hydrogen atoms in a certain arrangement R.

In the case of the property of being an earthquake, a similar approach does not seem to work. There is no particular arrangement of parts with properties which holds for every case of earthquakes. There are a vast variety of different ways in which there could be an earthquake. Moreover, the more we seek to characterize what might be common to all cases of earthquakes, the more the properties with which we characterize this commonality are unlikely to be narrowly physical property causes. The proposal thus faces a dilemma. If you go highly specific and seek to characterize everything in terms of narrowly physical property causes, then we need some other way of characterizing why the property of being an earthquake, as opposed to instances of the property of being an earthquake, is a broadly physical property. On the other hand, if you go highly general, you might be able to capture what is common to all cases of the property of being an earthquake (let us suppose), but now you open up a gap between the narrowly physical property causes and the more general specification.

It might be argued that the property of being an earthquake is a broadly physical property because all of its instances involve narrowly physical properties. However, this will not do as an answer by itself. If you claim that two properties have the same instance, then nothing has been established as to whether the properties in question are physical. Instead, the instance may involve both physical and non-physical properties. If it is claimed that instances don't combine physical and non-physical properties, we need to know why. Whatever answer is given at this stage is the answer to our question of what is involved in 'nothing more' and not the claim that the putative broadly physical property only has instances that are physical. For example, if it is said that two properties do not have the same instance unless one of them metaphysically necessitated the other, then that is the answer to our question. Unconstrained talk of identity of instances cannot yield the right results. Suppose, for example, that narrowly physical property causes nomically necessitate certain emergent properties. We could not establish that the emergent properties were physical rather than non-physical simply by saying that the same instance has these nomically related elements. Our defence of the claim that they have the same instance requires discussion of the relationship between the properties themselves and whether this relationship is compatible with them sharing instances. Certainly, if it is not metaphysically possible for there to be an instance of the property of being an earthquake without it being an instance of an arrangement of narrowly physical property causes, then this shows something about the nature of the property of being an earthquake. But now we simply have an assertion of the relation of supervenience which was being eschewed. For these reasons, O'Connor and Wong's approach doesn't seem to avoid the problem to which the appeal to a metaphysically necessary supervenience relation was the solution.

Once appeal to metaphysical necessity is in play, it is reasonable to consider whether we need anything else. We would have captured the idea that emergent properties are, in some sense, inexplicable by saying that their instantiation requires a fundamental physico-psychological law. The idea that there is a metaphysically necessary connection between arrangements of narrowly physical properties and broadly physical properties is a placemarker for various kinds of explication that might be provided of the connection. There seems no reason to limit ourselves to certain types or even claim that we will be able to grasp the kind of explication that is at the basis of the metaphysical necessity—for instance, we might be cognitively closed to it (McGinn 1989).

I conclude that the distinction between broadly physical properties and a strong kind of emergence is best made out by a difference in the strength of the second necessity operator in the formulation of supervenience and that no other account promises to do better without appealing to the same materials.

2. A COUNTERFACTUAL ANALYSIS OF PROPERTY INSTANCE CAUSATION

The way in which I have distinguished broadly physical from non-physical emergent properties has consequences for the development of a counterfactual analysis of property instance causation and, thereby, of property causation. For the purposes of the present discussion, I am going to make two simplifying assumptions. The first is that, subject to a successful response to the objection below,

> p_1 is a cause of p_2 if (1) if p_1 were not instantiated, then p_2 would not be instantiated and (2) if p_1 were instantiated, p_2 would be instantiated (where 'p_1' and 'p_2' are property instances and they are distinct existences in a sense to be made clearer below).

So I am setting aside issues of indeterminism for which appeal to chance raising counterfactuals would be necessary and, by only providing a sufficient condition, not engaging with the question of redundant causation (see, e.g., Noordhof 1999c).

The second simplifying assumption is that, for our purposes, Lewis's analysis of, and similarity weighting for, counterfactuals is correct. Thus

> A counterfactual 'If it were that A, then it would be that C' is (non-vacuously) true if and only if some (accessible) world where both A and C are true is more similar to our actual world, overall, than is any world where A is true but C is false. (Lewis 1986a [1979]: 41)

with similarity of worlds measured by the following conditions:

> (A) It is of the first importance to avoid big, widespread, diverse violations of law.
> (B) It is of the second importance to maximize the spatio-temporal region throughout which perfect match of particular fact prevails.
> (C) It is of the third importance to avoid even small, localized, simple violations of law.
> (D) It is of little or no importance to secure approximate similarity of particular fact, even in matters which concern us greatly. (Lewis 1986a [1979]: 47–8)

In fact, the similarity weighting needs some reform but the ways in which it needs reform should not touch the discussion to follow.

With these assumptions in place, we can see the problem for a counterfactual analysis. Let ep_1 be an instance of an emergent property and $A(p_1, p_2, p_3 \ldots)$ be an arrangement of narrowly physical property causes from which ep_1 emerges. It seems possible that ep_1 is not a cause of e but $A(p_1, p_2, p_3 \ldots)$ is. The case seems to be one in which a cause, $A(p_1, p_2, p_3 \ldots)$, has two effects ep_1 and e without it following that one of the effects is a cause of the other. The standard treatment of this in a counterfactual analysis turns on whether a backtracking conditional between ep_1 and $A(p_1, p_2, p_3 \ldots)$ holds. There seems no more reason to allow for the possibility of backtracking counterfactuals being true in the particular case of emergent properties and their N-supervenience-bases than in standard cases of two effects. We obtain more perfect match by retaining $A(p_1, p_2, p_3 \ldots)$ and violating the law to ep_1 than by removing $A(p_1, p_2, p_3 \ldots)$, keeping the law, and securing additional perfect match by covering up the traces of $A(p_1, p_2, p_3 \ldots)$ in the future. So, by the similarity weighting, the closest worlds will not be worlds in which, when ep_1 is absent, $A(p_1, p_2, p_3 \ldots)$ is absent. There seems to be no possibility of the counterfactual analysis of property instance causation falsely concluding that there is causation when dealing with emergent properties.

The situation is different for broadly physical properties. Let bp_1 be an instance of a broadly physical property, and $A(p_1, p_2, p_3 \ldots)$ the arrangement of narrowly physical property causes which constitutes its M-supervenience-base. At its most basic, this may just be the instantiation of a single narrowly physical property—the formulation is just for generality. There are two types of broadly physical properties we need to consider: occupant attributing properties and other broadly physical properties which are putative occupants of a causal role. For each, the challenge breaks down into two components. First, can there be cases in which $A(p_1, p_2, p_3 \ldots)$ is efficacious and bp_1 not? Second, if so, can a counterfactual theory capture this? I will consider these questions in reverse order since it will usefully narrow the discussion.

Before I do this, though, I will make a third simplifying assumption which has the effect of ruling out one possible line of escape for the counterfactual theorist. When we consider what would be the case were bp_1 to be absent, one way of retaining the presence of e is to suppose that there might be some replacement property(ies) instantiated, $bp_1{}^*$ perhaps or simply some $A(p_1, p_2, p_3 \ldots)^*$, which still causes e. We don't have a case of redundant causation exactly because these replacement properties are not actually instantiated. Nevertheless, they would be in the envisaged counterfactual circumstances. Call this *close-world redundant causation*. There are ways to deal with this within the counterfactual framework (see, e.g., Ganeri, Noordhof and Ramachandran 1996, 1998; Ramachandran 1997; Noordhof 1999c). The important point is that they should offer no succour to the counterfactual theorist with regard to the present problem.

With the assumption in place, we should conclude that if the occupant attributing properties were not instantiated, then e would not occur either. That would be so, even if $A(p_1, p_2, p_3 \ldots)$ were still present. The changes in the laws required for the absence of the occupant attributing property would also make $A(p_1, p_2, p_3 \ldots)$ not a cause of e. So the challenge to the counterfactual analysis of property instance causation is direct. If occupant attributing properties are inefficacious with regard to e, the counterfactual analysis will not capture this verdict. For other broadly physical properties which don't have, as part of their supervenience-base, laws relating narrowly physical properties—macro-properties might be an example—their absence would mean that $A(p_1, p_2, p_3 \ldots)$ is absent. According to my account of broadly physical properties, there is a metaphysically necessary relationship between instances of $A(p_1, p_2, p_3 \ldots)$ and these other broadly physical properties. So, once more, e would not occur. This time because its cause, $A(p_1, p_2, p_3 \ldots)$, is absent.

It seems that the counterfactual theory of property instance causation cannot allow that broadly physical properties of either sort fail to have efficacy in the circumstances envisaged. Yet these verdicts are, to put it mildly, contested. Here is how Jaegwon Kim expresses the worry (e.g. Kim 1998). I adapt it for the particular account of the relationship between broadly physical properties and narrowly physical property causes I defended in the previous section. First, it is assumed that physics is complete, in our terminology the sufficient causes of all instances of narrowly physical property causes are other narrowly physical property causes. Suppose that $A_1(p_1, p_2, p_3 \ldots)$ is causally sufficient for $A_2(p_{100}, p_{101}, p_{102} \ldots)$ and that the first is the supervenience-base for bp_1 and the second is the supervenience-base for bp_2. Assume, for the sake of argument, that bp_1 is a cause of bp_2; bp_1 causes bp_2 either directly or by causing $A_2(p_{100}, p_{101}, p_{102} \ldots)$. If it causes bp_2 directly, then either $A_1(p_1, p_2, p_3 \ldots)$ is insufficient for bp_2 by causing $A_2(p_{100}, p_{101}, p_{102} \ldots)$ or bp_1 is an overdetermining cause. If bp_1 causes bp_2 indirectly by causing $A_2(p_{100}, p_{101}, p_{102} \ldots)$, then the same choice holds regarding $A_2(p_{100}, p_{101}, p_{102} \ldots)$. It is implausible to suppose that there should be systematic—since the reasoning is entirely general and the situation envisaged widespread—overdetermination in either of these ways. Therefore, bp_1 is inefficacious.

At this point, the rejection of all kinds of interlevel causation can seem a tempting defence for the counterfactualist. A key idea is that causation must involve the right level of generalization and not simply determination (e.g. see Gibbons 2006). A sophisticated development of this strategy holds that different levels are associated with different patterns of close-world redundancy (Menzies this volume). Thus, if we consider what would be the case if a certain instance of a mental property were absent, we must suppose that all of its M-supervenience-bases are absent. Whereas, if we consider what would be the case if a certain

M-supervenience-base is absent, then (arguably) another M-supervenience-base of the property would be there instead. So we might be led to conclude that the mental property is efficacious and its M-supervenience-base is not.

It seems to me both that there are difficulties in implementing this strategy and problems with it as a dialectical move. On the implementation side, the kind of properties that distinguish a level often have their character specified in terms of a causal role. For instance, actions are partly characterized in terms of their causal history, part of which will include an intention. The standard response to the claim that entities so characterized cannot stand in causal relations, because they fail to be distinct from each other, is that they can because of the existence of lower level entities not specified in that fashion. Thus, intention and action are related as cause to effect because of their narrowly physical properties (e.g. Davidson 1963: 14–17). If the efficacy of the lower level is not allowed to transmit to the higher level, we have no explanation of why this lower-level causation makes it true to say that intentions cause actions. If it is allowed to transmit, then it is unclear why the lower-level physical properties cannot be counted as causes of actions as well, so giving rise to the familiar concern.

As a dialectical move, the worry about the denial of interlevel causation is that, even if causation is not simply a matter of determination, this is a necessary condition for causation. Suppose we give the term 'cause' to the deniers of interlevel causation and, thereby, accept that narrowly physical property instances are not causes of higher-level properties and vice versa. Deniers of interlevel causation must still accept that there is a competition of determination. If it can be shown—and this is what Kim's argument purports to do—that determination resides at the level of narrowly physical property instances, then it is still not plausible that the higher-level properties count as causes of entities at their own level. If, on the other hand, determination transmits, then there will be no problem. But in those circumstances, we could go straight to GO with a determination account of causality, and property instance causation in particular.

A proponent of workers' rights might also protest that the honest toil of the narrowly physical properties should not be impugned because, if the work had not been done by one of them, it would have been done by another, and should not accrue to the mental property just because it would be there whichever narrowly physical property was in play. According to such a picture, it would seem that something would only be credited with the work it did if there were nothing else on hand to do it in its place.

The problem that my preferred way of distinguishing between non-reductive physicalism and emergent property dualism presents, then, may be put like this. The counterfactual theory of property instance causation and Kim's argument are in conflict over the cases. The metaphysically necessary relationship between the supervenience-base of broadly physical properties and these broadly physical properties is a plausible case of counterfactual dependence spuriously indicating causal dependence. Successful defence of a counterfactual theory of property instance causation requires a treatment of this concern together with a diagnosis of what is wrong with Kim's argument. A similar challenge would be faced by conditional chance-raising accounts of causation (see, e.g., Sober 1999: 145–8).

3. AN ACCOUNT OF PROPERTY CAUSATION

As I have already noted, an account of property causation breaks down into two components: an account of property instance causation and the element of generality which grounds the claim that it is the property and not just a particular instance of it that is efficacious. If a counterfactual analysis of property instance causation is appropriate for the first component, we need an explanation of how, with some qualifications, if the M-supervenience-base properties of a broadly physical property are efficacious, then the broadly physical property is efficacious. As we saw in the previous section, the counterfactual analysis will pronounce most of them to be so. Without such an explanation, its credibility is undermined.

To this end, I wish to defend a qualified version of the principle that

(TC) If an instance of $A(p_i, p_{i+1}, p_{i+2}\ldots)$ is a cause of e and $\Box_m(x)(A(p_i, p_{i+1}, p_{i+2}\ldots)x$ 6 $bp)$, then the instantiation of bp is a cause of e.

I have labelled it (TC) for transmission of causality principle since it is a response to the intuition that there is no doubting the efficacy of narrowly physical property causes. The question is whether there is any further property causation. In fact, I question whether this is the appropriate way to look at things. The more neutral way of posing the issue is whether there can be harmony regarding the attribution of efficacy at different levels. Since this does not alter the character of the argument to follow, I have suppressed such qualms.

I have a two part defence of the principle. The first is a generalization from intuitively efficacious broadly physical properties. The second is an accusation of asymmetry.

If we begin with mental properties and are convinced that they are broadly physical because we are physicalists, then the combination of Kim's argument plus residual worries about how mental properties fit with the narrowly physical world can convince us that they are inefficacious. They just don't seem the right kind of thing to have an impact upon the instantiation of narrowly physical properties. If we begin instead with earthquakes, rivers or glaciers, match lightings and other events characterized by macro-properties, Kim's argument seems much less compelling. Instantiations of all of these properties are metaphysically necessitated by some arrangement of narrowly physical property causes and yet they are efficacious. This provides some support for the transmission of causality principle.

Simple cases of determinable and determinate properties provide us with additional support. If my eleven-stone weight caused the chair to break, it seems churlish to deny that my having some weight or other did so, too. Obviously, in this case, the former seems more adequate than the latter (a matter I will come back to in a moment). There are examples in which the more determinable property captures what is required where a determinate seems superfluous. According to the myth, bulls are enraged by red capes. Presumably, then, the fact that this particular cape is scarlet is beside the point. If there is no univocal proclamation in favour of the determinate, then it is plausible that determinable properties may still be efficacious in circumstances in which they are inadequate but necessary. If there is no univocal proclamation in favour of the determinable, then it is plausible that determinate properties may be efficacious while still having superfluous features.

One response to this line of reasoning is to reject it on the grounds that focus on mental properties has simply helped us to question an unjustified orthodoxy with regard to the other broadly physical properties mentioned, being an earthquake, a river, or a match lighting. This is not a response that Kim emphasizes. Instead, he seeks to introduce a block to the generality of the argument, so isolating mental properties from their unproblematic broadly physical potential cousins. There are some cases in which supervening properties are inefficacious because efficacy resides in their supervenience-base and there are some which are efficacious: the micro-based properties. These he characterizes as follows.

> P is a micro-based property just in case P is the property of being completely decomposable into nonoverlapping proper parts, $a_1, a_2, \ldots a_n$ such that $P_1a_1, P_2a_2, .. P_na_n$ and $R(a_1, a_2, \ldots a_n)$. (Kim 1998: 84)

Intuitively efficacious broadly physical properties turn out to be micro-based, mental properties sadly do not.

Kim's response has caused puzzlement. The argument at the end of the previous section seemed to rely upon the following facts. First, the supervenience-base properties necessitated broadly physical properties. Second, instantiations of supervenience-base properties were sufficient causes of other supervenience-base properties. These facts hold just as much for the relationship between arrangements of narrowly physical property causes and micro-based properties as they do between the former and other supervening properties (Noordhof 1999b: 109–14). I take it that an object O's possession of a micro-based property is distinct from there being entities $a_1, a_2, \ldots a_n$ such that $P_1a_1, P_2a_2, .. P_na_n$ and $R(a_1, a_2, \ldots a_n)$ for two reasons. First, the micro-based property is a property of the object rather than a series of properties of the entities $a_1, a_2, \ldots a_n$, together with a relation between them. Second, micro-based properties may be variably realized: the property of being an earthquake would be a good example. Because of this, I don't know what to make of Kim's response that his argument does not apply to micro-based properties because the idea of a micro-based property having a micro-base is obscure and there is no relation of determination (as opposed, perhaps, to identity) between there being entities $a_1, a_2, \ldots a_n$ such that $P_1a_1, P_2a_2, .. P_na_n$ and $R(a_1, a_2, \ldots a_n)$ and the micro-based property (see Kim 1999a: 116–17).

Without a block, it seems to me that we are more confident that properties such as being an earthquake are efficacious than we are about the principles behind Kim's argument. One way of supporting this position is to argue that we need to refine our notion of the causal completeness of physics to allow that it is not impugned by certain other non-narrowly physical property instances being counted as causes. For instance, we might say that physics is complete partly because its proprietary properties necessitate broadly physical properties. The efficacy of bp accrues to $A(p_i, p_{i+1}, p_{i+2} \ldots)$ because of this fact. Even if physics provides complete coverage and focuses on the very small (both contentious claims), there is no reason to insist that this should mean that efficacy is always located first and foremost at the narrowly physical level and then must be spread outwards as it were. Alternatively, we may argue that instances of broadly physical properties just identify a component of the causal relationship between narrowly physical property instances or have as parts instances of narrowly physical causally related properties. Regarding the first of these two possibilities, there is no reason to suppose that every such component should be identified by physics with it own distinctive explanatory aims. All of these options involve explanatory burdens that are not easy to meet but not of the character that threatens the plausibility of this line of response. We shouldn't abandon a philosophical strategy, anymore than we should abandon the search for a cure of a disease, because it is difficult.

Another related way of supporting the recommended position focuses on refining our understanding of overdetermination. Overdetermination is not simply a matter of there being two distinct causes of certain target effect. Two property instances overdetermine an effect only if they are independent causes where this is understood to mean that either one would cause the effect without the other. Two property instances would be genuine causal competitors only if they are independent causes in this sense (Bennett 2003: 476–92; Noordhof 1999a: 305). My characterization of broadly physical properties in terms of M-supervenience rules them out as independent causes of the effects caused by arrangements of narrowly physical properties and hence denies that there is genuine causal competition.

It is not open to opponents simply to resist this refinement of overdetermination and stick with the original one. Everybody agrees that two links in the same causal chain are not competitors and don't overdetermine a certain effect. So we need some characterization of why there is no overdetermination in this case. A flat-footed response is to insist that they are not overdetermining causes because they bring about the effect by the same causal chain. However, this is not right. The earlier property instance in a causal chain involves connecting causal elements that the later does not.

The next move is to say that the two causal chains should not have wholly overlapping parts (presumably at the section of the chain leading up to the effect). However, two events may cause an effect by two chains with a wholly overlapping part leading up to the effect if either event would not have been able to cause the effect without the other. Imagine that the effect requires vigorous stimulation which can only be brought about by the events in concert. The idea we need is that one chain leading from a cause to the effect could have brought about the effect in the absence of the other, that is, that they are independent causes. However, this is precisely what we don't have in the case of narrowly physical property instances and the broadly physical properties which supervene upon them. Thus a necessary condition for genuine overdetermination is not met.

The second part of my strategy regarding the defence of the transmission of causality principle was the charge of unmotivated asymmetry. The charge concerns the apparently *initially* different attitudes taken to, for instance, macro-causal relations and other macro-properties. To begin with, the existence of instances of macro-properties is not supposed to be in doubt. They are taken to supervene upon arrangements of instances of narrowly physical micro-properties. Afterwards, when their efficacy is impugned, the existence of instances of macro-properties may be denied but it is important to recognize that this is not the initial situation. It is agreed on all sides that instances of micro-causal relations exist. In which case, the instances of macro-causal relations should supervene upon an arrangement of them in the same way. To deny this seems an unmotivated asymmetry. What is so strange about the macro-causal that its instances should not be constituted from instances of micro-causal relations in the same way that instances of macro-properties in general are constituted from instances of narrowly physical micro-properties? The assumption seems to be that *if* the micro-causal relations fix what must happen, then the macro-causal relations have nothing further to contribute. Yet if instances of micro-causal relations do constitute an instance of a macro-causal relation, then the macro-causal relation has already made its contribution and to demand it again is to be unnecessarily insistent.

We have seen that there is one major class of cases in which the transmission of causality principle is defensible. The efficacy of determinable properties—my other illustration of the intuitive force of the transmission principle—has been questioned. The main charge is that determinate properties have all the causal powers attributed to determinable properties and hence there is no reason to take the latter to exist. It appeals to what has been called the 'subset' account of realization. An object that has the *determinate* property of being a sphere of a certain size has all the causal powers that accrue to it as a result of that property. Obviously if the object is a sphere of a certain size, it is also simply a sphere. Yet, as the reasoning goes, the latter property is not required by the object to confer the causal powers distinctive of being simply spherical. These causal powers are simply a subset of the causal powers of being a spherical object of a certain size. What confers no causal powers does not exist (Gillett and Rives 2005: 490–1).

The reasoning appears questionable in two respects. First, it seems tacitly to rule out the possibility that the instantiation of a determinate property has some of its causal powers *because* it involves the instantiation of a determinable property. This seems to stem from an antecedent commitment to the non-existence of the determinable. In which case, the argument can hardly hope to establish in non-question-begging fashion that determinables do not exist. If instantiations of determinate properties have the determinable's subset of powers by involving their instantiation, then there will be no double-counting of causal powers (a charge made by Gillett and Rives (2005: 486–7)). It is common ground that there is a subset of causal powers and, by being a subset of causal powers, we have a genuine resemblance between objects which possess that subset. That seems a good basis to conclude that determinable powers exist even if the subset view is correct.

Second, in fact, it is not clear that instances of determinable properties have causal powers which are simply a subset of an instance of the determinate property that realizes them. Suppose that there is a circular hole of 4 cm width. A sphere 4 cm or less can go through it, a cube would have to be around 2.8 cm or under to go through. The property of being a sphere has causal powers, then, that the property of being a 5 cm sphere does not, namely passing through the hole for a range of sizes at or below 4 cm.

The point may be made even more plausibly with regard to particular ways in which the property of being spherical may be realized, which are not appropriately characterized as determinates of the determinable being spherical. For example, suppose a spherical rock travels through a window at speed leaving a spherical hole in the glass (along with cracks spreading outwards). A spherical piece of cotton wool has no such effects. The property of being spherical has causal consequences the property of being a spherical piece of cotton wool does not, namely that the former is a cause of the spherical character of the hole. Such cases only appear problematic if you assume that the powers of a determinable must be derived from a determinate rather than simply be due to the determinables in their own right.

A natural response is to say that, although a determinable property doesn't have a subset of the causal powers of its determinate realizing property on a particular occasion, it can be thought of as having a disjunction of subsets of the causal powers, one subset for each of its determinates (Gillett and Rives 2005: 502, fn. 10). That means that, for a particular instantiation, the determinable property does not have the causal powers it has when instantiated with another of its determinates.

This is fine as a position but it cannot constitute an *argument* against the existence of determinables and, in particular, not a causal argument against their existence. It presumes that there is no reality to a determinable other than its instantiation by a particular determinate of it. Standardly, an object or property has its causal powers regardless of whether it is in circumstances conducive to their manifestation. Applying this thought to the case in hand, the property of being spherical has the power to go through spherical holes for a certain range of sizes or make spherical holes in windows regardless of whether, on a particular occasion, it cannot because it has too great a size or is a characteristic of cotton wool. The claim can only be dismissed if it is thought that the existence of the determinable on a particular occasion is simply the determinate in which it is instantiated. Naturally enough, with this assumption, it will seem obvious that determinables have no distinctive powers free of their determinates and hence that determinables don't exist.

Of course, determinables require some determinate or other to be instantiated. This does not mean that, given all the determinates of a world, we have explained all the causal powers in that world. A world does not involve instantiations of all a determinable's determinates. Even if it did, we would still lack an explanation of how, for a particular instantiation of a determinable, it has causal powers which outstrip those of its determinate. The fact that there are other determinates with different causal powers seems to have no impact upon the causal powers of the particular determinate in question.

If determinables don't exist, I do not pretend to have shown they do. The argument of the last few paragraphs rested upon the assumption that they did exist and considered what followed from that. Nor, then, would we have a counter-example to the principle of the transmission of causality. Efficacy does not transmit to entities that do not exist (of course). My claim is simply that if there is a metaphysically necessary relationship of the required type between instantiations of properties (each of which exist), then the necessitated will be efficacious if the necessitating are.

The support I have produced for the causal transmission principle does not have the upshot that all occupant attributing properties are efficacious. The proper treatment of this type of case divides into two, depending upon whether or not the occupant attributing properties are part of a powers ontology. Suppose, first, that we are not considering properties for which a powers ontology holds. The narrowly physical properties have a causal role to play, settled by the laws that hold. These laws don't M-supervene upon which properties are instantiated (though the pattern of properties that are instantiated may determine which laws hold) (for further discussion, see Noordhof 1997). In those circumstances, the occupant attributing properties are not M-supervenient upon the narrowly physical properties (or particular spatio-temporal arrangements of them). Of course, they would M-supervene upon narrowly physical properties plus the laws (or the particular patterns of properties, which determines that the laws hold if a regularity view of laws is adhered to). However, we should simply exclude this characterization of the M-supervenience-base. So, one qualification to the transmission of causality principle is that the m-supervenience-base should include neither laws so understood nor the patterns of properties which settle which laws hold.

Suppose now we consider what we should say about occupant attributing properties in a powers ontology. There are at least two options to consider here. If the laws M-supervene upon particular narrowly physical properties (say) then they will, in effect, be part of the supervenience-base for occupant attributing properties. However, this should not threaten the efficacy of such occupant attributing properties. In such circumstances, the transmission of causality principle simply reflects how combinations of causal role at one level—the narrowly physical—relate to the instantiation of causal roles at another level—the mental. So the qualification I made earlier about how the laws should not be part of the M-supervenience-base should be understood to concern only laws which themselves don't M-supervene upon which properties are instantiated. The principle so formulated provides a constraint upon how we should understand the idea that causes and effects must be distinct existences.

In the previous section, I explained how, even if one is committed to a powers ontology, one need not suppose that, in every world, if a property is instantiated, a particular causal role (and hence particular laws) are instantiated. In that case, even according to the powers ontology, laws do not M-supervene upon which properties are instantiated. However, in that case, the occupant attributing properties do not M-supervene upon narrowly physical properties either. Putting these ideas together, we have the following general treatment. Occupant attributing properties are not efficacious if they M-supervene upon a supervenience-base which includes laws which, themselves, fail to M-supervene upon which properties are instantiated.

Another qualification to the transmission of causality principle,

(TC) If an instance of $A(p_i, p_{i+1}, p_{i+2}\ldots)$ is a cause of e and $\Box_m(x)(A(p_i, p_{i+1}, p_{i+2}\ldots)x \rightarrow bp)$, then the instantiation of bp is a cause of e,

is needed. It arises because the simple relation of metaphysical necessitation does not ensure that the bp has anything to do with the efficacious part of the supervenience-base property. Suppose that $A(p_i, p_{i+1}, p_{i+2}\ldots)$ is a cause of e, that $A(p_k, p_{k+1}, p_{k+2}\ldots)$ (where k i) is not and further that $A(p_k, p_{k+1}, p_{k+2}\ldots)$ metaphysically necessitates bp* (where bp* bp). Then $A(p_i, p_{i+1}, p_{i+2}\ldots)$ & $A(p_k, p_{k+1}, p_{k+2}\ldots)$ metaphysically necessitates bp*. Nevertheless, we would not want to conclude that bp* was efficacious. Intuitively, the efficacy of the supervenience-base had nothing to do with what was required to instantiate bp*.

For this reason, we must impose a condition upon $A(p_i, p_{i+1}, p_{i+2}\ldots)$ for it to transmit its efficacy, namely that it should be the minimal supervenience-base for the instantiation of the property to which the efficacy is being transmitted. As a first stab at this condition, we might hold that B is not the minimal supervenience-base of a property P if P is still instantiated if not B but B^- (where B^- is B without the instantiation of certain properties and no additional properties are instantiated except those which are metaphysically necessitated by the non-instantiation of these other properties) (Noordhof 1999a: 307). No doubt there may be problems with this proposal as it stands, but the general idea, it seems to me, holds good. Even if an analysis is not possible and we just have to take the idea of a minimal supervenience-base as a primitive, this does not seem to be particularly damaging. For instance, it doesn't seem determined by facts about when the supervening property is efficacious.

With these qualifications in place, we now have a defence of the verdicts that the counterfactual analysis of property instance causation provided for broadly physical properties. However, this does not complete the story. We need to explain how this may be turned into an account, not of property instance causation, but property causation. Elsewhere, and in response to the rejection of interlevel causation, I have explained why it is not productive to consider property causation to be a more refined version of property instance causation in which the putative property causes are somehow more appropriate/adequate for the effects (e.g. Noodhof 1999a: 303–7). Instead, as I have already remarked, an element of generality is needed. This is reflected by the fact that, for the narrowly physical properties, appeal to laws was required. When we turn to properties that supervene upon narrowly physical properties, I think this requirement needs to be relaxed. In its place, I have recommended the following

> F is a property cause of G only if each minimal M-supervenience base of F is such that all its instantiations would cause (or in the case of indeterminism, raise the probability of) an instantiation of one of the minimal M-supervenience-bases of G if they were in some causal circumstances C—where C may vary for each kind of supervenience-base. (Noordhof 1999a: 307)

Thus if F has three types of minimal supervenience-bases then each of these should always cause Gs with a certain minimal supervenience-base in a specified set of circumstances. It is plausible to claim that the property is a cause because the identified relationship holds for all the supervenience-bases. Thus there is no reason for supplanting such property causation by property causes which are simply the minimal supervenience-bases in question.

If my characterization of emergent properties is correct, then the condition for property causation in the case of broadly physical properties will not apply to them. Instead, there will be fundamental laws which relate the emergent properties to a particular type of effect. The situation is no different than for narrowly physical property causes. Although the emergent properties supervene in some sense upon narrowly physical property causes, they do not supervene in the way that I require for the condition to apply. Thus the requirement for emergent property causation is commensurately stronger. It does not allow that there is emergent property causation if the properties upon which emergent properties N-supervene satisfy the corresponding condition formulated in terms of N-supervenience-bases. The intuitive reason for this is that the question of whether a particular emergent property is efficacious is independent of the conditions for its instantiation (although these may be part of the causal conditions in which the emergent property operates).

Some may hold that there is a need for further refinement motivated by the thought that property causation requires not simply generality (the property of being a hammer always causes nails to go in) but also precision (it's not really being a hammer but being an object of a certain weight and resistance that is required for the nails to go in). I have defended the approach above against this line of objection elsewhere so I won't go into it here (see Noordhof 1999a, 2006).

4. EMERGENT CAUSATION

I have distinguished emergent properties from broadly physical properties and explained how a counterfactual analysis of property instance causation can be defended, and developed into an account of property causation. The basic idea is that the favoured analysis should be plugged in where mention of cause occurs in the generality condition supplied at the end of the last section. One way of understanding emergent causation would then be to say that emergent causation is simply causation by emergent properties. However, matters are not so straightforward.

A second, and perhaps central, way to understand emergent causation is as causation which is in some way novel or unpredictable when we consider the causal interactions from which it emerges. Making this more precise is difficult. It should not be thought of as what would strike us as novel because there are possible cases of emergent causation which would fail to do so.

There are two ways in which $A(p_i, p_{i+1}, p_{i+2} \ldots)$ may have consequences which fail to be a result of the laws that govern the instances of narrowly physical properties which make up the arrangement. The first way is if the arrangement has an effect that is neither the result of all the individual causal relations which the instances of properties would have outside of that arrangement, nor the result of the same laws that govern the individual causal relations applying to that arrangement (for further discussion see Noordhof 2003: 90–3). There is no puzzle, for instance, about the baseball's breaking of a window just because the breakage would not occur as a result of all the individual causal relations between elements of the baseball and elements of the glass. The same laws are at work. Here, there seems a relatively straightforward connection between apparent novelty and emergent causation. There is emergent causation because there are *emergent consequences* indicating the operation of different laws.

The second way in which there may be emergent causation is if the arrangement *fails* to have an effect which, taking into account all the individual causal relations and/or the laws concerning them, we would expect it to have. Here we don't so much have a case of emergence as *submergence*. If an arrangement has a submerged consequence, it does not follow that the expected effect fails to occur. An emergent property of the arrangement may cause it to occur. In such circumstances, we might formulate law statements concerning narrowly physical properties oblivious of the fact that arrangements have submerged consequences for which the presence of emerged properties compensates. If $A(p_i, p_{i+1}, p_{i+2}\ldots)$ is sufficient for the emergent property, then there may be no obvious need to mention the presence of the emergent property. The occurrence of the effect subsequent to the instantiation of $A(p_i, p_{i+1}, p_{i+2}\ldots)$ may seem in no way surprising. Nevertheless, we would have a case of emergent causation in the first sense. In these circumstances, the arrangement would not so much fail to have an effect as fail to have it in the way we assume.

The point has application against Kim's argument that such emergent properties will be epiphenomenal. He claims that, if $A(p_i, p_{i+1}, p_{i+2}\ldots)$ is sufficient for an emergent property to be instantiated, it is sufficient for anything which the instance of the emergent property causes. Hence, he concludes, the emergent property has no causal work to do (Kim 1999b: 32–3; see also Sober 1999: 139). The conclusion does not follow. The efficacy of the emergent property is revealed by the fact that, if the emergent property were not present, then the target effect would not occur even though $A(p_i, p_{i+1}, p_{i+2}\ldots)$ is still present. This is quite compatible with allowing that, given that $A(p_i, p_{i+1}, p_{i+2}\ldots)$ is sufficient for the instantiation of the emergent property, it will also be sufficient for the target effect. This is simply a generalization of a familiar point, that the sufficiency of preceding links in the causal chain does not threaten the efficacy of subsequent links, to non-symmetric nomologically simultaneous relations of the envisaged kind. The familiar point is something for which Kim allowed in earlier work (Kim 1993 [1989]: 252). I can see no reason for rejecting the generalization to this case.

The point just made is also compatible with, indeed supports, the following possibility that may be behind Kim's reasoning. Consider a world in which emergent property dualism is true. There will be a possible world in which the same pattern of narrowly physical property causes over time will be instantiated without the emergent properties being present. This doesn't threaten the efficacy of emergent properties in the first world, however. The world that does not need emergent properties for the causal relations to hold will be a world with different laws to the laws that hold if emergent property dualism is the case. What proves to be insufficient in some worlds in virtue of the laws may be sufficient in others. Nor does this point rest upon rejecting the idea that a powers ontology is true (the laws being fixed simply by the kinds of properties that are instantiated), given the points made in section 1.

Emergent causation in the second sense identified—that involving either emergent or submergent consequences—does not seem to require the existence of emergent properties as causes. Indeed, given that emergent properties, themselves, are emergent consequences, it appears that there must be at least some emergent causation in which there are no emergent causes.

The suggested characterization of emergent causation in the second sense is a relatively straightforward application of the account of the distinction between the broadly physical and the emergent outlined in section 1.

xAy *strongly supervenes* on xBy just in case, necessarily, for each x and each causal relation in A, if xFy, then there is a pattern of (micro-)causal relations G such that xGy and, necessarily, if any x, y stand in G, then they stand in F i.e. □ (x)(y)(xFy & F *e* A □ (G) (G *e* B & xGy & □ (u)(v)(uGv □ xFy))).

A-relations are complex characterizations of exactly the way in which two things are causally related and so two things may stand in many such causal relations. The B-relations are characterizations of patterns of causal relations between x and y's proper parts, for example, if x and y are property instances, then they hold between parts of their minimal supervenience-bases. Thus x stands in G to y if (say) x has parts p_{x1}, p_{x2}, p_{x3} . . .p_{xn} which are causes of the respective parts of y, p_{y1}, p_{y2}, p_{y3} . . .p_{yn}. Suppose that, in a wide range of cases, perhaps in all, if a part of the supervenience-base of x is a cause of p_{yi}, then x is a cause of p_{yi} (in virtue of that part). It would not follow that an emergent causal relation does not hold between x and y. There may be a causal relation between the parts and also an emergent causal relation between x and y taken as a whole. Hence the importance of not thinking of the A-relations as simply whether or not a causal relation holds.

The second modal operator is once more to be understood as that of metaphysical necessity. If a case of non-emergent causation is just an arrangement of causal relations between instances of narrowly physical properties, then metaphysically necessarily, with this arrangement, non-emergent causation will be present. On the assumption that part of the supervenience-base for the existence of causal relations are the laws which cover them (when they do), then laws implicated in the existence of causal relations between instances of narrowly physical properties will be part of the legitimate supervenience-base of causal relations between instances of broadly physical properties. If these laws imply that instances of broadly physical properties will have causal relations which are not simply an arrangement of the causal relations of instances of narrowly physical properties, then, once more, these causal relations will hold in all possible worlds in which the supervenience-base is realized. Emergent causation will occur when neither of these two circumstances is met, most specifically, when the laws which explain how an arrangement of narrowly physical properties has certain causal consequences are not those which are implicated in the causal relations between instances of narrowly physical property instances alone. Since these laws may fail to hold while the laws concerning narrowly physical properties still hold, the second modal operator would then only be that of nomological necessity.

One concern about this characterization of emergent causation is that it is too strong. There may be emergent causation if laws covering narrowly physical properties have distinctive consequences when applied to their arrangements. There are two possibilities here. Either the laws that cover the instances of the narrowly physical properties of the arrangement have distinctive consequences for the arrangement or laws that cover other instances of narrowly physical properties have distinctive consequences for an arrangement which does not involve these other instances. The latter would happen if properties of the arrangement were not shared by the property instances arranged. To give a toy illustration, laws covering square things may apply to square particles and square arrangements of round particles.

Either way, this would make emergent causation extremely ubiquitous and weak. Even if the law covering narrowly physical properties is additive, the result may yield distinctive causal consequences—for example, the smashing of a window by a baseball—which cannot be described simply in terms of the addition of the various causal contributions of the instances of narrowly physical properties but is partly the result of the nature of the thing to which these contributions are applied. Perhaps such cases might be excluded—and the notion of emergence thereby become stronger—if it is not so much a matter of whether there are distinct causal consequences as a result of the nature of the interacted upon but rather simply whether the laws are additive or not. Additive laws—such as additive force laws—are of particular significance if there are unit values for the determinable properties the law concerns. In that case, a law concerning the determinable properties can be reduced to laws about determinate properties with these unit values. So there is certainly a type of emergence to recognize here. It is a version of the weak emergence characterized in the previous section applied to causal relations. Carl Gillett has argued that this was the position of the pre-eminent British Emergentist, Samuel Alexander (Gillett 2006: 275–6, 285–6, and references to Alexander's work therein; Alexander 1920). By contrast, Brian McLaughlin holds that Samuel Alexander, and others, supposed that there were fundamental forces which were only exerted by types of configurations of particles (McLaughlin 1992: 52, 66–7).

A second concern about emergent causation (understood in the way I have recommended) is that any causal powers which are putatively possessed by the emergent property seem legitimately attributed to narrowly physical properties instead (Ginet suggested this, as reported by O'Connor (1994: 89–99). Suppose that $A(p_i, p_{i+1}, p_{i+2} \ldots)$ is sufficient for the instantiation of a particular emergent property, E_1 and that, in circumstances C, E_1 putatively causes E_2. Then p_i has the following power, in circumstances $A(\ldots, p_{i+1}, p_{i+2} \ldots)$ plus C, p_i is a cause of E_2. In which case, any candidate case of emergent causation would turn out to involve causal relations between instances of narrowly physical properties alone, so satisfying my analysis of non-emergent causation (see also Shoemaker 2002).

Considerations of theoretical simplicity are unlikely to speak in favour of postulating an emergent property. Although we would need only to attribute the disposition to instantiate E_1 in circumstances C to p_i, the number of dispositions we are attributing overall would be greater. There would be all the dispositions attributed as a result of the instantiation of E_1 together with p_i's disposition to instantiate E_1 (as O'Connor and Wong acknowledge (2005: 672)).[3]

Special arrangements of instances of narrowly physical properties, those with distinct causal powers, are emergent phenomena. The point is not that, if we recognize special arrangements of such properties, we can avoid recognizing an emergent phenomenon (see O'Connor 2000a: 113–14). The issue is rather whether we need to recognize emergent properties in addition to emergent causation. There does not seem to be any reason to do so. Of course, if a causal theory of properties is true, then the distinct causal powers attributed to the arrangement of instances of narrowly physical properties, rather than the instances of narrowly physical properties themselves, will be sufficient to establish the existence of an emergent property. Nevertheless, it is important to recognize that this additional premise is required.

The correct response to the second concern is that emergent causation is present, not because causal relations between instances of narrowly physical properties fail to necessitate it, but because the laws envisaged attribute fundamental forces to narrowly physical properties when they occur as part of a certain kind of arrangement or, in some way, seem isolated from the other laws that hold. Such laws should not figure as part of the supervenience-base in the account given earlier. This is more important than to which property the causal power should be attributed. Nevertheless, there can be reason to postulate the existence of emergent property, either because of the pattern of causal relations we observe or because we have other, non-causal reasons for believing in its existence, for example to characterize the nature of our experience. In the last section of this chapter, I summarize the various kinds of emergence I have distinguished and relate them to two possible cases of emergent phenomena.

5. CONCLUDING REMARKS AND POSSIBLE CASES OF EMERGENCE

I have argued that strongly emergent properties are to be distinguished from broadly physical properties by the fact that the strongly emergent properties

[3] Unfortunately, for reasons of space, I cannot discuss O'Connor and Wong's argument that without postulating emergent properties we would be left with action at a distance (see O'Connor and Wong 2005: 672–3).

are only related to their supervenience-base by nomological necessity. Weakly emergent properties are a subcategory of broadly physical properties in which the properties are in some way hard to infer from arrangements of narrowly physical properties. Causation by strongly emergent properties is a species of emergent causation. Nevertheless, broadly physical properties may also stand in emergent causal relations. These also come in strong or weak versions. According to the strong version, emergent causal relations are related only by nomological necessity to patterns of causal relations between instances of narrowly physical properties. Weak emergent causation just insists that narrowly physical property instances should have different powers when taken together in a particular arrangement than they would taken in isolation. Physical laws will reflect this.

Quantum mechanics has been thought to support emergent physicalism, that is, narrowly physical properties which are emergent from other narrowly physical properties. Suppose that emergent physicalism is true. Is there any reason to be an emergent non-physicalist due to mental phenomena?

Phenomenal consciousness does not seem to be captured by recognizing emergent causation by non-emergent properties. We have little evidence for the existence of emergent causation as a result of phenomenal consciousness unless we take seriously the idea that phenomenal consciousness can only be captured by non-physical phenomenal properties. Intercranial causal relationships seem to be nothing out of the ordinary. Indeed, that's partly why the causal completeness of the physical realm seems so plausible. Of course, if there are emergent phenomenal properties, then there will be emergent causation in both senses. For that reason, from the point of view of the philosophy of mind, emergent physicalism seems to me to be an unattractive doctrine. It postulates emergent causation for no reason and yet denies that phenomenal properties are broadly non-physical. This reservation does not apply to Gillett's emergent physicalism only involving weakly emergent causation (Gillett 2003, 2006).

Should we be forced to accept the existence of non-physical emergent phenomenal properties, this won't just be because it is difficult to attribute a predicate to anything other than a system of entities or because it is difficult or impossible to attribute a predicate on the basis of the knowledge of boundary conditions plus dynamic laws (see Heard (2006: 58–61), who attributes these commitments to the emergentist). If it is simply difficult but not impossible to attribute a predicate, or only epistemically impossible for McGinn-style closure reasons, then there are no emergent properties in the sense we have identified. No sensible emergentist is going to accept that there is a straightforward inference from the fact that something can only be predicated of a whole system, and that the development of a system can be understood in terms of laws that make no reference to the activity of elements, to the claim that there exist emergent properties. Rather, this proposal will be assessed in the light of competing considerations of which the observation just mentioned is only one. Deciding that there are emergent non-physical phenomenal properties will be the complex result of balancing causal arguments for physicalism against different possible accounts of the explanatory gap and our introspective experience. Thus O'Connor emphasizes the apparent non-structurability and subjectivity of phenomenal properties, and Kim that they apparently cannot be functionalized (O'Connor 2000a: 116–17; Kim 2005).

Libertarian accounts of free will provide another potential case of emergence. According to some, agents have a property in virtue of the fact that they are able to cause events to happen in their bodies (see, e.g., O'Connor 2000a: 121–3). A certain way of understanding libertarian free will may seem to threaten the claim that emergent properties even N-supervene upon narrowly physical properties. Suppose that $A_1(p_1, p_2, p_3 \ldots)$ indeterministically causes $A_2(p_{100}, p_{101}, p_{102} \ldots)$ and in one subject, when in $A_1(p_1, p_2, p_3 \ldots)$, the subject considers whether to imagine sensuously purple, decides to and, when in $A_2(p_{100}, p_{101}, p_{102} \ldots)$, he or she does. In another subject, his or her decision may go the other way. If you insist that $A_2(p_{100}, p_{101}, p_{102} \ldots)$ will still be caused by $A_1(p_1, p_2, p_3 \ldots)$, then this can't be even the N-supervenience-base for imagining sensuously purple. The argument can presumably generalize (see O'Connor 2000b: 111–12).

Two points should be recognized. The first is that this way of understanding libertarian free will is not required. We can keep the connection between $A_2(p_{100}, p_{101}, p_{102} \ldots)$ and imagining sensuously purple so long as the decision to imagine another colour instead indeterministically causes a different arrangement of narrowly physical properties $A_3(p_{1000}, p_{1001}, p_{1002} \ldots)$. There is no reason to suppose that the physico-psycho law is indeterministic. Indeed, there is no need to suppose that libertarian free will involves additional emergent properties. We can just suppose that there is emergent causation. If the world is indeterministic, there will be a range of possible events which might occur in agents' brains. Which one does on any occasion is not just by chance nor by law, but directly influenced by an agent. Nor need such a picture involve a revision to the laws of physics since the range of alternatives are set by them. All we would have is one unlikely sequence rather than another actualized by the agent. If the causation in question is a case of agent causation, then it will be emergent not because of the application of some new law but because of the activity of an agent.

The second point is that, if there is interlevel indeterminism (and why not), then the supervenience claim should be understood correspondingly. Thus N-supervenience will postulate an indeterministic but still nomic relationship between narrowly physical properties and mental properties. It will no longer be appropriate to say that if two subjects are identical in their narrowly physical properties, then they should be identical in their mental properties. Rather, the claim will be that the probabilities of various mental properties will be the same.

However, as I have already indicated, this dimension of the position is not mandatory.

Thus, we see that the distinction between emergent properties and emergent causation drawn in this chapter enables us to differentiate between emergent approaches to the mind that are a challenge to physicalism and those which are not.

REFERENCES

Alexander, S. 1920. *Space, Time and Deity: The Gifford Lectures 1916–1918*, 2 vols. London: Macmillan.

Bedau, M. A. 1997. 'Weak Emergence'. In J. E. Tomberlin (ed.), *Philosophical Perspectives* 11: 375–99.

Bennett, K. 2003. 'Why the Exclusion Problem Seems Intractable, and How, Just Maybe, to Tract It'. *Noûs* 37.3: 471–97.

Chalmers, D. 2006. 'Strong and Weak Emergence'. In P. Clayton and P. Davies (eds), *The Re-Emergence of Emergence*. Oxford: Oxford University Press.

Davidson D. 1980 [1963], 'Actions, Reasons and Causes'. *Journal of Philosophy* 60, reprinted in his *Essays on Actions and Events* (Oxford, Oxford University Press), pp. 3–19.

Ganeri, J., Noordhof, P. and Ramachandran, M. 1996. 'Counterfactuals and Preemptive Causation'. *Analysis* 56: 216–25.

—— 1998. 'For a (revised) PCA Analysis' *Analysis* 58.1: 45–7.

Gibbons, J. 2006. 'Mental Causation without Downward Causation'. *The Philosophical Review*, 115.1: 79–103.

Gillett, C. 2003. 'Non-Reductive Realization and Non-Reductive Identity'. In S. Walter and H.-D. Heckmann (eds), *Physicalism and Mental Causation*. Charlottesville, VA: Imprint Academic, 31–57.

—— 2006. 'Samuel Alexander's Emergentism: Or, Higher Order Causation for Physicalists'. *Synthese* 153: 261–96.

—— and Rives, B. 2005. 'The Non-Existence of Determinables: Or, a World of Absolute Determinates as Default Hypothesis' *Noûs* 39.3: 483–504.

Heard, D. 2006. 'A New Problem for Ontological Emergence'. *Philosophical Quarterly* 56.222: 55–62.

Heller, M. 1998. 'Property Counterparts in Ersatz Worlds'. *Journal of Philosophy* 95.6: 293–316.

Hüttemann, A. and Papineau, D. 2005. 'Physicalism Decomposed'. *Analysis* 65: 33–9.

Jackson, F. 1998. *From Metaphysics to Ethics*. Oxford: Oxford University Press.

Kim, J. 1993 [1984]. 'Concepts of Supervenience'. In idem, *Supervenience and Mind*. Cambridge: Cambridge University Press, 53–78.

—— 1993 [1989], 'Mechanism, Purpose or Explanatory Exclusion'. In *Supervenience and Mind*. Cambridge: Cambridge University Press, 237–264.

Kim, J. 1998. *Mind in a Physical World*. Cambridge, MA: The MIT Press.
—— 1999a. 'Supervenient Properties and Micro-Based Properties: A Reply to Noordhof'. *Proceedings of the Aristotelian Society* 99: 115–18.
—— 1999b. 'Making Sense of Emergence'. *Philosophical Studies* 95.1–2: 3–36.
—— 2005. *Physicalism, or Something Near Enough*. Princeton, NJ: Princeton University Press.
Lewis, D. 1986a [1979], 'Counterfactual Dependence and Time's Arrow'. *Philosophical Papers*, vol. 2. Oxford: Oxford University Press, 32–52.
—— 1986b. *On the Plurality of Worlds*. Oxford: Basil Blackwell.
McGinn, C. 1991 [1989], 'Can we solve the Mind-Body Problem?' *Mind* 98.391: 349-66, repr. in his *The Problem of Consciousness* (Oxford: Basil Blackwell, 1991), 1–22.
McLaughlin, B. 1992. 'The Rise and Fall of British Emergentism'. In A. Beckermann, H. Flohr, and J. Kim (eds), *Emergence or Reduction? Essays on the Prospects of Nonreductive Physicalism*. New York: Walter de Gruyter, 49–93.
Noordhof, P. 1997. 'Making the Change: the Functionalist's Way'. *British Journal for the Philosophy of Science* 48: 233–50.
—— 1999a. 'Causation by Content?'. *Mind and Language* 14.3: 291–320.
—— 1999b. 'Micro-based Properties and the Supervenience Argument: a Response to Kim'. *Proceedings of the Aristotelian Society* 99: 109–14.
—— 1999c. 'Probabilistic Causation, Preemption and Counterfactuals'. *Mind* 108.429: 95–125.
—— 2003. 'Not Old . . . But Not That New Either: Explicability, Emergence and the Characterization of Materialism'. In S. Walter and H.-D. Heckmann (eds), *Physicalism and Mental Causation*. Charlottesville, VA: Imprint Academic, 85–108.
—— 2006. 'Environment-Dependent Content and the Virtues of Causal Explanation'. *Synthese* 149: 441–575.
O'Connor, T. 1994. 'Emergent Properties'. *American Philosophical Quarterly* 31.2: 91–104.
—— 2000a. *Persons and Causes*. Oxford: Oxford University Press.
—— 2000b. 'Causality, Mind and Free Will'. In J. E. Tomberlin (ed.), *Philosophical Perspectives* 14. Boston MA and Oxford: Blackwell, 105–17.
—— and Wong, Hong Yu. 2005. 'The Metaphysics of Emergence'. *Noûs* 39.4: 658–78.
Papineau, David (forthcoming). 'Must a Physicalist be a Microphysicalist?'.
Ramachandran, M. 1997. 'A Counterfactual Analysis of Causation'. *Mind* 106: 263–77.
Shoemaker, S. 1980, 'Causality and Properties'. In P. Van Inwagen (ed.), *Time and Change* (Dordrecht: D. Reidel) and repr. in his *Identity, Cause and Mind* (Cambridge: Cambridge University Press, 1984), 206–33.
—— 1981, 'Some Varieties of Functionalism'. *Philosophical Topics* 12.1: 83–118, repr. in his (2003), *Identity, Cause and Mind: Expanded Condition* (Oxford: Oxford University Press), 261–86.
—— 2002. 'Kim on Emergence'. *Philosophical Studies* 108: 53–63.
Silberstein, M. 2006. 'In Defence of Ontological Emergence and Mental Causation'. In Philip Clayton and Paul Davies (eds), *The Re-Emergence of Emergence*. Oxford: Oxford University Press, 203–26.
Sober, E. 1999. 'Physicalism from a Probabilistic Point of View'. *Philosophical Studies* 95.1–2: 135–74.

Van Cleve, J. 1990. 'Mind-Dust or Magic? Panpsychism Versus Emergence'. In J. E. Tomberlin (ed.), *Philosophical Perspectives*, vol. 4: *Action Theory and Philosophy of Mind*. Atascadero, CA: Ridgeview, 215–26.
Wilson, J. 2005. 'Supervenience-based Formulations of Physicalism'. *Noûs* 39.3: 426–59.
Yablo, S. 1992. 'Mental Causation'. *The Philosophical Review* 101.2: 245–80.

7

Emergence: Laws and Properties: Comments on Noordhof

Simone Gozzano

Any philosopher who wants to avoid reductive physicalism, while maintaining a form of physicalism, is committed to defend the following three theses:

(1) Mental properties are different from physical properties (hence not reducible to them).
(2) Mental properties are causally efficacious (so not epiphenomenal).
(3) There is no systematic causal overdetermination between mental and physical properties.

However, holding these three theses has proved to be difficult, because:

(A) 1 *and* 2 run against the so-called Principle of Causal Closure of the physical domain, according to which if a physical event has a cause at time *t* it has a physical cause at time *t*, and
(B) 1 *and* 2 *and* 3 may condemn physical properties to epiphenomenalism since if on a given type of causal relation mental properties are efficacious (by 2) and different from the physical properties occurring at that type of causal relation (by 1) and there is no systematic overdetermination in that type of causal relation (by 3), then physical properties are causally inert on that type of causal relation.

One way to solve at least problem A, consists in construing mental properties as emerging from physical properties. Emergent properties have to be causally efficacious and different from the properties from which they emerge. This is a very minimal requirement on emergent properties, so it is important to provide further characterizations of them, and one of the main goals of the chapter by Noordhof is how to individuate emergent properties, an issue that originates from the three theses stated at the outset of this comment. My aim here is to provide a sketch of Noordhof's strategy (section 1), to evaluate how it fares with other emergentists' approaches (section 2) and to set some sceptical remarks on its viability (section 3).

1. NOORDHOF'S MAIN POINTS

To begin with, Noordhof distinguishes between narrowly physical properties, those identified by physical sciences and mentioned in strict laws of the form (x) (Fx $\rightarrow$ Gx), and broadly physical properties that stand in some relation to the narrow ones. Emergent properties could be broadly physical, with respect to some set of narrow physical properties, or may fail to supervene on the narrowly physical. However, emergent properties will be determined by the narrowly physical ones, and the central question is whether the properties so determined are *new* in any sense or not. Emergentists think they are, while non-reductive materialists deny this, and Noordhof thinks that the contrast between these two views can be traced in different construals of his preferred formulation of Strong Supervenience (thesis **S**).[1]

Let's say that A-properties strongly supervene on B-properties just in case:

(S): $\Box$ (x) (F) (Fx *and* F $\in$ A $\rightarrow$ ($\exists$G) (Gx *and* G $\in$ B) *and* $\Box$ (y) (Gy $\rightarrow$ Fy))

That is: necessarily, if x has F and F belongs to properties of level A, then there is a property G, belonging to level B, such that x has G and necessarily if anything has G it has F. Now, the crucial problem is how to interpret the modal operators of necessity, that is, if one has to read them in the nomological ($\Box_n$) or in the metaphysical ($\Box_m$) reading. If one is taking the following reading of the modal operator one is committed to non-reductive physicalism:

(nRP-S): $\Box_n$ (x) (F) (Fx *and* F $\in$ A $\rightarrow$ (G) (Gx *and* G $\in$ B) *and* $\Box_m$ (y) (Gy $\rightarrow$ Fy)).

If one is taking the following reading one is committed to emergent (property) dualism:

(ED-S): $\Box_n$ (x) (F) (Fx *and* F $\in$ A $\rightarrow$ ($\exists$G) (Gx *and* G $\in$ B) *and* $\Box_n$ (y) (Gy $\rightarrow$ Fy)).

Both nRP-S and ED-S are ways to make explicit thesis 1. However, holding **S** in either form, but in particular in the ED reading, must be confronted with some objections, such as the view that properties are individuated through causal roles and the causal theory of properties. Noordhof thinks he can resist the attacks, but another problem that lurks behind **S** is how it fares with the counterfactual theory of property instance causation. Noordhof admits that it seems that emergent properties fare better than broadly physical properties.

The counterfactual theory, though, is construed for instances of property causation. If it can be generalized, it would show that it is properties as such,

[1] This way of expressing strong supervenience is due to Kim (1984), but see Horgan (1982) for a similar formulation.

and not only their instances, that are efficacious (meeting thesis 2 above). If a property is a supervenient one (not narrow), then its efficacy must be preserved in the passage from token to type. To this end the Transmission of Causality principle must hold:

(**TC**): if an instance of $A(p_i, p_{i+1}, p_{i+2} \ldots)$ is a cause of *e* and $\square_m$ (x) $(A(p_i, p_{i+1}, p_{i+2} \ldots)x \rightarrow bp)$ then the instantiation of bp is a cause of *e*.

TC is supported by consideration involving micro-macro relations and, to some extent, determinables-determinate relations with respect to property causation. As with thesis **S**, Noordhof provides a defence of the import such a relation has on causation in general. The purpose of **TC**, as it is defended in the paper, is to support the view that determinates' efficacy is derived from that of their determinables (Noordhof this volume: 83). However, the relation is tighter than this, so Noordhof qualifies it with a requirement that has to preserve efficacy while allowing for generality. This is the minimal base requirement: in a nutshell, a property F causes a property G just in case the minimal supervenience-base of F causes the minimal supervenience-base of G in some causal circumstances C. If such a requirement is met, then the flowing of causal efficacy is secured at each occasion by a different minimal supervenience-base, so the only way to preserve property causation is to consider the macro-property (determinable), instead of the micro-realizer of it (determinate property).

As a final point, Noordhof argues that emergent properties and emergent causation are partially independent of each other: if there are emergent properties in causal relation then there is emergent causation; however, if there is emergent causation there should not necessarily be emergent properties.

2. EMERGENTISM

Noordhof's main concern is to qualify the modal operators of the supervenience thesis, a crucial step of clarification, as has already been noted by Lewis (1986). Kim recognizes that specifying such point is an important hallmark of any supervenience thesis: 'different reading of the modal terms will generate different supervenience theses' (1984: 166). On the same score Stalnaker (1996) stresses that the force of the concept of necessity has direct consequences on the force of the reductionist thesis. The difference between nRP-**S** and ED-**S** reduces to different views of the second modal operator: metaphysical or nomological necessity? Let's consider ED-**S**. Since in it both operators have the same strength, it is possible to operate the following derivation, made explicit by Kim:

If A strongly supervenes on B, then for each property F in A there is a property G in B such that necessarily (x) $(Gx \leftrightarrow Fx)$, that is, every A-property has a *necessary coextension* in B. (Kim 1984: 170)

Here the necessity is nomological, so every A-property has a nomologically necessary coextension in B. Some emergentists are willing to accept such a consequence. Beckermann, for instance, notes that Broad was one of them: 'according to Broad, emergent properties must strongly supervene on microstructural properties. For otherwise the presence of such properties could in no way be explained by reference to the corresponding microstructure' (Beckermann 1992: 103) but this does not entail that emergent properties can be deduced from the complete knowledge of the microstructure, and this is the crucial feature of emergent properties according to Broad. So, if **S** is accepted for this reason, emergence would be nothing but an epistemical feature, and the ontological difference between mental and physical properties is lost. One may wonder on the appropriateness of applying Kim's derivation on Noordhof's formulation of supervenience: the former results in a biconditional where the latter has only a conditional. Kim's point, however, is that if the properties included in the supervenient-base are individuated via their causal roles, then the supervenience thesis comes to this much more strong formulation. If, on the contrary, one allows laws into the supervenience-base, as Noordhof is willing to do, then the conclusion drawn by Kim is not secured anymore. But on the viability of having laws included in the supervenience-base I will cast my doubts in the next section. So, specifying emergentism in terms of supervenience may prove not to be an easy task.[2]

At the opposite end of the emergentists' spectrum there is Humphreys (1997) who takes emergent properties to be the product of physical fusion processes, which cannot be captured by any supervenience relation. His positive example is quantum entanglement, where the state of the compound system determines the states of the constituents. However, it must be said, Humphreys makes explicit appeal to our ignorance, because he cannot be sure that such a case is relevant for the main reason emergent properties are discussed: that is, the mental (perhaps Penrose could use such a strategy).

Others, O'Connor and Wong (2005), who figure among the polemical targets of Noordhof, isolate emergent properties as those properties that are wholly nonstructural, that is, that cannot be analysed in decompositional terms. These properties are basic properties of composite individuals. Their strategy is somewhat analogous to that of Humphreys, in that both deny that emergent properties strongly supervene on basic properties. They diverge in that O'Connor and Wong accept global supervenience, whose autonomy from strong supervenience has been proved by Paull and Sider (1992), while Humphreys denies it. So, some emergentists reject Noordhof's strategy from the beginning. It is one of Noordhof's tasks to show that the best way of construing emergentism is in terms of strong supervenience in the nomological reading.

[2] I wish to thank Graham Macdonald for having raised this issue.

3. LAWS AND PROPERTIES

In the case of nRP reading of thesis **S**, the derivation presented by Kim should be strengthened, if metaphysical necessity entails nomological necessity (something I would like to leave open). In that case, in fact, there is nomologically necessary coextension *and* the supervenient properties are the metaphysically necessary condition for the subvenient ones (remember, the second part of thesis **S** states that (y) (Gy → Fy)).

This construal of the supervenience relation is, in Noordhof's view, consistent with the inclusion, in the supervenience-base, of laws concerning narrow causal properties, a qualification imposed by objections concerning ED-**S** with respect to the supervenient base. As Noordhof says: 'The supervenience base should not just include narrowly physical properties causes but also any laws concerning them alone' (this volume: 73). It should be kept in mind that this is the weaker reading of **S**, so such qualification holds a fortiori for the stronger case.

The relation between properties and laws is a crucial one in deciding whether there are emergent properties, because the individuation conditions of properties can be, in some interpretation of them, interdependent on scientific laws. In particular, I would like to cast some doubts on Noordhof's idea of having laws in the supervenience-base. As you may remember, the second conjunct of strong supervenience states that there cannot be variations in the supervenient properties without variations in the subvenient ones. Suppose there is a supervenient variation and we accept Noordhof's qualification: then either there are variations in the subvenient properties, or in the subvenient laws, or in both. The only way in which there can be interesting variations in the subvenient laws is by having modifications in what they state (taking laws to be universally quantified expressions of the form (x) (Px → Qx)). But if there are such modifications, then the supervenience base is changed and the supervenient relation is not secured any more. For, suppose that the mass of an object has changed. Then, there must be changes in the masses of its parts, or in their relations or, as Noordhof adds, in one of the laws concerning mass. However, it is in virtue of what these laws state that we can figure it out whether there have been changes in the mass of the whole object or in any of its parts. So, if laws are in the supervenience-base these cannot be taken for granted any more, and the supervenience relation is lost.

The relation between laws and properties surfaces again, as an annoying thorn in Noordhof's reasoning, when he defends his demarcation against the causal theory of properties by appealing to counterpart relations. For instance, he seems to think that as the shape of a particular is intrinsic to it, even if it is accidental, given that that very particular may have different shapes in other worlds, so by the same token we can allow that the laws are intrinsic to a property, while the causal role for which they are responsible is accidental, because it is counterparts to that property which possess different causal roles in different worlds (cf. Noordhof this volume:

73). Now, I think that we can accept the intrinsic/accidental distinction with respect to particulars only on the background of a more robust view of laws.

Consider an object O: it has spherical shape in virtue of its atomic or molecular structure. So, it is intrinsic to it to have a shape, even if it is accidentally spherical. However, if causal roles are somewhat disconnected from the laws that intrinsically apply to a particular, then it is possible that in other possible worlds the laws governing the causal roles of the atoms and molecules that compose object O are such to determine the causal role typical of gases. So shape cannot be counted as one of O's intrinsic properties anymore, and the distinction between O's intrinsic and accidental properties is lost, given that a property is intrinsic if it is invariant under possible worlds' transformations.

The importance of this point with respect to the general issue of emergentism can be made clear in some other ways. One of the common intuitions behind the idea of emergence, is the view that a property is emergent if it is somewhat *new*, an idea that Noordhof himself embraces. For instance, when my daughter was born, she was the bearer of many new properties: her DNA token has never been manifested before. Hardly any emergentist would be satisfied by considering her never expressed DNA as an interesting instance of an emergent property. Now, suppose she, for the first time in human kind, was born with the left iris partially blue and partially yellow. This bicoloured iris has some specific causal powers: it reflects light thus and so, causes her some excitement in sunny days and bad moods in the cloudy ones. Such a property would be a new type of property, but it would not satisfy the emergentist, I dare to say, for two reasons. First, even if new, it would be the result of the combination of known properties, not an entirely new manifestation. Second, having a bicoloured iris is an unstable property, one that we do not know how to have again, a result by chance, so to say. These two features can be considered as prominent in the analysis of property emergence: an emergence property is a *stable* and *new* property, but this is not enough.

One possibility is to conceive novelty in terms of unpredictability, but this cannot be the case. Lottery results are typically unpredictable, though not unexpectable given that, say, the winning number is included in the set ranging from 1 to 90. Another option, advanced by Chalmers, insists on the notion of deducibility: 'We can say that a high-level phenomenon is *strongly emergent* with respect to a low-level domain when the high-level phenomenon arises from the low-level domain, but truths concerning that phenomenon are not *deducible* even in principle from truths in the low-level domain' (Chalmers 2006: 244). Here it is quite important to have the notion of level clearly formulated, as otherwise there are important counter-examples to such a view. The reason why it is necessary to clarify such a notion is that Chalmers characterizes weak emergence, as opposed to strong emergence, just as the unexpectability of high-level phenomena given the principles governing the low-level domain from which they arise. Clearly, as the case of my daughter's left iris shows, weak emergence does not entail strong emergence, as Chalmers himself notes. So, the sense in which a phenomenon is weakly or strongly emergent is equivalent to it not being expectable or deducible, respectively, given the contrast between low-level domain and high-level phenomena. How should we consider levels here?

Consider what happened when Neptune was discovered. Newtonian laws of motion plus the initial conditions of the heavenly bodies whose presence was recorded at the time did not allow the deduction of the orbit of Uranus. It needed further facts, that is, the hypothesis concerning the presence of another planet behind it, Neptune. Clearly, we are facing phenomena that are at the same level, in a very intuitive sense, but not everybody would agree with such a notion.[3] This seems to force toward a revision in the definition of emergence in terms of deduction, along the following lines: not deducible from a *complete* knowledge of all the relevant facts on a given level. So, when Chalmers (2006: 245) says: 'if there are phenomena whose existence is not deducible from the facts about the exact distribution of particles and fields throughout space and time (along with the laws of physics), then this suggests that new fundamental laws of nature are needed to explain these phenomena', he says something incomplete. The non-deducibility must be from *all* the *relevant* facts. Whether such a condition entails new laws is a further matter.

The classical view of reduction holds that a theory (A) can be reduced to another theory (B) if it can be deduced from such theory plus bridge laws connecting the terms of the reducing theory (B) to those of the to-be-reduced one (A). How can a phenomenon be declared not deducible with respect to the laws appearing at a lower level if no bridge connection between the phenomenon and the terms comprised in the reducing law can be established? So, we must suppose that such a bridge could be established, in pain of making the non-deducibility trivial.[4] This would make the above definition of strong emergence stronger indeed. In fact, the truths concerning the higher-level phenomenon would be non-deducible from all the relevant truths concerning the lower-level domain, notwithstanding the presence of conditional or (even stronger) biconditional statements linking the two domains with respect to the phenomenon. This way of strengthening the definition may prove fatal for the emergence relation.

In the history of science, in fact, the introduction of new fundamental laws is ubiquitous, so would be emergence. For instance, Mendel and subsequently genetic laws are fundamental given the problem they provide an explanation to, but they cannot be derived from any physical laws plus knowledge concerning genetic facts. Moreover, very often new fundamental laws are formulated at lower level than those of the phenomena they have to explain. In order to explain

[3] For instance, not Kim (1998).

[4] The presence of bridge laws is a necessary requirement for evaluating the deducibility of 'a fact' as Chalmers put the problem, or a proposition. I am indebted for this point to Ausonio Marras, whom I thank.

the behaviour of electrons, new fundamental entities, such as quarks and other particles, and laws are introduced at a lower level. The same holds for biological sciences: isolating new diseases triggers the quest for a bio-molecular search. Science, so to say, passes from higher-level phenomena, whatever these are, to lower level laws. None of the previous cases, though, has been considered as a positive example of emergence.

All these considerations apply to the supervenience reading provided in ED-S. In fact, as we saw, from that construal it was possible to derive that every higher-level property has a necessary nomological coextension in a lower-level property. If this is the case, then there is no ontological novelty in the passage from one level to the next one, at most an epistemological novelty. I wonder whether any emergentists would be satisfied by such a result.

REFERENCES

Beckermann, A. 1992. 'Supervenience, Emergence, and Reduction', in A. Beckermann, H. Flohr, and J. Kim (eds), *Emergence or Reduction?* Berlin: De Gruyter, 94–118.

Chalmers, D. 2006. 'Strong and Weak Emergence', in P. Clayton and P. Davies (eds), *The Re-Emergence of Emergence*. Oxford: Oxford University Press, 244–55.

Horgan, T. 1982. 'Supervenience and Microphysics'. *Pacific Philosophical Quarterly* 63: 29–43.

Humphreys P. 1997. 'How Properties Emerge'. *Philosophy of Science* 64: 1–17.

Kim, J. 1984. 'Concepts of Supervenience'. *Philosophy and Phenomenological Research* 45: 153–76.

—— 1998. *Mind in a Physical World*. Cambridge, MA: MIT Press.

Lewis, D. 1986. *On the Plurality of Worlds*. Oxford: Blackwell.

O'Connor, T. and Wong, H. Y. 2005. 'The Metaphysics of Emergence'. *Noûs* 39: 658–78.

Paull, C. and Sider, T. 1992. 'In Defense of Global Supervenience'. *Philosophy and Phenomenological Research* 52: 833–54.

Stalnaker, R. 1996. 'Varieties of Supervenience'. *Philosophical Perspectives* 10: 221–41.

8

The Causal Autonomy of the Special Sciences

*Peter Menzies and Christian List**

1. INTRODUCTION

The systems studied in the special sciences are often said to be causally autonomous, in the sense that their higher-level properties have causal powers that are independent of those of their more basic physical properties. This view was espoused by the British emergentists, who claimed that systems achieving a certain level of organizational complexity have distinctive causal powers that emerge from their constituent elements but do not derive from them.[1] More recently, non-reductive physicalists have espoused a similar view about the causal autonomy of special-science properties. They argue that since these properties can typically have multiple physical realizations, they are not identical to physical properties, and further they possess causal powers that differ from those of their physical realizers.[2]

Despite the orthodoxy of this view, it is hard to find a clear exposition of its meaning or a defence of it in terms of a well-motivated account of causation. In this paper, we aim to address this gap in the literature by clarifying what is implied by the doctrine of the causal autonomy of special-science properties and by defending the doctrine using a prominent theory of causation from the philosophy of science.

The theory of causation we employ is a simplified version of an 'interventionist' theory advanced by James Woodward (2003, 2008a, 2008b), according to which a cause makes a counterfactual difference to its effects. In terms of this theory, it is possible to show that a special-science property can make a difference to some effect while the physical property that realizes it does not. Although other philosophers have also used counterfactual analyses of causation to argue for

* This chapter presents an application of earlier technical results in List and Menzies (forthcoming).

[1] For a very clear statement of this view see Broad (1925); and for historical background see Kim (1992) and McLaughlin (1992).

[2] An important statement of the non-reductive physicalist position can be found in Fodor (1974).

the causal autonomy of special-science properties,[3] the theory of causation we employ is able to establish this with an unprecedented level of precision. It permits us to identify necessary and sufficient conditions for the causal autonomy of a higher-level property, and to show that these are satisfied when causal claims about higher-level properties have a special feature we call *realization-insensitivity*. This feature consists in the fact that the relevant claims are true regardless of the way the higher-level properties they describe are physically realized. Our findings are consistent with those of other philosophers, for example Alan Garfinkel (1981), who have noted the realization-insensitivity of higher-level causal relations as a distinctive feature of the special sciences and have suggested that this feature ensures their independence from lower-level causal relations.

Our discussion proceeds as follows. In section 2, we clarify what it means to say that the causal powers of special-science properties are independent of those of their underlying physical properties. In section 3, we present a simplified version of the theory of causation as counterfactual difference-making. In section 4, we employ this theory to specify the conditions under which an instance of a higher-level, special-science property can have causal powers not possessed by the instance of the physical property that realizes it. In section 5, we compare our results with Garfinkel's discussion of the indispensability of higher-level causal explanations in the special sciences, and argue that his insights can be systematized in our framework.

We discuss the causal autonomy of the special sciences in the context of non-reductive physicalism rather than British emergentism. The emergentists' views are difficult to interpret because of their unfamiliar terminology and philosophical preoccupations; non-reductive physicalism is a more familiar framework for our purposes. To be sure, this framework has been criticized by several philosophers, most notably Jaegwon Kim (1998, 2005), who has argued that its commitment to the causal efficacy of higher-level properties makes it inherently unstable. A defence of non-reductive physicalism against these arguments is not the topic of this paper. We have discussed it elsewhere (List and Menzies forthcoming; Menzies 2008), and the technical results stated in section 4 below draw on this work.

2. CAUSAL AUTONOMY

Non-reductive physicalists believe that even if the higher-level properties of special-science systems are not identical to lower-level physical properties, they

[3] See, for example, Crane 2001; Horgan 1989; Le Pore and Loewer 1987. In a recent unpublished paper, Raatikainen (2006) has independently developed a similar analysis of mental causation in terms of Woodward's interventionist theory of causation.

nonetheless supervene on them, meaning that there can be no difference in a special-science property without an accompanying difference in some physical property.[4] For convenience, we focus on a concrete instance of such a supervenience claim, namely the relationship of mental or psychological properties to physical properties, though the morals of this paper can be generalized to other special-science properties. The non-reductive physicalist about the mind maintains that mental properties are not identical to physical properties—notably because of multiple realizability—but they nonetheless supervene on physical properties. The relevant physical properties vary from one supervenience claim to another. In the case of mental properties, they are usually taken to be neurophysiological properties.[5] Accordingly, the non-reductive physicalist holds that any two individuals, actual or possible, who are duplicates with respect to their neurophysiological properties, will be duplicates with respect to their mental properties. When an individual instantiates a particular mental property by virtue of instantiating a subvenient physical property, it is customary to say that the instance of the physical property *realizes* the instance of the mental property.[6]

What does it mean to say that special-science properties, in particular mental properties, have causal powers that are independent of those of their physical realizers? To answer this question, we must clarify two terminological issues. First, in discussing a property's *causal powers*, one might discuss its forward-looking powers to cause certain states, or its backward-looking powers to be caused by certain states, or a combination of both. We focus on forward-looking powers, as they are most relevant to our present concerns. Second, in talking about the causal powers of properties, one might refer to *properties* or *property-instances*. Which is appropriate depends on whether one is discussing type-causation or token-causation. Since we are concerned with token-causation, we focus on the causal powers of property-instances or states. We say that the state Fa, an instance of the property F, has the *forward-looking causal power* to produce the state Gb, an instance of the property G, just in case Fa can in suitable conditions

[4] More precisely, the supervenience thesis should be understood as a contingent, global supervenience thesis: any world that is a minimal physical duplicate of the actual world is a duplicate with respect to special-science properties and relations. A *minimal physical duplicate* of the actual world is a world that has the actual world's physical entities, physical properties, and laws, *and nothing else*.

[5] Intentional mental properties are often thought to have wide content in the sense that their content implicates the existence of objects and properties outside the skin of the subjects of those properties. If this is so, intentional properties do not supervene on neurophysiological properties. Here it is necessary to bracket the issue about wide content because of limitations of space. Thus we focus on non-intentional properties and on intentional properties whose contents can be specified, perhaps somewhat artificially, in a narrow way.

[6] Throughout this paper we understand an *instance of a property* to consist in an object's instantiating the property at a certain time. (Usually, the reference to time will be tacit.) We also refer to a property-instance as a *state*. We understand the identity conditions of property-instances or states to be such that two property-instances or states are identical only if the corresponding properties are identical. We do not construe property-instances as tropes or abstract particulars.

cause Gb. The causal powers of properties can then be understood in terms of generalizations about the causal powers of their instances.[7]

What is involved in affirming or denying the claim that certain special-science properties, say mental properties, are autonomous and independent of those of their physical realizers? It is easier to begin with what is involved in the denial of such claims. The following thesis, we believe, captures the view of many philosophers who deny the causal autonomy of mental properties:

The Physical Determination of the Causal Powers of Mental States: For all mental properties M and physical properties P, if an instance of property M is realized by an instance of property P, then the causal powers of the M-instance are a *subset* of the causal powers of the P-instance.

The formulation of this thesis in terms of subsets allows for the special case in which the causal powers of the mental state are identical to those of its realizing physical state. So, for example, the instances of M and P may have the same causal powers to produce in identical conditions the behavioural effects B_1 and B_2. But equally, the causal powers of the instance of M may be a proper subset of those of the corresponding instance of P: perhaps the instance of M has the power to produce B_1 and B_2 under certain conditions, while the instance of P has the power to produce B_1, B_2 and B_3 under the same conditions.

Now the assertion of the causal autonomy of mental properties is best viewed in terms of the denial of this determination thesis. The autonomy thesis can thus be stated as follows:

The Causal Autonomy of Mental States: For some mental property M and physical property P, where an instance of property M is realized by an instance of property P, the causal powers of the M-instance are *not* a subset of those of the P-instance.

If the thesis is true, as we seek to show, some mental states have causal powers that are not causal powers of their realizing physical states. This claim is controversial and denied by many philosophers. Jaegwon Kim (1998), for example, affirms a version of the Physical Determination of Mental States for functionally defined mental properties. His Causal Inheritance Principle states that if a mental state is functionally defined in terms of its causal role then its causal powers must be identical with those of the physical state that realizes that causal role. Similarly, Sydney Shoemaker (2001) argues for something like the Physical Determination of Mental States in connection with functionally defined properties. Indeed, he defines the notion of realization in terms of this thesis. He writes: 'In general, then, property X realizes property Y just in case the conditional powers bestowed by Y are a subset of the conditional powers bestowed by X . . . Where the realized property is multiply realizable, the conditional powers bestowed by it will be a proper subset of the sets bestowed by each of the realizer

[7] For example, the statement that the property F has the forward-looking power to cause G can be understood as meaning that instances of F can in suitable conditions cause instances of G.

properties' (2001: 78–9). Although Kim's and Shoemaker's theses are restricted to functionally defined properties, our arguments below refute not only the unrestricted Physical Determination thesis but also the restricted ones accepted by Kim and Shoemaker.

3. DIFFERENCE-MAKING CAUSATION

Philosophical discussions of the causal autonomy of the special sciences often invoke intuitive principles about causation. Since intuitions can be misleading, especially when chosen selectively, a better procedure is to base the discussion on a fully developed, well-motivated theory of causation. For this purpose, we turn to the interventionist account of causation developed recently by James Woodward (2003, forthcoming a, b). This theory has gained increasing support from philosophers of science as providing an instructive account of causal concepts in science. More generally, the interventionist framework forms the basis of a productive research programme for studying causation in philosophy, computer science, and psychology.[8]

Many theories of causation link the concept of causation with that of making a difference. Woodward's interventionist theory falls within this tradition. On his theory, the causal relata are variables, and causation relates changes in one variable to those in another. The simplest version of the theory states that variable X *causes* variable Y just in case if the value of X were to change as a result of an intervention, then the value of Y would change too.[9] Although Woodward presents this simple definition as an account of type-causation, we shall also use it for analysing token-causation, setting aside those situations that require Woodward's more involved account of token-causation. When applied to token-causation, the theory uses variables whose values represent the occurrence or non-occurrence of an event, or the instantiation or non-instantiation of a property by an object at a time.

It is crucial that the changes in the causally related variables occur by virtue of a hypothetical, if not actual, intervention on the cause variable. Changes in one variable may accidentally be correlated with changes in another without any causal relation between them. For example, decreases in barometer readings are correlated with onsets of storms, but this correlation is due to these phenomena being the effects of a common cause—drops in atmospheric pressure. However, if the changes in the barometer reading were brought about by an intervention,

[8] For other works that employ this interventionist framework, see Gopnik and Schulz 2007; Hitchcock 2001; Pearl 2000; Spirtes, Glymour, and Scheines 2000 [1993].

[9] This simple version of the theory assumes causation to be deterministic and so does not cover probabilistic causation. It is also not intended to cover more complicated cases of pre-emption and overdetermination. Moreover, Woodward defines two causal concepts: the concept of a total cause and of a direct cause. The two concepts coincide in the simple cases we discuss here.

say by an experimenter fixing the reading of the barometer, the correlation with the onset of a storm would disappear. This is the central difference between correlations and causal relations: a genuine causal relation is robust under interventions that change the values of the cause variable.

Woodward's interventionist theory bears some resemblance to manipulability theories that state that a causal relationship exists when a human agent can bring about one event by manipulating another. A crucial difference, however, lies in Woodward's definition of an intervention. Very roughly, an *intervention* on one variable X with respect to another variable Y is an idealized experimental manipulation that causes X to change its value in such a way that all other variables that previously were causally relevant to X no longer influence it; and in such a way that any change in Y can only come about through the change in X. Thus an intervention could be the result not only of a human action, but also of a 'natural experiment'.[10]

Further, on the interventionist theory, causal claims have an implicit contrastive structure, which can be made explicit using 'rather than' constructions. So the standard form of a causal claim might be represented: the contrast between X's taking the value x rather than x' causes the contrast between Y's taking the value y rather than y'. When the variables involved are many-valued, it can be indeterminate which of the possible values of X and Y are being contrasted with their actual values, but since our focus is on cases involving binary variables, the relevant contrast is always clear.[11] We suggest that in the binary case the difference-making condition for causation can be adequately expressed in terms of a pair of counterfactuals, where x, x' and y, y' are the possible values of the variables X and Y, respectively:

Truth conditions for difference-making causation: $X = x$ makes a difference to $Y = y$ if and only if (a) $X = x \mathrel{\Box\!\!\rightarrow} Y = y$; and (b) $X = x' \mathrel{\Box\!\!\rightarrow} Y = y'$.

These counterfactuals must be understood according to an interventionist, non-backtracking reading. A backtracking counterfactual involves reasoning from an outcome to earlier events and then forwards again, as, for example, when one

[10] This notion of an intervention does not burden the theory with the anthropocentric implications of manipulability theories. For instance, the theory implies that a causal relation can exist between two variables independently of whether any human agent does or could carry out an intervention on the variables. Another difference from orthodox manipulability theories is that it does not attempt to reduce causal concepts to non-causal ones. The notion of an intervention is defined in terms of causal concepts, which means that the definition of causation in terms of interventions is non-reductive. Woodward argues persuasively that the definition is nonetheless informative in virtue of having many non-trivial implications regarding causal relationships.

[11] To be sure, this is an artificial restriction because many causal variables, even in everyday life, are best seen as many-valued. But the simplifying assumption that causal statements involve just binary variables makes the questions we discuss more easily tractable. More generally, for X to be a cause of Y, there must exist at least two different values of X, x and x', and two different values of Y, y and y', such that under an intervention that changes X from x to x', the value of Y changes from y to y'.

reasons that if the barometer reading were low, then this would mean that the atmospheric pressure is low, which in turn would mean that the storm is going to occur. Evaluated as a backtracking conditional, the counterfactual 'If the barometer reading were low, then the storm would occur' is true. By contrast, a non-backtracking counterfactual is evaluated by supposing that its antecedent is made true by an intervention, which breaks any existing relationship between the antecedent and its causes. When the barometer reading is set to some value through an intervention, one cannot infer back from this value to the value that the atmospheric pressure must have had, and thus the counterfactual 'If the barometer reading were low, then the storm would occur' is false under the non-backtracking reading.

Generally, the counterfactuals are to be understood according to the standard possible-worlds semantics, developed by Lewis (1973), which defines their truth conditions in terms of a similarity relation between possible worlds. The similarity relation is represented by an assignment to each possible world w of a system of spheres of worlds centred on w, subject to standard constraints.[12] The idea is that the smaller a sphere is around w, the more similar to w are the worlds in it. Now a counterfactual conditional P □→ Q is true in world w if and only if Q is true in all the closest P-worlds to w. Figure 8.1 shows a situation in which the counterfactual P □→ Q is true in the world w at the centre of the system of spheres. The set of P-worlds is represented by the region with diagonal lines, the set of Q-worlds by the larger region that includes the set of P-worlds.

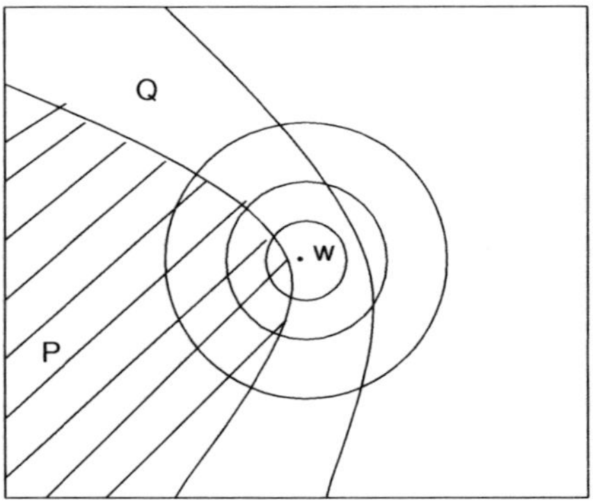

Figure 8.1.

[12] *Nestedness*: For any two spheres S and T, either S is included in T or T is included in S. *Weak centring*: w is contained in every sphere. *Exhaustiveness*: There is a largest sphere containing all relevant possible worlds. *Limit assumption*: For any world w and any proposition P, there is a smallest sphere around w containing some P-world, called the *smallest P-permitting sphere around* w. The *closest P-worlds to* w are defined as the P-worlds within the smallest P-permitting sphere around w.

As discussed in more detail in List and Menzies (forthcoming), our semantic framework diverges from Lewis's in one respect: we adopt a weaker centring requirement than Lewis by allowing the smallest sphere around w to contain more than one world. Lewis, by contrast, requires it to contain only w. A relaxation of Lewis's strong centring requirement is essential for our purposes. Strong centring implies that whenever P and Q are true in some world so is P $\Box\!\!\rightarrow$ Q. So, whenever X=x and Y=y are true in the actual world, clause (a) of our truth-conditions for difference-making causation would automatically hold as well, which would trivialize this condition. For the difference-making account of causation to work, clause (a) must be non-trivial: it must rule out insufficiently specific candidate causes. In particular, the counterfactual formulation must allow that even if the candidate cause and effect are both exemplified in the actual world, the smallest sphere around it also contains some other worlds exemplifying the candidate cause.

4. PROPORTIONAL CAUSATION AND REALIZATION-INSENSITIVE CAUSATION

In this section we employ the difference-making account of causation to determine whether the thesis of the Physical Determination of the Causal Powers of Mental States is true, drawing on earlier work in List and Menzies (forthcoming). Recall that the thesis states that if an instance of a mental property M is realized by an instance of a physical property P, then the causal powers of the M-instance are either identical to those of the P-instance or a proper subset of them. If this thesis is true, then its negation, the thesis of the Causal Autonomy of Mental States, must be false. Correspondingly, if the first thesis is false, then the second must be true.

Many philosophers of mind believe that the Physical Determination thesis is true. Their acceptance of the thesis, we think, stems from a more general preference for lower-level, physical causal variables over higher-level, special-science variables: it is the physical properties and states that do all the causal work, it is assumed, and properties and states supervening on them derive whatever causal efficacy they have from the underlying physical ones.[13] Moreover, these philosophers seem to assume that the Physical Determination thesis is an a priori truth. Kim (2005), for example, says it is an analytic truth. But this is mistaken, as we show in this section. If it is a fact about the world that every mental state derives its causal powers from its realizing physical properties then it is an

[13] A striking manifestation of this general preference is the Exclusion Principle, used by Kim in his Exclusion Argument for the conclusion that non-reductive physicalism is committed to epiphenomenalism about mental properties and states. One version of the principle states that if a physical cause exists for a physical effect, that excludes any mental cause for the same effect. For an evaluation of this argument, see Menzies (2008) and List and Menzies (forthcoming).

empirical fact about the world. Notwithstanding this point, a priori conceptual knowledge can shed light on this issue. Knowing what the concept of causation entails, we are in a better position to understand the precise meaning of the Physical Determination thesis, and so in a better position to determine whether it is true or false in the light of our empirical knowledge about the world.

So we propose to interpret the Physical Determination thesis in terms of the difference-making account of causation. It is convenient to evaluate the thesis by considering a schematic example. Suppose one of us—say Peter—has an intention to signal a taxi (an instance of mental property M) and he waves his arm (an instance of behavioural property B); and suppose that his intention (the M-instance) is realized by some neural state (an instance of neural property N_1), but it could also have been realized by other neural states (say, instances of the neural properties $N_2, \ldots, N_n$). What is the cause of Peter's waving his arm—the mental state Ma (where 'a' refers to Peter) or the neural one N_1a? The difference-making account of causation permits it to be the case that the mental state Ma, and not the neural state N_1a, is the cause of the behavioural outcome Ba. Thus it may be the case that both counterfactuals (1a) and (1b) are true, but not both counterfactuals (2a) and (2b) are true:

(1a) $Ma \mathrel{\Box\!\!\rightarrow} Ba$.
(1b) $\sim Ma \mathrel{\Box\!\!\rightarrow} \sim Ba$.

(2a) $N_1a \mathrel{\Box\!\!\rightarrow} Ba$.
(2b) $\sim N_1a \mathrel{\Box\!\!\rightarrow} \sim Ba$.

This situation might be described by saying that Peter's waving his arm rather than not waving it was caused by his having the mental property M rather than not having it, and not by his having the neural property N_1 rather than not having it.

It is not surprising that this situation could obtain. There are many common situations in which a supervenient state has causal powers not possessed by the subvenient state that realizes it. Consider a familiar example from the philosophy of action. Imagine you have feuded with an irascible neighbour for some time but you decide to break the ice by greeting him. Your neighbour is startled by your saying 'Hello' unexpectedly. As it happens, your greeting is rather abrupt and loud. But your neighbour is startled, not because you say 'Hello' loudly, but because you simply say it. Here the relationship between your saying 'Hello' and saying 'Hello' loudly is analogous to the relationship between Peter's having the intention to wave his arm and his being in neural state N_1: the first state of each pair supervenes on the second. However, it is the supervenient state, not the subvenient one, that does the causal work. The difference in the states' causal status is reflected in terms of the difference in the truth values of the following pairs of counterfactuals:

(3a) You say 'Hello' $\mathrel{\Box\!\!\rightarrow}$ your neighbour startles.

(3b) You don't say 'Hello' $\square\!\!\rightarrow$ your neighbour doesn't startle.

(4a) You say 'Hello' loudly $\square\!\!\rightarrow$ your neighbour startles.
(4b) You don't say 'Hello' loudly $\square\!\!\rightarrow$ your neighbour doesn't startle.

Both counterfactuals (3a) and (3b) are true, whereas not both (4a) and (4b) are true. In particular, counterfactual (4b) is false because in some of the closest worlds in which you don't say 'Hello' loudly, such as those in which you say it normally, your neighbour still startles. The same point can be put in terms of contrastive focus: your neighbour startled rather than didn't startle because you said 'Hello' rather than didn't say it, and not because you said 'Hello' loudly rather than didn't say it so.

Stephen Yablo (1992) has argued for a similar conclusion on the basis of what he calls a proportionality constraint on causation. Yablo claims that causes must be *proportional* or *commensurate* with their effects in the sense that a cause must have the right degree of specificity to account for its effect—a cause cannot be underspecific or overly specific. So, citing Peter's having the neural property N_1 as the cause of his waving his arm, or your saying 'Hello' loudly as the cause of your neighbour's being startled, does not satisfy the proportionality constraint since these states are more specific than is required to account for their respective effects. They are overly specific precisely because they suggest erroneously that Peter would not have raised his arm if he had not had the neural property N_1, or that your neighbour would not have startled if you had not said 'Hello' loudly.

Yablo states his proportionality constraint in terms of a metaphysical framework of event essences. By contrast, we agree with those philosophers who suggest that the idea of causal proportionality is described more satisfactorily in terms of the contrastive character of causation (Craver 2007; Woodward 2008 a and b). In particular, we suggest that the proportionality constraint can be expressed in terms of a pair of counterfactuals having the structure of the (a) and (b) counterfactuals above. The function of the counterfactuals is to ensure that the candidate causes are of the right degree of specificity. The function of the (a) counterfactual is to rule out candidate causes that are not specific enough to account for the change in the effect variable, while the function of the (b) counterfactual is to rule out candidate causes that are too specific to account for this change.[14] In this way, the contrastive, counterfactual account of causation, proposed above, captures the idea of proportionality as well as that of difference-making.

Returning to the example about Peter's waving his arm with the intention of signalling a taxi, we can readily imagine conducting experiments the results of which confirm (1a) and (1b) and disconfirm (2b). Such experimental evidence

[14] For more details about the functions of the two counterfactuals, see List and Menzies (forthcoming).

would establish the falsity of the Physical Determination thesis, since it would establish that a mental state has a causal power not enjoyed by the physical state that realizes it. In demonstrating the falsity of the Physical Determination thesis, it would thereby demonstrate the truth of the Causal Autonomy thesis.

We emphasize again that whether some state satisfies the counterfactual conditions that constitute the difference-making causal relation is a completely empirical matter. Facts about the world determine which is the right level of causation. They determine which type of variable, a higher- or a lower-level one, is such that variation in its value can bring about a variation in an effect variable. In the example at hand, the higher-level variable is the source of causal influence. But in other circumstances lower-level variables can constitute the right level of causation. For example, suppose your interactions with your agitated neighbour have a different history. Suppose you have been getting along fine with him, but on one occasion you startle him because of the loudness of your greeting. In other words, he startles because you say 'Hello' loudly, not because you simply say 'Hello'. In these circumstances, the lower-level variable is the proportional cause of the effect. Likewise, suppose that what is to be explained in the example of Peter's waving his arm is a change of fine-grained motor control rather than coarse-grained behaviour. This change might be explicable only in terms of a variation in his neural states and not in terms of a variation in his mental states. In this case a neural state would be the proportional difference-making cause of the effect. Generally, the right level of causation is determined by the contrast to be explained and by the empirical facts about which variables can be varied in such a way as to account for the given contrast.

As we have just seen, the Physical Determination thesis is not *generally* true. One might wonder, nonetheless, whether it is not true for the most part, or true more often than not. One benefit of formulating the difference-making conception of causation in terms of counterfactuals is that it makes this question logically tractable.

Returning to the case of the mental state Ma, the neural state N_1a and the behaviour Ba, one can prove that the causal powers of Ma are a subset of those of N_1a—i.e., that if Ma causes Ba, then N_1a causes Ba—*only under very special conditions*. To state this result, call a causal relation between Ma and Ba *realization-sensitive* if Ba fails to hold in all those Ma-worlds that are closest $\sim N_1a$-worlds (i.e., where Ma has a different realizer from the actual one). The result is the following:

Entailment Result (List and Menzies forthcoming): If Ma causes Ba, then N_1a causes Ba if and only if the causal relation between Ma and Ba is realization-sensitive.

Rather than prove this result here, it is more instructive to describe a situation that exemplifies the result. So consider the situation represented in Figure 8.2. As before, the concentric spheres represent sets of more and more similar worlds to the actual world; the innermost sphere contains the actual world, labelled w,

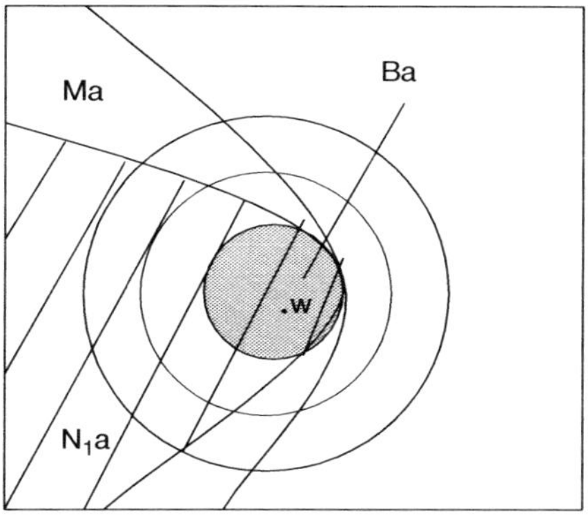

Figure 8.2.

and the other worlds deemed maximally similar to it. The set of N_1a-worlds is represented by the region with diagonal lines, the set of Ma-worlds by the larger region that includes the set of N_1a-worlds. The shaded region represents the set of Ba-worlds. In this situation, it is easy to see that Ma causes Ba. First, since Ma holds throughout the innermost sphere, that sphere picks out the closest Ma-worlds, and since Ba also holds in it, counterfactual (1a) is true. Second, since Ba does not hold in any ~Ma-worlds, it fails to hold in all the closest ~Ma-worlds and thus counterfactual (1b) is true. Further, the causal relation between Ma and Ba is realization-sensitive: since Ba does not hold in any ~N_1a-worlds, it follows a fortiori that it does not hold in any of the closest ~N_1a-worlds that are Ma-worlds. And finally, N_1a does indeed cause Ba: counterfactuals (2a) and (2b) can easily be verified to be true.

In view of this result, it is open to the defender of the Physical Determination thesis to argue that the thesis is sometimes true even if not always true. It is important to note, however, that the conditions under which the counterfactual pair (1a)–(1b) implies the pair (2a)–(2b) are very special. Figure 8.2 illustrates this point nicely. Although both Ma and its actual realizing state N_1a are difference-making causes of Ba here, the realization-sensitivity of the causal relation between Ma and Ba means that small perturbations in the way in which Ma is realized would result in the absence of Ba. In other words, if Ma were realized by any neural state other than N_1a (such as N_2a, N_3a, and so on), then Ba would cease to hold. When might we expect these conditions to obtain? If the mental property M were identical to the neural property N_1, then we would certainly expect instances of M to stand in realization-sensitive causal relations with respect to instances of N_1. The fact that M-instances had certain effects when and only when N_1-instances are present would simply reflect the identity of the properties. What other explanations could there be for the realization-sensitivity of higher-level causal relations? It is hard to think of any explanation other than the identity of the properties. But this explanation will not be available if we assume, in keeping with our overarching presupposition, that the higher-level properties are multiply realized by physical properties, and so not identical with them.

At this point it is useful to consider a logically equivalent formulation of the Entailment Result that shows more directly that the Physical Determination thesis is not generally true. In analogy with the earlier definition, call a causal relation between Ma and Ba *realization-insensitive* if Ba holds in some Ma-worlds that are closest $\sim N_1a$-worlds (i.e., where Ma has a different realizer from the actual one). The following proposition is an immediate corollary of the Entailment Result:

Downwards Exclusion Result (List and Menzies forthcoming): If Ma causes Ba, then N_1a does not cause Ba if and only if the causal relation between Ma and Ba is realization-insensitive.

Again let us consider a schematic example that exemplifies this proposition, focusing on the situation represented in Figure 8.3. As before, the system of spheres represents sets of worlds with greater or lesser degrees of similarity to the actual world, labelled w. The set of N_1a-worlds is represented by the region with diagonal lines, and the set of Ma-worlds by the larger region that includes the set of N_1a-worlds. The shaded region represents the set of Ba-worlds. This figure shows that Ma causes Ba, since Ba holds in all the closest Ma-worlds and fails to hold in all the closest $\sim$Ma-worlds, i.e., counterfactuals (1a) and (1b) are both true. It is also easy to see that this causal relation is realization-insensitive: Ba continues to hold in some, indeed all, of the Ma-worlds that are closest $\sim N_1a$-worlds. Finally, it is easy to see that N_1a does not cause Ba: the counterfactual (2b) is false, since Ba holds in all the closest $\sim N_1a$-worlds.

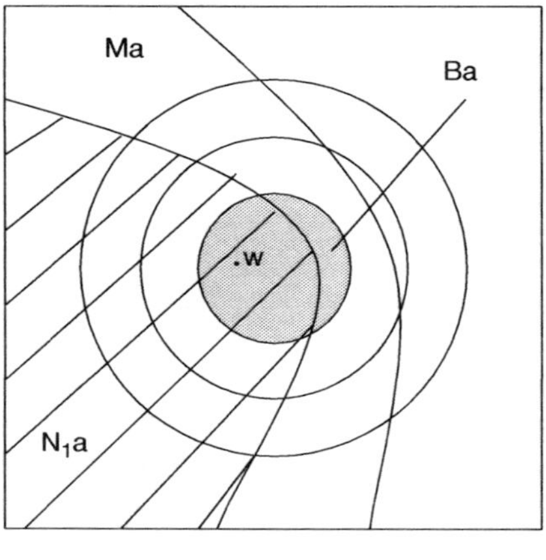

Figure 8.3.

This result is of great significance with respect to the question of whether the Physical Determination thesis or the Causal Autonomy thesis is true. If some mental state Ma stands in a realization-insensitive causal relation to another state Ba, then this mental state has a causal power to bring about a certain effect that does not belong to the set of causal powers of its physical realizing state. Hence, the Physical Determination thesis is false and the Causal Autonomy thesis true in this situation. These logical inferences are depicted in Figure 8.4, where the proposition in each box logically implies the proposition in the box below it.

What is the upshot of this discussion? If we have reason to believe that a mental state stands in a realization-insensitive causal relation to some other state, then we are entitled to think that this higher-level causal relation is independent of any lower-level causal relation enjoyed by the neural realizer of the mental state. We have plenty of reason to believe that mental states do indeed stand in realization-insensitive causal relations to other states. Given that a mental state is typically realized in many different ways, we can expect that whatever causal powers it has, it has them independently of the particular way it is realized. In other words, we can expect that a mental state's causal powers do not depend on which of its possible realizers happens to be the actual one.

More generally, there is reason to think that most higher-level causal relations are realization-insensitive in ways that ensure their autonomy. Several

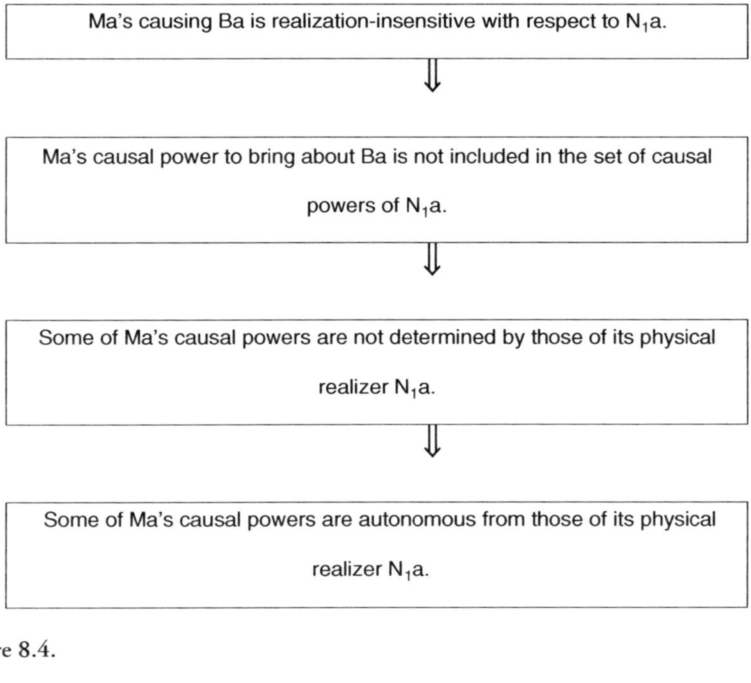

Figure 8.4.

philosophers have noted that we intuitively require causal relations to be 'insensitive' in the sense that they would continue to hold under perturbations of the actual circumstances. Lewis (1986) was the first to make this observation, but Woodward (2006) has developed the point in greatest detail. He says that a causal statement is insensitive to the degree that a pair of counterfactuals such as the (a) and (b) counterfactuals above would continue to hold even if the actual circumstances were varied in certain admissible ways; and that the more insensitive the counterfactuals are, especially the (a) counterfactual, the more willing we are to count the corresponding causal statement as true. As to what counts as admissible variations to the actual circumstances, Woodward says that the changes should not be too unlikely or far-fetched; but more generally, the answer is context-sensitive, depending partly on discipline-specific considerations. We suggest that in the special sciences causal relations are typically required to be invariant under changes to the physical realization of the higher-level properties involved. And we suggest that the required realization-insensitivity of higher-level causal relations is an instance of the more general phenomenon noted by Lewis and Woodward.[15]

5. GARFINKEL ON THE INDISPENSABILITY OF HIGHER-LEVEL EXPLANATIONS

In this section we compare the conclusions of the last section with Alan Garfinkel's views about reductionism (1981). Garfinkel argues for very similar conclusions about the indispensability of higher-level causal explanations in the special sciences, although his arguments proceed without the aid of a systematic account of causal explanation. We hope to show that Garfinkel's interesting and original insights have an internal coherence that is explicable in terms of the framework we have developed. Many of his insights follow straightforwardly as consequences from the account of difference-making causation and the formal results described above. This fact indicates a two-way relationship of confirmation: the systematizability of Garfinkel's insights in terms of our framework provides some independent confirmation of them, but the antecedent plausibility of his insights also offers some evidence in support of our framework.

Garfinkel was one of the first philosophers to emphasize the contrastive character of explanation and its importance for many issues in the philosophy of science. He advanced now-familiar arguments for the claim that explanation is relative to a contrast space. Although he focused on explanation, his observations apply equally to causation, as we have seen. Of particular interest is Garfinkel's appeal to the contrastive character of explanation to criticize reductionist claims, such as the claim that psychology is reducible to neurophysiology, the claim

[15] For further discussion of this issue see List and Menzies (forthcoming).

that thermodynamics is reducible to statistical mechanics, or the claim that social laws are reducible to principles about the actions of individuals. By *reductionism*, Garfinkel means any doctrine according to which the phenomena in the explanatory domain of one theory are best explained in terms of a lower-level theory. Reductionism is sometimes thought, he says, as an ideal, which is possible 'in theory' but not 'in practice'. In contrast, he argues for the view that reductionism is often impossible even in theory, since the explanations of lower-level theories are simply not good enough to replace the explanations of higher-level theories.

Garfinkel discusses an example of a purported microreduction in which an explanation in terms of the macrostates of a system is eliminated in favour of an explanation in terms of its microstates (1981: 53 ff.). He asks us to imagine an ecological system composed of foxes and rabbits where the population levels of the two species periodically fluctuate: 'The explanation [of the fluctuations] turns out to be that the foxes eat the rabbits to such a point that there are too few rabbits left to sustain the fox population, so the foxes begin dying off. After a while, this takes the pressure off the rabbits, who then begin to multiply until there is plenty of food for the foxes, who begin to multiply, killing more rabbits, and so forth' (1981: 53).

Now suppose that a particular rabbit r has been killed. What is the explanation of this? It is plausible to say that the cause of the rabbit's death was that the fox population was high. Can such an explanation be replaced by an explanation in terms of the underlying microstate of the system? The microstate will be an enormously complex state, specified in terms of the number and location of all the foxes and rabbits, their interactions, and perhaps their physiological states such as their reaction times. Garfinkel argues that an explanation in terms of this microstate is not satisfactory because it does not provide an answer to the question that is implicitly being asked. This is the contrastive question: Why was the rabbit r eaten rather than not eaten? The explanation in terms of the microstate contains a great deal of information that is not relevant to this question and does not really answer it. At best, the microexplanation answers the different question: Why did the rabbit get eaten by fox f at place p and time t rather than by some other fox at some other place and time? In other words, the explanations have completely different objects; and so it is inappropriate to try to match one explanation with the object of the other.

Why doesn't the explanation in terms of the complex microstate of the system provide a satisfactory answer to the question of why the particular rabbit was eaten rather than not eaten? Garfinkel says that microexplanation does not work because it does not say how things would have to be different in order for the rabbit not to get eaten. For example, if the microstate, specified in terms of the number and location of foxes and rabbits and their interactions at the particular time, had been slightly different, the rabbit would not have been eaten by fox f, but probably would have been eaten by another fox, given that the fox population was so high. An explanation should ideally tell us, Garfinkel argues, how the outcome is sensitive to changes in the conditions. But the microexplanation does not do this, since perturbations in the microstate still leave us with the same outcome.

More generally, a satisfactory causal explanation of the fact that the rabbit was eaten rather than not eaten must, according to Garfinkel, provide some account of 'the sensitive aspects of the causal connection'. He explains this as follows:

> We can imagine the space of the substratum as underlying the whole process. We have a complete set of microstates and a principle of microexplanation, V, which explains the microstate Y_0 in terms of X_0:
>
> $$X_0 \rightarrow_V Y_0.$$
>
> The rabbit was eaten by fox f (= Y_0) because it was at a certain place, time and so on (= X_0). For most X_0, this evolution is smooth; small changes in X_0 do not make for qualitative changes. But at certain critical points, small perturbations do make a difference and will result in the rabbit's wandering out of the capture space of the fox. These critical points mark the boundaries of the regions of smooth change. They partition the underlying space into equivalence classes within which the map is stable. The crucial thing we want to know is how this set of critical points is embedded in the substratum space, for that will tell us what is really relevant and what is not. Therefore, what is necessary for a true explanation is an account of how the underlying space is partitioned into basins of irrelevant differences, separated by ridge lines of critical points. (Garfinkel 1981: 63–4)

This account of causal explanation employs ideas and terminology taken from catastrophe theory. But the basic ideas are simple. Suppose we want to explain a contrast that can be represented as the difference in the values of a variable. For convenience, let us call these the *explanandum values*. In the example, the explanandum values are the rabbit's being eaten and its not being eaten. To explain this contrast, Garfinkel tells us that we have to partition the states of the system into equivalence classes. It is implicit in the passage and his subsequent examples that the partition must satisfy certain conditions. One is that the resulting equivalence classes must be such that the laws of the system map all the members of the same class onto the same explanandum value. Accordingly, perturbing the system to change it from one state to another in the same equivalence class will not make for qualitative changes in outcome. Another condition is that the equivalence classes must be such that the laws of the system map members of different classes onto different explanandum values. Hence perturbing the system to change it from a state belonging to one class to a state belonging to another will make for qualitative changes in outcomes. These changes will mark, in his terms, the boundaries of the regions of smooth transition. In sum, the explanatory partition must be such that 'the underlying space is partitioned into equivalence classes within which differences do not make a difference but across which differences *do* make a difference' (Garfinkel 1981: 65).

In terms of this account of causal explanation, it is easy to understand Garfinkel's remarks about what counts as a satisfactory explanation of the fact that rabbit r was eaten rather than not eaten. Partitioning the states of the system into the class of states in which the fox population is high and the class of states in which it is not high satisfies, or approximately satisfies, the two conditions.[16] First, the laws of the system map all states in a given equivalence class onto the same explanandum value; and second, the laws map states in different equivalence classes onto different explanandum values. Accordingly, the transition from a state in which the fox population is high to a state in which it is not crosses the boundary between regions of smooth transition. By contrast, the partition of the states into a class consisting of a single microstate X_0 and a class consisting of all other microstates will not satisfy the conditions on an explanatory partition. The states in the second class are not all equivalent from the point of view of the laws of the system. For the laws map the states in this class onto different explanandum values and, indeed, many of these states will be mapped onto the same explanandum value as the state X_0, reflecting the fact that even if rabbit r was not eaten by fox f it may have been eaten by some other fox.

How does Garfinkel's account of causal explanation relate to the difference-making account of causation? It is not too difficult to see the structural parallels between the two accounts. Suppose that the property-instance Fa is a difference-making cause of the property-instance Gb. Then the following pair of counterfactuals must be true:

(5a) Fa $\Box\!\!\rightarrow$ Gb
(5b) ~Fa $\Box\!\!\rightarrow$ ~Gb

The truth of these counterfactuals entails that every closest Fa-world is a Gb-world and that every closest ~Fa-world is a ~Gb-world. Notice that the set of closest Fa-worlds and the set of closest ~Fa-worlds need not belong to the same sphere of worlds in the system of spheres. Indeed, the closest Fa-worlds may be a subset of one sphere and the closest ~Fa-worlds a subset of a different sphere. But if we focus on the special case in which they belong to the same sphere, we can see that the truth of these counterfactuals implies the existence of a partition on the common sphere that satisfies Garfinkel's two conditions on an explanatory partition. The possible worlds in this sphere fall into two equivalence classes: one class consists of worlds whose laws map the state Fa onto the state Ga, and the other class consists of worlds whose laws map the state ~Fa onto the state ~Gb.[17]

[16] It follows from the fact that the explanandum consists of a contrast between two values that the explanatory partition must consist of two equivalence classes. But this need not generally be the case.

[17] If the closest Fa-worlds and the closest ~Fa-worlds do not belong to a common sphere of worlds, the difference-making account of causation is informationally richer, but it is still possible to construct a partition of a suitable set of worlds that satisfies Garfinkel's conditions on an explanatory partition, though the construction is not so intuitively natural.

So we have an argument that if a state is a difference-making cause of another, then the first state will be a good causal explanation of the second on Garfinkel's account. The argument in the other direction is even more straightforward. If in all relevantly similar systems, the laws map Fa-states onto Gb-states and ~Fa-states onto ~Gb-states, then it is a simple matter to show the pair of counterfactuals above will be true. In sum, the difference-making account of causation and Garfinkel's account of causal explanation are structurally similar to each other.

This is not the only parallel between our framework and Garfinkel's. Garfinkel argues that many macroexplanations in the special sciences cannot be eliminated and replaced by microexplanations. His justification of this claim is based on the fact that an explanation must have a certain amount of stability under perturbations of its conditions; and that certain structural features of special-science systems ensure the stability of macroexplanations. For example, the successful causal explanation of the rabbit's being eaten in terms of the high fox population rests on a certain stability or resilience of the causal processes in this system: the rabbit was eaten by fox f, but if it had not been eaten by this fox it would have been eaten by another fox. So the causality with which the effect is produced has a strong resilience: 'The very fact that the rabbit did not wander into the capture space of fox f makes it likely that it will be eaten by another fox' (Garfinkel 1981: 57). When this is true of a system, Garfinkel says, we have a case of 'redundant causation'. He writes:

> Systems exhibiting redundant causality have, for every consequent Q, a bundle of antecedents (P_i) such that:
>
> 1. If any one of the P_i is true, so will be Q.
> 2. If one P_i should not be the case, some other will.
>
> Obviously, in any system with redundant causality, citing the actual P_i that caused Q will be defective as an explanation. This will apply to many cases in which P_i is the microexplanation. (Garfinkel 1981: 58)

We think that it is, strictly speaking, misleading in this context to use the term 'redundant causation', which is normally used to describe cases of pre-emption and overdetermination. The situation described here is essentially different from one in which multiple causes lead to the same effect, such as when a victim is hit by multiple bullets. We suggest that the phenomenon described by Garfinkel is actually the realization-insensitivity of causal relations. Consider the causal relation between P and Q, in Garfinkel's notation, where P is a higher-level macrostate that is actually realized by a microstate P_i but could have been realized by any of the microstates $P_1, \ldots, P_n$. This causal relation is realization-insensitive just in case Q is true in some of the closest $\sim P_i$-worlds that are still P-worlds, i.e., Q remains true in some of the worlds in which P is realized differently. It is easy to see that the conditions that Garfinkel stipulates for 'redundant causation' ensure that the causal relation between P and Q is realization-insensitive in this sense. His second condition ensures that among the closest $\sim P_i$-worlds there are some that are P-worlds; and his first condition ensures that all these closest $\sim P_i$-worlds that are P-worlds are ones in which Q is true.

What is the point of these remarks? It is that Garfinkel has arrived at essentially the same conclusion reached in the last section: the key to understanding the ineliminability of macrostate causal explanations is the realization-insensitivity of the causal links they invoke. In the last section, we argued that when an upper-level macroexplanation rests on a realization-insensitive causal relation, it cannot be replaced by a lower-level microstate explanation. Garfinkel says essentially the same thing. The causal explanation of the death of rabbit r in terms of the high fox population cannot be replaced by an explanation in terms of the actual microstate in which it is eaten by fox f because even if the high fox population were realized by some other microstate, it would still be true that the rabbit would have been eaten, if not by fox f then by some other fox. The Downwards Exclusion Result, described in the last section, implies that the macroexplanation conveys essential contrastive, difference-making information that is not conveyed by the microstate explanation.

In conclusion, the results of this chapter recapitulate many of the conclusions reached less formally by Garfinkel. There are, indeed, some striking similarities between our approaches. First, both approaches attach special significance to the contrastive character of causation or causal explanation in establishing the right level of causation or causal explanation. Second, both approaches provide accounts of causation or causal explanation in terms of a partition of states or possible worlds into classes satisfying certain constraints concerning the lawful mappings of one state onto another. Finally, both approaches justify and explain the ineliminability of higher-level causation or causal explanations in terms of the fact that they are realization-insensitive. The framework developed in this chapter, involving the difference-making account of causation and the Entailment and Downwards Exclusion Results, is especially advantageous in showing how these common ideas can be systematized in a coherent way.

REFERENCES

Broad, C. D. 1925. *The Mind and Its Place in Nature*. London: Routledge & Kegan Paul.

Crane, T. 2001. *The Elements of Mind*. Oxford: Oxford University Press.

Craver, C. 2007. *Explaining the Brain*. New York: Oxford University Press.

Fodor, J. 1974. 'Special Sciences, or the Disunity of Sciences as a Working Hypothesis'. *Synthese* 28: 97–115.

Garfinkel, A. 1981. *Forms of Explanation*. New Haven: Yale University Press.

Gopnik, A. and Schulz, L. 2007. *Causal Learning: Psychology, Philosophy and Computation*. New York: Oxford University Press

Hitchcock, C. 2001. 'The Intransitivity of Causation Revealed in Equations and Graphs', *Journal of Philosophy* 98: 273–99.

Horgan, T. 1989. 'Mental Quausation'. *Philosophical Perspectives* 3: 47–76.
Kim, J. 1992. 'Downward Causation'. In A. Beckermann, H. Flohr, and J. Kim (eds), *Emergence or Reduction*. New York and Berlin: De Gruyter.
—— 1998. *Mind in a Physical World: An Essay on the Mind-Body Problem and Mental Causation*. Cambridge, MA: MIT Press.
—— 2005. *Physicalism or Something Near Enough*. Princeton, NJ: Princeton University Press.
Le Pore, E. and Loewer, B. 1987. 'Mind Matters'. *Journal of Philosophy* 84: 630–42.
Lewis, D. 1973. *Counterfactuals*. Oxford: Blackwell.
—— 1986. 'Postscripts to Causation'. *Philosophical Papers*, vol. 2. Oxford: Oxford University Press.
List, C. and Menzies, P. forthcoming. 'Non-reductive Physicalism and the Limits of the Exclusion Principle'. *Journal of Philosophy*.
MacLaughlin, B. 1992. 'The Rise and Fall of British Emergentism'. In A. Beckermann, H. Flohr, and J. Kim (eds), *Emergence or Reduction: Essays on the Prospects of Nonreductive Physicalism*. New York and Berlin: W. de Gruyter, 49–93.
Menzies, P. 2008. 'The Exclusion Problem, the Determination Relation, and Contrastive Causation'. In J. Hohwy and J. Kallestrup (eds), *Being Reduced: New Essays on Reductive Explanation and Special Science Causation*. Oxford: Oxford University Press.
Pearl, J. 2000. *Causality: Models, Reasoning, and Inference*. Cambridge: Cambridge University Press.
Raatikainen, P. 2006. 'Mental Causation, Interventions, and Contrasts'. Unpublished manuscript, University of Helsinki.
Shoemaker, S. 2001. 'Realization and Mental Causation'. In C. Gillett and B. Loewer (eds), *Physicalism and Its Discontents*. Cambridge: Cambridge University Press.
Spirtes, P., Glymour, C. and Scheines, R. 2000. *Causation, Prediction and Search*. Cambridge, MA: MIT Press.
Woodward, J. 2003. *Making Things Happen: A Theory of Causal Explanation*. Oxford: Oxford University Press.
—— 2006. 'Sensitive and Insensitive Causation'. *Philosophical Review* 115: 1–50.
—— 2008a. 'Cause and Explanation in Psychiatry: An Interventionist Perspective'. In K. Kendler and J. Parnas (eds), *Philosophical Issues in Psychiatry: Explanation, Phenomenology, and Nosology*. Baltimore: Johns Hopkins University Press.
—— 2008b. 'Mental Causation and Neural Mechanisms'. In J. Hohwy and J. Kallestrup (eds), *Being Reduced: New Essays on Reductive Explanation and Special Science Causation*. Oxford: Oxford University Press, 218–62.
Yablo, S. 1992. 'Mental Causation'. *Philosophical Review* 101: 245–80.

9

Causal and Explanatory Autonomy: Comments on Menzies and List

Ausonio Marras and Juhani Yli-Vakkuri

The chapter by Menzies and List offers a refreshing and much needed naturalistic perspective on the mental causation debate, and the debate over causation in the domains of the special sciences more generally. Though the literature they are responding to is concerned mainly with mental causation, Menzies and List are surely right that the answers they propose, if correct, generalize to any of the other presumably causal processes studied in the special sciences. By calling Menzies and List's perspective 'naturalistic', we mean to say that it represents the kind of philosophy that is sensitive to empirical science, its results and practices. It is a puzzling fact that 'naturalism' in this sense, though it has influenced—and, we think, improved—just about every other area of philosophy, has been so late in coming to the mental causation literature. But now we are confronted nearly with an embarrassment of riches: Raatikainen (2006), Shapiro and Sober (2007), the present chapter, as well as Menzies (2008), and Woodward (2008) all apply some version of the interventionist theory of causation to the problem of mental causation—a theory that has as good a claim as any to being an accurate account of the concept of causation at work in empirical science. These authors reach some interesting conclusions, some of which we think have a pretty good chance of also being true.

What's true, without a doubt, is that there is mental causation: mental events sometimes cause (mental and) other events, and even do so in virtue of being instances of mental properties. The contemporary *problem* of mental causation is that there are arguments, the most famous ones being due to Jaegwon Kim, that purport to show that if non-reductive physicalism is true, then there is no mental causation because—the reasons given for this vary—it cannot be that both mental events (or properties) and their physical realizers are causally efficacious, and our physicalist commitments somehow tell us that the physical realizers are the efficacious ones. Let us call these 'causal exclusion arguments'. The solution to the problem is to figure out what's wrong with the causal exclusion arguments.

The answer is, no doubt: *a lot*. Anyone who feels like challenging their ancillary premises—metaphysical platitudes, according to their advocates—has a lot to choose from. In our paper (2008), we laid out what we thought were the *non*-platitudinous assumptions in Kim's (2005) causal exclusion argument and singled out for criticism one that pertained to the identity conditions of events. However, this piecemeal approach of find-a-weak-premise-and-explain-why-it's-weak has never been successful at winning new converts to non-reductive physicalism, and in the concluding 'polemical remarks' in our paper (§10) we expressed some dissatisfaction with the way the entire mental causation game was being played: the concept of causation deployed by the disputants seemed utterly divorced from that at work either in common sense or science. A philosopher with a broadly naturalist outlook should show some interest—a lot of interest, in fact—in what scientists have to say about causation. When one does turn one's attention to scientific practice, we suggested, one will find weightier reasons to doubt the reality of *physical* causation than of mental or other higher-level causation: it is only in the special sciences, after all, that scientists explicitly take themselves to be investigating causal relationships (does smoking cause lung cancer?, etc.). Even a principle as supposedly fundamental as the causal closure of the physical domain, then, could not be taken for granted on a naturalistic approach. In n. 40 (Marras and Yli-Vakkuri 2008: 129) we suggested, however, that some form of the interventionist theory of causation could be used to vindicate the idea that there is causation going on, even at the 'bottom' level of fundamental physics. Indeed, it now seems to us that the correct interventionist theory of causation, supposing it is something along the lines of Woodward (2003), would, jointly with the assumption that mental and other higher-level properties are realized by physical properties, imply that every event that has a cause at all has a simultaneous physical cause—a principle even stronger than that which Kim (2005: 43) calls 'Closure'. Others, however, have reached different conclusions on the basis of different versions of interventionism.

In their contribution, Menzies and List seek to use a version of the interventionist theory to redefine the assumptions of the mental causation debate. While we think this is just what the debate needs, we find some of the conclusions Menzies and List reach questionable. In the following we will try to articulate our main concerns about their argument for, as well as attribution of, the 'causal autonomy' thesis, and the responses to the causal exclusion arguments that are implicit in their contribution.

NON-REDUCTIVE PHYSICALISM AND CAUSAL AUTONOMY?

Menzies and List attribute to non-reductive physicalists the view that the systems studied in the special sciences 'have causal powers that are independent of those of their more basic physical properties' (this volume: 108). No doubt this claim is true of the British emergentists of the early twentieth century, but is it true of contemporary non-reductive physicalists? Menzies and List cite Jerry Fodor as an example of a contemporary non-reductive physicalist who holds this view. But to attribute this view to Fodor is, to say the least, surprising: didn't Fodor (1985: 42) famously claim that 'if mind/brain supervenience goes, the intelligibility of mental causation goes with it'? The point of Fodor's remark was that the systems studied by psychology, and, by extension, the special sciences in general, would have no causal powers *at all* unless their properties supervened on the underlying physical properties of their constituents. For Fodor, any non-reductive physicalist who does not believe in magic must accept that the causal powers of special science systems are *dependent* on, and *determined* by, the causal powers of their more basic properties. No doubt every non-reductive physicalist must insist that mental—or, in general, higher-level—properties are *distinct* from the lower-level properties that realize them, and that, consequently, their causal powers must also be distinct (assuming, as most now do, that 'real' properties are individuated by their causal powers); but to assert the *distinctness* of the former properties and causal powers from the latter properties and causal powers is not the same as to assert their *independence*. When Fodor argued for the *autonomy* of the special sciences he was arguing for their *explanatory* autonomy, not for the *causal* autonomy of the systems which they study. The main concern of Fodor (1974)—the *locus classicus* of Fodor's exposition and defence of the 'autonomy thesis'—was to deny the *reducibility* of the special sciences in a sense of 'reduction' that would require the coextension of each special science predicate with a predicate of physics, while insisting that the real and legitimate aim of reduction should be to 'explicate the physical mechanisms whereby events conform to the laws of the special sciences' (1974: 27). Quite clearly, the point of explicating such physical mechanisms is precisely to show *how* special science systems and their properties manage to be causally efficacious: they do so by way of the physical mechanisms which implement them. To suppose otherwise would have been, from Fodor's point of view, to believe in occult powers.

Now it is a separate question whether Fodor was *right* to link the possibility of mental causation with psychophysical supervenience as he did—a matter to which we will return. However, the conception of higher-level causation as 'working through' underlying *physical* causal processes, which is presupposed in Fodor (1974), is widely shared by physicalists, of both the reductivist and non-reductivist variety. That being the case, the basis for attributing to the latter the view that the causal powers of higher-level properties are independent of those of physical properties is not clear.

But what do Menzies and List mean by saying that 'the higher-level properties of . . . systems [studied in the special sciences] have causal powers that are independent of those of their more basic physical properties'? The sense in which they think this is true turns out to be a bit surprising. While one might have thought that an affirmation of an 'independence' thesis is equivalent to the denial of a supervenience thesis (so that two systems might differ with respect to the causal powers of their higher-level properties while being indiscernible with respect to the causal powers of their physical properties), Menzies and List, apparently, simply mean to say that some higher-level properties have causal powers which are not causal powers of any of their physical realizers. Though more modest than one might have expected, the thesis is certainly interesting. The examples by which they attempt to establish this thesis involve higher-level properties whose physical realizers are 'too specific' to count as causes of their effects. In Menzies and List's argument, both a higher-level property and its physical realizer may turn out to be, so to speak, 'causally excluded'—which is excluded depends on contingent, empirical facts about each case.

This argument is based on their version of the interventionist theory of causation. In effect, Menzies and List use interventionist ideas to motivate a principle that is very nearly equivalent (see note 1) to the 'proportionality constraint' of Yablo (1992). The former does seem like an improvement on the latter in at least one respect: while Yablo's proposal was deeply involved in—some might say marred by—intuition-driven essentialist metaphysics, Menzies and List's proposal is presented as being in accord with good scientific practice. Nonetheless, the outcome is very similar, and it bears asking whether the outcome is true.

It is worth pausing to consider just why Yablo's and their proposals assign the same truth values to the causal judgements Menzies and List consider. They do so for very nearly the same reason. Consider, for example, the two rival judgements about Yablo's (1992: 258) pigeon example:

(Red) The triangle's being red caused the pigeon to peck.

(Crimson) The triangle's being crimson caused the pigeon to peck.

Given that the pigeon had been trained to peck at red things, (Red) satisfies Yablo's proportionality constraint but (Crimson) does not. (Red) satisfies the constraint because both of the counterfactuals, 'Had the triangle been red, the pigeon would have pecked' and 'Had the triangle not been red, the pigeon would not have pecked', are true, whereas (Crimson) does not satisfy it because only the first of the two counterfactuals, 'Had the triangle been crimson, the pigeon would have pecked' and 'Had the triangle not been crimson, the pigeon would not have pecked', is true. The latter is false because in one of the nearest possible worlds in which the triangle is not crimson, the triangle is some other shade of red, resulting in the pigeon pecking. So on Yablo's proportionality account (Red) is assigned True and (Crimson) False; i.e. the triangle's being red, not crimson, was the cause of the pigeon's pecking. But the very same truth value assignments, and essentially for the same reasons, would fall out of Menzies and List's account; the only difference is that the antecedents of the relevant counterfactuals are understood as being made true by an intervention.[1] The reason (Crimson) would be false, on their view, is that there is an intervention I—namely one that changes the triangle from crimson to another shade of red—such that if the triangle had been non-crimson as a result of I, the pigeon would have pecked regardless.

Applying similar reasoning to claims that ascribe causal efficacy to higher-level properties (instead of redness) versus their physical realizers (instead of crimson) yields the desired conclusion. Given Menzies and List's version of the interventionist theory of causation, it is overwhelmingly likely that there are *some* cases in which the higher-level property but not the realizer will turn out to be a cause—but this will be, as they correctly point out, an empirical question.

Is this reasoning sound? We have our doubts. The weakest part of the case for the 'causal autonomy' thesis is Menzies and List's version of the interventionist theory, which they present as a 'simplified' version of Woodward's. Let us consider the differences between the two theories. Menzies and List are (roughly[2]) committed to the following.

(ML) A causes (or, as Menzies and List say 'makes a difference to') B iff (i) $A \mathbin{\Box\!\!\rightarrow} B$; and (ii) $\sim A \mathbin{\Box\!\!\rightarrow} \sim B$.

These are 'interventionist counterfactuals', in which the antecedent is to be understood as being made true by an intervention in an ideal experiment. This stipulation invalidates the rule $A, B/(A \mathbin{\Box\!\!\rightarrow} B)$, which is valid in Lewis's (1973) counterfactual logic since, clearly, the bare truth of A and B does not guarantee that B would still be true if A were made true in an ideal experiment. However, apart from a minor technical revision made to accommodate this fact, Menzies and List's counterfactuals are understood as having the familiar Lewisian semantics. It follows, then, that if $A \mathbin{\Box\!\!\rightarrow} B$ is true in a world w, then B is true in *every* world w' that resembles w as much as A's being made true by an intervention in w' will allow.

[1] This, of course, makes a difference to truth conditions, so Yablo's and Menzies and List's proposals do not assign exactly the same truth conditions to causal claims (for example, it appears that on Yablo's proposal a barometer reading might qualify as a cause of a storm). However the truth *values* they assign in the cases that interest us are the same.

[2] This is essentially a notational variant of Menzies and List's account. To take into account the 'contrastive' character that causal claims have according to interventionism, they use a more cumbersome notation which quantifies over (possibly many-valued) variables and values assigned to them. Since they themselves focus on simple cases involving binary variables representing the occurrence or non-occurrence of an event, and since in nearly all discussions of mental causation the examples concern causal relations obtaining between *token* events (e.g., Jones's desire for water causes him to reach for the glass), which are naturally reported using sentence nominalizations ('Jones desiring water', 'Jones reaching for the glass') that take one of two values, we will for simplicity's sake mostly use 'A' and '~A' in place of 'A = 1', 'A = 0', and the like. Sometimes, however, it will be more natural to speak of a binary variable 'A' 'changing' its value, so we will alternate between the two idioms. Notably, (W) below is stated using the idiom of variables and values.

Woodward, on the other hand, is (roughly) committed to this.

(W) *X* causes *Y* iff [there is an intervention *I* on *X* such that if *I* were to change the value of *X*, then the value of *Y* would also change].

This too is a simplification,[3] but it preserves an essential feature of Woodward's theory which is not present in Menzies and List's: namely, that the right-hand side of (W) has the form of an *existential generalization* over interventions. The right-hand side of (ML), on the other hand, is a conjunction of two claims which are, *in effect*, universal generalizations over interventions.

Here's what we mean by the 'in effect'. Let us abbreviate '*A* is made true by an intervention' as '**I**(*A*)'. Since Menzies and List assume, *mutatis mutandis*, the usual Lewis semantics, the right-hand side of (ML) is true iff [*every nearest* **I**(*A*)*-world is a B-world* and *every nearest* **I**(∼*A*)*-world is a* ∼*B-world*]. In other words, a claim is being made about *all* interventions that bring about *A* (or ∼*A*) and which occur in worlds resembling the actual world as much as the truth of **I**(*A*) (or **I**(∼A)) will permit.

On Woodward's theory, on the other hand, the existence of even *one* intervention on *X* that would alter *Y* implies that *X* is a cause of *Y*. It is clear that (ML) and (W) deliver different verdicts about (Red) and (Crimson). According to (W), both (Red) and (Crimson) are true, since there is an intervention that 'changes' the redness of the triangle (i.e., makes it non-red) under which the pigeon's pecking would be 'changed' (i.e., the pigeon would not peck), as well as one that 'changes' the scarlet-ness of the triangle (making it non-scarlet by making it a non-red colour), under which the pecking would be 'changed'. One can reason similarly about mental properties and their physical realizers.

Woodward uses up a few pages explaining why 'Causal Claims [Tell] Us What Happens Under Some (Not All) Interventions' (2003: 65), and we will not repeat what he has to say here. Rather, we will say what seems to us correct about Menzies and List's version of the proportionality constraint. If we replaced the words 'causes' with 'causally explains' and the 'iff' with 'only if' in (ML), we would have, we think, a plausible claim—call the resulting claim (ML*). A good causal explanation obviously does more than make a true causal claim—for example, 'The cause of lung cancer causes lung cancer' may be true but explains nothing. 'Smoking causes lung cancer' is more enlightening but still leaves something to be desired: in particular, the latter claim does not tell us *which* interventions on smoking would affect lung cancer and how, but only that some would. Good causal explanations specify the relationship between two variables *X* and *Y*—say, by means of an equation $Y = F(X)$—in a way that tells us just how *Y* would change under interventions on *X*. When what is to be explained is a single token event, which can be viewed as the taking of a particular value (1 or 0) by a binary variable according to whether the event occurs (1) or doesn't occur (0),

[3] The full account is given in Woodward (2003: 59), where it is labelled '(**M**)'.

it seems plausible that the explanation should specify a variable by means of the manipulation of which the event could be made to both occur and not occur. If the explaining variable is also binary, then the right-hand side of (ML*) seems a plausible condition for adequacy of the explanation. (But (ML*) would only state a necessary condition, as it does not rule out claims like 'That the cause of the fire occurred causally explains why the fire occurred'.)

It is an important point, made by Batterman (2002) and Woodward (2003: 231f), among others, that often lower-level explanations of phenomena are simply inappropriate in science—for example, explanations that cite the positions of each of the 9×10^{70} molecules that compose a thermodynamic system do not adequately answer questions like 'Why is the pressure of the gas *P*?' This is arguably, as Woodward does argue, because they do not provide us with information that we could use for manipulation of the explanandum (moving the individual molecules around is not a very good strategy for altering the pressure of the gas). This is another sense, besides Fodor's, in which the special sciences have *explanatory* autonomy, but the case for the *causal* autonomy of the systems they study remains to be made.

CAUSAL EXCLUSION ARGUMENTS

What is the reply to the causal exclusion arguments, in particular to Kim's (2005: ch. 2), that is implicit in Menzies and List's chapter? It is evident that this would entail rejecting Closure—the assumption that, according to Kim, guarantees the result that the physical cause will 'win' whenever mental properties and their physical realizers 'compete' for causal efficacy. The very same considerations that militate in favour of Menzies and List's 'causal autonomy' thesis, *if* their (ML) is assumed, will militate against Closure. That an event *E* has a mental cause *M* occurring at *t* is no guarantee that it will have a physical cause also occurring at *t*, for any putative physical cause *P* occurring at *t* may be ruled out by (ML) as 'too specific', i.e., *P* may fail to satisfy (ML)(ii). Both Raatikainen (2006) and Menzies (2008) respond to the causal exclusion arguments on the basis of a similar understanding of the interventionist theory of causation.

There are two problems with this line of response. The first is that, as we have argued (Marras and Yli-Vakkuri 2008: 111), Closure is redundant to Kim's causal exclusion argument. Kim can make his case that non-reductive physicalism implies epiphenomenalism without that assumption (if he can make it at all). Menzies and List might, however, raise another objection to Kim's argument: two of Kim's implicit assumptions concerning how supervenience relates to causation—labelled 'SC I' and 'SC II' in our paper (2008: 106f)—turn out to be no more tenable than Closure, if (ML) is assumed. SC I says that an event *C* can only cause a (higher-level) event *E* by causing *E*'s supervenience base, and

SC II says that a (higher-level) event C can cause E only if C's supervenience base causes E. Given the proportionality constraint encoded in (ML), however, we have no reason to expect this to be the case: a higher-level event that causes E may be too non-specific to qualify as a cause of E's supervenience base, and the supervenience base of a higher-level event that causes E may be too specific to qualify as a cause of E.

The second problem is that, again, (ML) itself looks untenable, and if (ML) is false, the objections to Closure, SC I, and SC II we just sketched are unsound. If, as we suppose, (W), not (ML), is a (more nearly) correct account of causation, we can, in fact, give arguments for all of Closure, SC I, and SC II. To illustrate with Closure: suppose a higher-level event H causes another event E; then by (W) there is an intervention I that sets $H = 0$ such that if I were carried out, it would be the case that $E = 0$. Supposing H is realized by a physical event P_i, and that $P_1, \ldots, P_k$ are all the possible realizers of M, then there is an intervention on P_i that would set $E = 0$, namely one that sets $P_j = 0$ for each $1 \leq j \leq k$. (Why is it guaranteed that there is such an intervention? Because the intervention I that sets $H = 0$ itself is such that it sets $P_j = 0$ for each $1 \leq j \leq k$.) It follows by (W) that P_i also causes E. So, if an event has a higher-level cause, then it has a physical cause—this principle is, in fact, stronger than Kim's Closure.

What, then, is wrong with the causal exclusion arguments? We suggest that, if (W) is correct, the culprit is the causal exclusion principle itself, which is, in one form or another, common to all the arguments: in Kim's (2005: 42) version, it is the principle that 'No single event can have more than one sufficient cause occurring at any given time—unless it is a genuine case of overdetermination'. Again, it seems plausible that both a higher-level event H and its physical realizer P can be intervened on in such a way as to bring about the non-occurrence of some putative effect E of H, showing both H and P to be causes of E.[4]

CONCLUDING REMARKS

Menzies and List are surely right that whether special science properties are causally autonomous or not is an empirical matter, and likewise the question of whether special science properties, if distinct from physical properties, are ever causally efficacious. We also agree with them that the interventionist theory, broadly construed, is a promising framework for answering these questions. However, within this framework, metaphysical questions concerning the truth

[4] We assume here Kim's 'fine-grained' conception of events, on which each event is an instance of exactly one property. If this assumption is not made, a different reply, which we outline in 'The "Supervenience Argument" ' (Marras and Yli-Vakkuri 2008), is available.

of causal claims, and epistemological questions concerning the adequacy of explanations, can and must be kept apart, and it seems to us that Menzies and List's attempts to both defend the causal efficacy of special science properties and argue for their causal autonomy founder on their conflation of these two types of questions. Mental and other higher-level causation is no less defensible for that, but there is, as far as we can see, no case for the causal autonomy of higher-level properties that does not rest on a conflation of explanatory adequacy with causal efficacy.

Finally, we would like to return to Fodor's claim that the possibility of mental causation depends on the truth of psychophysical supervenience. One surprising result that becomes evident as soon as we consider the question of mental causation within the interventionist framework is that Fodor was wrong about this. If interventionism is right, then mental causation is real just in virtue of the fact that there are relationships between mental and other properties that we can exploit for manipulation—nothing further is required. Fodor's conception of mental and other higher-level causation as 'working through' physical causal mechanisms is an *empirical hypothesis*; it is not an account of the *nature* of causation. On this we are, we presume, in agreement with Menzies and List, though we perhaps part company with them in tentatively accepting Fodor's hypothesis.

REFERENCES

Batterman, R. 2002. *The Devil in the Details: Asymptotic Reasoning in Explanation, Reduction, and Emergence*. Oxford: Oxford University Press.

Fodor, J. 1974. 'Special Sciences, or the Disunity of Science as a Working Hypothesis'. *Synthese* 28: 97–115.

—— 1985. *Psychosemantics: The Problem of Meaning in the Philosophy of Mind*. Cambridge, MA: MIT Press.

Kim, J. 2005. *Physicalism, or Something Near Enough*. Princeton, NJ: Princeton University Press.

Lewis, D. 1973. *Counterfactuals*. Cambridge, MA: Harvard University Press.

Marras, A. and Yli-Vakkuri, J. 2008. 'The "Supervenience Argument": Kim's Challenge to Nonreductive Physicalism'. In F. Orilia and S. Gozzano (eds), *Tropes, Universals, and the Philosophy of Mind*. Frankfurt: Ontos Verlag.

Menzies, P. 2008. 'The Exclusion Problem, the Determination Relation, and Contrastive Causation'. In J. Howhy and J. Kallestrup (eds), *Being Reduced: New Essays on Reductive Explanation and Special Science Causation*. Oxford: Oxford University Press.

Raatikainen, P. 2006. 'Mental Causation, Intervention, and Contrasts.' Unpublished. Available at www.mv.helsinki.fi/home/praatika/ (accessed 9 January 2008).

Shapiro, L. and Sober, E. 2007. 'Epiphenomenalism—the Do's and the Don't's'. In G. Wolters and P. Machamer (eds), *Studies in Causality: Historical and Contemporary*. Pittsburgh: University of Pittsburgh Press.

Woodward, J. 2003. *Making Things Happen: A Theory of Causal Explanation*. Oxford: Oxford University Press.
—— 2008. 'Mental Causation and Neural Mechanisms'. In J. Howhy and J. Kallestrup (eds), *Being Reduced: New Essays on Reductive Explanation and Special Science Causation*. Oxford: Oxford University Press, 218–62.
Yablo, S. 1992. 'Mental Causation'. *Philosophical Review* 101: 245–80.

10

Emergence and Downward Causation

*Cynthia Macdonald and Graham Macdonald**

A plethora of recent publications has announced the presence in the world of a large variety of emergent phenomena.[1] According to these, emergence is ubiquitous. According to others, though, it is rather like the Scarlet Pimpernel: we seek it here, we seek it there, we seek it everywhere, but it remains just as elusive. Worse still, there does not seem to be much agreement on what it is that is ubiquitous, or what exactly we are missing when we do not find it.

The fact is that there are a variety of ways in which emergence has been, and can be, conceptualized, and equally, a variety of views on whether it manifests itself in the world (and if so, to what extent).Our aim in this chapter is to outline a particular way of developing an emergentist view of the nature of mind, one that goes under the name of non-reductive monism, and to defend it against some recent objections voiced by those who propose a different way of developing that emergentist view. To this end, we begin in section 1 by characterizing some important versions of the doctrine of emergentism in order to identify some core commitments shared by them and eventually to settle on a version that we will take as 'the' doctrine of emergentism for the purposes of our discussion (a version of what is known as 'strong' emergence). Section 2 develops the challenge anti-emergentists set for advocates of strong emergence and in particular for proponents of non-reductive monism, that of demonstrating how emergent properties can be causally effective. In section 3 we set out our own proposal for dealing with the challenge. Then, in section 4 we outline and develop a principled argument against non-reductive monism based on its emergentist commitments (Kim 2003, 2005), and dismantle the argument by appealing to the metaphysics of our own version of the position. Section 5 defends our position against objections from opponents and argues against the opposing strategy.

The principal charge against the emergentist account of the nature of mind, including non-reductive monism, is that it leads to incoherence because it is

* This work has been supported by a grant from the Royal Society of New Zealand Marsden Fund.

[1] See, for example, Johnson 2002; Morowitz 2002; Clayton and Davies 2006.

committed to 'downward' causation (Kim 1999, 2003, 2005), and it is this charge that we here aim to defeat by appeal to a specific metaphysics of mental causation (Macdonald and Macdonald 2006: Macdonald 2005). Since that metaphysics has come under attack, part of our defence is to address the opposing view. A secondary charge against emergentist accounts is that they cannot explain how emergent properties can have 'new' distinctive causal powers. Though limitations of space makes it impossible for us to address this charge here, elsewhere (Macdonald and Macdonald 1995, 2006) we do so in considerable detail.

1. CONCEPTIONS OF EMERGENCE

Two types of doctrines of emergence have been prevalent in the history of the literature on it. One type, which we shall call 'complex systems emergence', seems to be what many contemporary theorists take to be a doctrine of 'weak emergence'. The other type, which we shall call 'holism emergence,' seems to be what contemporary theorists consider to be a doctrine of 'strong emergence'. Beginning with the distinction between complex systems emergence and holism emergence, we shall work our way toward the distinction between weak and strong emergence.

According to many versions of the doctrine, *complexity* is the key to emergence, the idea being that complex systems exhibit behaviour that is unexpected, or inexplicable, given knowledge only of the simpler parts of the complex whole. Here is one recent characterization of this view:

> complexity emerges at the higher levels of the hierarchy of structure on the basis of the underlying physics, leading to emergent behaviours that cannot be reduced to a description at any lower level. (Ellis 2006: 79)

This usefully incorporates a number of themes relevant to debates about emergence: the notion of hierarchical levels, the thought that underlying the hierarchy is a fundamental physical 'level', and that descriptions of the behaviour of entities at higher levels cannot be *reduced* to descriptions of the behaviour of lower-level entities. Connected to the theme of non-reductibility is *unpredictability* which, in turn, is spoken of in terms of *inexplicability* and causal autonomy, and sometimes this phenomenon of emergence is 'explained' as a result of the *spontaneous* appearance of unprecedented order, such spontaneity itself being said to be the product of the *self-organizing* capacities of the underlying parts.

Given this generous assemblage of characteristics that systems or properties or entities can possess in order to qualify as emergent, it is no surprise to find emergence everywhere, from the behaviour of multicellular slime molds to the operation of financial markets. The price of the coffee one buys in Starbucks will be emergent with respect to the underlying behaviour of the individuals comprising the production and consumption 'base' for coffee.

A second approach to emergence is associated with those who are impressed by what they see as something like a categorical difference between different aspects of reality. This view takes the physical, or the lower level, to be categorized by mechanist, efficient, causality, whereas other levels exhibit some teleological, or final, causality. Here is a quote from one enthusiast, Jan Chistiaan Smuts, in an article on 'holism' in science:

> [Holism] regards natural objects as wholes . . . It looks upon nature as consisting of discrete, concrete bodies and things . . . which are not entirely resolvable into parts; and . . . which are more than the sum of their parts, and the mechanical putting together of their parts will not produce them or account for their characters and behaviour. (Smuts 1929: 640)

Although Smuts says holism views nature as consisting of 'discrete' bodies and events, some contemporary descendants of the view stress the relational nature of the properties mentioned in special scientific explanations, in contrast to the supposedly 'atomistic', or individualistically individuated, entities at the lower level. Sometimes, as in some forms of social holism, the distinction between levels is denied *because* the putatively 'higher' level is said to be essential for the individuation of the entities at the lower level. So individual people are said to be essentially socially constituted; their psychological characteristics, it is said, cannot be understood except in relation to the social system of which they are a part. (Marx is a famous exemplar of this view: 'human essence is no abstraction inherent in each single individual. In its reality it is the ensemble of social relations' [Marx 1972: 109]).

If one were to represent these two views of emergence as ideal types, one could say the former, complex systems approach stresses the *ordinary workings* of nature as resulting in surprising behaviour of complex systems; the distinction between levels is characterized as a distinction between larger and smaller, or between mereologically composed wholes and their parts, with an emphasis on mundane processes, complicated only because they contain elements of feedback, producing 'resultant' emergence. This contains both a diachronic and synchronic aspect: the later, evolved system contains more than its simpler predecessors, *and* the behaviour of the later system is more complex with respect to its simpler parts and cannot be reduced to the behaviour of those parts.

The second, holistic and relational view is less concerned with what may be called 'unadorned complexity'. In the past, adherents of holism were impressed with the difference that life makes, and so were apt to stress the intricate interconnectedness and interdependence of what may be called 'vital' properties. On this view the difference of levels is *not only* one of larger complexes having properties not reducible to the properties of their simpler parts. Although such irreducibility is endorsed, there is an attempt to make it more substantial, or more principled. In the nineteenth century this produced the doctrine of 'vitalism', a view insisting on the *uniqueness* of the properties that constituted the essence of living forms. The difference of levels was seen as a difference induced not just by the emergence of new properties, but of a new *type* of property. As a consequence some adherents of this view were less inclined to explain this 'emergence' as being the result of simple processes working in unforeseen ways; such an explanation would have been seen as unduly materialist, relying on mechanistic causality which, it was claimed, could never produce the inbuilt *telos* of vital properties and processes. In general this approach emphasized synchronic emergence, the temporal origin of uniqueness being somewhat difficult to explain without invoking a *deus ex machina*.

Perhaps for this reason the holistic view of emergence is not particularly popular today. Even before Darwin and Wallace naturalized biological complexity, and teleology, via natural selection there was a group of scientists in Germany, known as the Munich materialists, who did their best to discredit what they took to be an anti-scientific lobby constituted by Driesch and other vitalists, and did so by insisting that one could create life in a test tube by mechanical means. Combined with the growing influence of Darwin, the more extreme versions of biological holism gradually withered and died. Nevertheless, we want to keep this view on the table, since it contains, we think, elements of our topic that are crucial to the conception of emergence we want, and think appropriate here, to pursue.

In what follows, we won't have much to say about what we call the 'complexity thesis', mainly because many of the issues here seem to us to be empirical. Our suspicion is that many of the enthusiasts for emergence conflate surprise, unpredictability, and (present) inexplicability with 'in principle' inexplicability. It also seems, to our sceptical eyes, that not even present inexplicability is deemed necessary for this type of emergence to exist. The behaviour of many complex systems is not inexplicable, though it can be surprising. The discovery that the cells composing multicellular slime molds do not have a 'leader', a cell assuming the organizing role, was surprising, given that the behaviour of the multitude of cells seemed to call for such an organizer. But we do now have explanations of how the cells composing slime molds behave the way they do, and why the price of coffee in Starbucks is what it is, and our view is that this type of 'emergent' behaviour does not present any serious concern for the philosopher.

The difference between complex systems emergence and holistic emergence has been characterized by David Chalmers (2006) as a distinction between weak and strong emergence. This distinction has been remarked upon by a number of philosophers and scientists; Herbert Simon, for example, claims that weak emergence is the view that the parts of a complex system have relations that do not exist for the parts in isolation. For example, the template of an enzyme has no function

until it is placed in an environment of other molecules of a certain kind. Even though the template's function is 'emergent', having no meaning for the isolated enzyme molecule, the binding process, and the forces employed in it, can be given a wholly reductionist explanation in terms of the known physico-chemical properties of the molecules that participate in it. (Simon 1996: 170–1)

Like us, Chalmers thinks that weak emergence contains little of philosophical interest:

If one is given only the basic rules governing a cellular automaton, then the formation of complex high-level patterns . . . may well be unexpected, so these patterns are weakly emergent. But the formation of these patterns is straightforwardly deducible from the rules (and initial conditions), so these patterns are not strongly emergent. . . . The existence of unexpected phenomena in complex biological systems, for example, does not on its own threaten the completeness of the catalogue of fundamental laws found in physics. (Chalmers 2006: 245)

This leaves us with strong emergence, of which there have again been many characterizations. Chalmers has this to say about it:

We can say that a high-level phenomenon is *strongly emergent* with respect to a low-level domain . . . when truths concerning that [high-level] phenomenon are not *deducible* even in principle from truths in the low-level domain. (Chalmers 2006: 244)

This is clearly a characterization of a synchronic relation; it is not part of this view that if one cannot deduce truths about the later stages of a process from truths about the earlier stages, then the later stage is strongly emergent. Any output of a non-deterministic process would qualify as strongly emergent if we were to use the above criterion diachronically, and that is clearly not the intended interpretation of those wishing to espouse strong emergence.

Now, we think that this is a very strong version of strong emergence indeed: pack enough into descriptions of the lower domain, including its history, and, arguably, truths about biofunctional properties will be deducible, and so not emergent. It is unsurprising, then, that the only candidate for strong emergence countenanced by Chalmers is consciousness, the notorious explanatory gap ensuring non-deducibility. The rest of the psychological and social world, it is claimed, will be emergent with respect to the physical, but that emergence will be dependent on the physical facts plus consciousness, and so is not *intrinsically* emergent. Given the truths about the physical and the truths about consciousness, the truths of the social, say, will be deducible.

The plausibility of this extreme view will depend to some extent on how it is fleshed out; we need more detail about which truths are included as truths about consciousness in order to assess the claim that given all those truths plus the physical truths, all the rest of the truths about the world are deducible, including truths about the price of Starbucks coffee. We will have no chance of deducing truths about the latter type of fact unless truths about representational facts, about propositional attitudes, are included as truths about consciousness—or so we believe.

Further, there is another famous explanatory gap that is not considered by Chalmers in this context, and that is one constituted by the is–ought gap, or, more generally, the supposed fact–value dichotomy. If there is the notorious Humean gap here, it would appear that there are two options available with respect to moral truths: deny there are any, or admit another strongly emergent domain. One reason for denying there are any is important to our present topic, though it has not been discussed directly in relation to it. This concerns claims about the causal inertness of moral properties (Harman 1977, 1985, 1986); that it is not instances of moral properties, but rather, of our *beliefs* about moral truths, that cause us to behave the way we do, the moral property being at best epiphenomenal. The lack of any causal power is thought to be a sure sign that there is no *real* moral property doing any causal, and so any explanatory, work.

If this is right, then non-deducibility will need to be supplemented by a more substantial test for emergence, given that emergent properties are claimed to have genuine causal powers. The formulation of strong emergence in terms of the relation between the truths about different domains is what we would call a 'formal' notion of emergence. While the non-deducibility of the higher-level truths is clearly of interest, one would like to know more about *why* the non-deducibility holds. What is it about the *nature* of phenomena and/or properties in the relevant domains that makes for the in principle non-deducibility of the truths of one domain from the truths of the other? Leaving it as a brute fact is unsatisfactory; it invites mystery where there should be none. Our own preference is for a more straightforwardly metaphysical interpretation, one which talks of emergent properties, rather than non-deducible truths, but of course one then needs to say what it is about the properties that makes them emergent. And, again, the problem is that simply asserting their irreducibility to physical properties is unsatisfactory; one needs an argument as to why irreducibility holds.

We think that the argument will be different for different cases (Macdonald 1992; Macdonald and Macdonald 1995). If one wants to defend the irreducibility of biofunctional properties, then one will need to pay attention to what kind of properties these are, note that they have a historical dimension, and that two instances of the same physical-chemical property may be different with respect to whether they are also instances of a specific biofunctional property. If they are thus different, then that physical-chemical property cannot be identified with the biofunctional property with which it sometimes shares an instance, it being a condition on such an identity of properties that they necessarily share all their instances. If one wants to defend the irreducibility of mental properties, then one will need to mount a different argument, one specific to the nature of the mental; and it may need to be different with respect to different types of mental property as well. An argument to the effect that intentional properties are irreducible will be likely to take a different form from an argument to the effect that experiential properties are irreducible. And so on, for other cases, say, that of moral properties.

We do not propose to mount any of these arguments here,[2] since we think that prior work needs to be done: there are powerful arguments purporting to show that irreducibility has *principled* problems, and this is what the present discussion is aimed at tackling. These arguments hark back to our brief discussion of moral properties, and to the thought that, when faced with the non-deducibility of truths about a domain from the truths about a lower-level domain, when faced with an explanatory gap, one has a choice. To put it succinctly, one can be emergentist or eliminativist, be realist or irrealist with respect to the relevant property-type. Or one can deny irreducibility and non-deducibility. Which way one goes will be case-dependent, but as we have indicated and will shortly discuss in more detail, considerations about causality will be highly relevant.

Given that Chalmers' version of strong emergence is too strong for our taste, and given our preference for a more robustly metaphysical version of the doctrine that focuses on properties rather than on truths, here is our favoured version of strong emergence suitable for the likes of non-reductive monism (i.e., for a physicalist position on the nature of mind):

A property, *M*, is an emergent property of a (mereologically complex) entity/event, *e*, if and only if:

(1) *M* supervenes on the physical properties, *P*, of *e* (or *e*'s parts).
(2) *M* is not possessed by any of *e*'s parts.
(3) *M* is distinct from any structural property of *e*.
(4) *M* has a causal influence on the behaviour of *e*.[3]

A number of points about these conditions are in order. The first concerns the appeal to supervenience. Emergent properties are said to 'emerge' from other, physical properties of things and, as we have seen, many emergentists believe that the doctrine of emergence is compatible with physicalism. More specifically, non-reductive monism, being committed to the claim that mental properties are distinct from and irreducible to physical ones, claims to be a genuine form of physicalism. The appeal to the notion of supervenience is intended to ward off the charge that mental (and other) properties, being emergent and irreducible, are 'spooky' properties. The idea behind the appeal to supervenience is—to put it in our terms—that even though mental properties are non-physical, they are not worryingly non-physical either; that mental properties, and emergent properties more generally, aren't 'free-floating'. How best to characterize an appropriate relation of supervience for psychophysical and/or other cases is a thorny issue,

[2] For discussions relevant to the mental case see Macdonald (1992) and Macdonald and Macdonald (1995, 2006).

[3] This characterization of an emergent property is an adaptation of one advanced by O'Connor (1994).

as is well known: formulations of supervenience abound and we do not propose to enter into an extended discussion of them here.For our purposes, however, supervenience between the mental and the physical can be characterized as that relation which holds between a mental property or set of properties, *M*, and another, physical one, *P*, such that any two objects/events indiscernible with respect to *P* cannot diverge with respect to *M*. Following Kim (1978, 1984), let us distinguish weak from strong supervenience. Then we can define a relation of strong supervenience thus:

SS:*M*-properties strongly supervene on *P*-properties =*df*. For any possible worlds *w* and *w**, and any individuals *x* and *y*, if *x* in *w* is a *P*-twin of *y* in *w**, and the actual world's laws of physics hold in both, then *x* in *w* is an *M*-twin of *y* in *w**.[4]

A second clarificatory point concerns conditions (2) and (3).These record the fact that an emergent property is one that is not deducible from the properties of any of its parts; it is not 'derivable' from them. Condition (2) makes this clear by explicitly ruling out the possibility that the *M* property of *e* is possessed by any of *e*'s parts; (3) makes it clear by ruling out the possibility that *M* is a structural property. *M* is a structural property of *e* if and only if the proper parts of *e* have properties that are wholly distinct from *M*, and their having those properties is constitutive of *e*'s having *M*.[5] What is at stake in conditions (2) and (3) is the 'distinctive' and 'new' nature of the emergent property.

Condition (4) speaks for itself. An emergent property is one that has distinctive, new, causal powers, powers not possessed by any properties on which it supervenes, nor possessed by any property formed by Boolean operations on such properties. It is this claim that lies at the core of the doctrine of emergence, and which is the source of the charge that emergence leads to incoherence. With this in mind, we turn to one of the most powerful objections to the claim that mental properties are emergent properties.

[4] This is an adaptation of the definition of strong supervenience given by McLaughlin (1995). By *M*-properties (/*P*-properties) we mean the non-empty set, *M* (/*P*), of properties. We choose this version over Kim's principally because it is weaker than his, though his entails it. Kim's implies that it is necessarily the case that if something has an *M* property, then it has some *P* property. But SS could be true if twins had no *P* property at all. It thus allows for the possibility that there might be purely mental worlds. We think this consequence desirable, given that we take physicalism to be true and contingent, and given the variable realizability of mental properties.

[5] This is a version of O'Connor's (1994: 5) formulation, adapted from Armstrong (1978). The formulation allows for the possibility that a simple conjunctive property A&B of emergent properties A and B might be non-structural (though non-basic). Although O'Connor here speaks of the having by e's parts of certain properties being constitutive of e's having the structural property M, other comments in this work and in more recent work (O'Connor and Wong 2005) clearly indicate that by 'constitutive of' he means 'ontologically reducible to': "there is *nothing more* to having the structural property than being composed by parts having certain relations to one another—it is ontologically reducible" (2005: 10). This contrasts with Bigelow and Pargetter (1989), whose theory of structural properties takes 'constitutive of' to mean 'essential to but ontologically distinct from'.

2. ANTI-EMERGENCE

The principled argument against strongly emergent properties is one that stresses that emergent properties must come blessed with 'new' causal powers. One rationale for insisting on this lies with 'Alexander's dictum' (Kim 2005), that any real property has distinctive causal powers, so that if emergent properties are to be taken ontologically seriously, they had better have new causal powers.

Again, though, either triviality or falsehood dogs our inquiry. What exactly is required for a new property to have distinctive causal powers? One way of explicating this will be to say that emergent properties can have causal powers, but that these will be essentially dependent on, in the sense of being derivable from, the causal powers of the base properties from which the emergent properties emerge. So, for example, the causal power of a 10kg weight will be different from the causal power of any proper subpart of the weight, but this 'emergent' causal power will be readily derivable from the power of the weight's constituent parts. On this conception, a *causal inheritance* principle is respected: the causal powers of the emerging properties are the product of, or are inherited from, the causal powers of the 'basic' properties. This, the trivially true, version is to be contrasted with the stronger conception of emergent properties as having causal powers whose causal 'action is not detectable at the base level' (Di Francesco 2005) and so not readily derivable from the causal powers of the base properties. This latter view is essentially the one that we have identified as the more appropriate one for our purposes. It has been claimed, plausibly, that any such strong reading requires that the emergent property have powers to influence and control the direction of the lower processes. Such 'downward' causation has been deemed by Jaegwon Kim to be incoherent and this view of the causal powers of emergent properties false (Kim 1999, 2003, 2005).

Kim's formal characterization of emergence is:

> *Emergence*: Property M is emergent from properties $N_1, \ldots, N_n$ only if (1) M supervenes on $N_1, \ldots, N_n$, and (2) M is not functionally reducible with $N_1, \ldots, N_n$ as its realizers. (Kim 2006: 197)

And he claims that these two clauses capture the concept as it was introduced and intended by the classical emergentists such as Samuel Alexander and C. D. Broad.

> [*Emergence*] can serve as a useful benchmark; any deviation from it is a deviation from the classic conception, and new proposals can be analysed and compared with one another in terms of how far, and in what ways, they deviate from [*emergence*] as a starting point. (Kim 2006: 198)

The difficulty, as Kim sees it, lies with downward causation:

> There is no question that emergentists should want downward causation. Emergent properties must do some serious causal work, and this includes their capacity for projecting causal influence downward. (Kim 2006: 198)

Many advocates of emergence (Alexander 1920; Morgan 1931; Sperry 1980, 1987) require emergent properties to play a significant explanatory role in scientific theory, and epiphenomenal properties, properties that can have no causal influence, cannot play such a role.

As Kim sees it, the deep problem for emergent causal powers arises from the causally/explanatorily closed character of the physical domain, which can be encapsulated as follows:

> *Closure*: If a physical event has a cause, it has a physical cause. And if a physical event has an explanation, it has a physical explanation. (Kim 2006: 199)

The emergentist is thus faced with the challenge: either give compelling reasons for rejecting the closure principle, or demonstrate that downward causally emergent properties are compatible with that principle.

Some emergentists do reject the causal closure of the physical, but we do not intend to go down that route. The challenge, as we see it, is to defend the possibility of emergent causality, consistent with causal/explanatory closure of the physical domain. So our task is to rebut objections to the very possibility that there can be higher-level properties that have an 'independent' causal profile. In order to do this, we need to spend a bit of time outlining our proposed solution to the possibility of mental causation (Macdonald and Macdonald 1986, 1995, 2006).

The argument for non-reductive monism trades on the distinction between causality, which relates events in extension, and nomologicality, which relates events but only in virtue of certain of their properties. Thus, the first step towards solving the problem of mental causation involves dividing it up and conquering the parts separately. Needless to say, the solutions to the parts had better hang together, and ours not only do so but have the additional advantage of supporting one another. Here, we claim, we have a case of the plausibility-quotient of the whole exceeding the sum of the plausibility-quotients of the separate parts—an example of 'credibility emergentism'.

As we see it, 'the' problem of mental causation consists of three parts. The first concerns the causal efficacy of mental events, the second concerns what we have called the causal relevance of mental properties (the so-called *qua problem*), and the third deals with the compatibility of different levels of causation, the problem of downward causation.

As noted, the problem of efficacy concerns causality taken in extension; two events can be causally related even though the way in which that causal relation is described is explanatorily unilluminating. One of the clearest examples of this is provided by the description of the event that caused *b* as 'the cause of *b*'; such a description in a true causal statement will pick out a property of the cause, but will be explanatorily useless. Nevertheless, the event so described will indeed be the cause of *b*, and so will be causally efficacious in bringing about the *b* event.

Clearly, then, there can be true causal claims that yield nothing useful in the way of explanation. The second component of the 'problem', however—the problem of causal relevance of mental properties—does concern explanation; the challenge is to show that event *a* is causally relevant to event *b* by specifying *a* in a way that can explain why *b* occurred. In the language that has become mandatory, the solution to the second part of the problem is to show how *a* causes *b qua a*'s possessing property *M*. As has become clear, not just any property specification will do the required explanatory work, and it is a matter of ongoing debate how best to locate those properties in virtue of which an effect occurs. For present purposes we simply need to point out that successfully doing this is different from, because it involves more than, solving the difficulty concerning causal efficacy. The conflation of these two problems has been responsible for much of the confusion surrounding the problem of mental causal efficacy, or so we believe (cf. Yablo 1992; for more on causal relevance see Macdonald and Macdonald 2006).

The third part of any satisfactory account of mental causation is the one we have seen as especially relevant to any hope of defending a substantial doctrine of emergentism, the possibility of downward causation. The crucial 'downward causation' claim here is that mental events can cause physical effects. If mental properties are understood as being higher-level ones, then this can make it appear that that claim requires that there be downward causation.

In what follows we indicate first, as briefly as possible, how our solution overcomes problems of causal efficacy; we then move on to deal with downward causation, as that is most relevant in answering the critics of emergentism. We conclude by considering objections to our solution based on an opposing emergentist view of mental causation.

3. CAUSAL CONSIDERATIONS

Our solution to the problem of causal efficacy of mental events pays special attention to the distinction between properties, construed as abstract and universal, and their instancings or exemplifyings, understood as events, where such instancings are not to be understood as tropes. It appeals to a version of the property exemplification account (PEA), *developed in a particular way*. According to this, events, such as Jones's running now, not only have properties, such as the property of being a running event, but are the exemplifyings of properties, such as the property *runs*. That is to say, they are identical with exemplifyings of (*n*-adic) act-or event properties at (or during intervals of) times in objects.[6] The objects in which such exemplifyings occur are the subjects of those events. And the properties, whose exemplifyings in subjects just are events, are properties, not of events, but of their subjects. In our example, the event of Jones's running now just is the exemplifying in Jones of a property of Jones, the property, *runs*, now. Such properties are sometimes termed constitutive properties of events, and are so termed because they are the properties of subjects whose exemplifyings in those subjects just are events. So when it is said that events 'have' constitutive properties, this is not to be understood as the claim that they possess such properties.[7]

Events construed along these lines are sometimes referred to as 'structured particulars', and are so deemed because they have not only constitutive properties, but also constitutive objects (or subjects) and constitutive times. That is to say,

[6] In the terminology preferred by Kim, whose version of the account we describe and develop further here, events are *exemplifications* of properties by objects at times (see Kim 1976). But Kim himself, and many others who take a universalist rather than a trope view of properties, often use the term 'instance' as an alternative to the term 'exemplification' (and thus claim, for example, that a mental event is an instance of a property at a time in an object).We ourselves prefer 'exemplifyings' to 'exemplifications' (for reasons akin to those given by Lawrence Lombard [1986]), since it makes clear that events are fundamentally changes, whose 'constitutive' properties are dynamic rather than static, or its cognate term, 'instancings', since we think that failure to do so blurs the crucial distinction between a substance and an event (see Macdonald 1989, ch. 4)).Given the universalist (as contrasted with a tropist) view of properties, according to which an exemplification/instance of a property just is the thing that has it, we would have to say that *Jones* is the instance of the property, *runs*, since, according to the property exemplification account, as developed by Kim, this is a property of Jones, and so is a constitutive property of the event which is Jones's running. But although Kim wants to say that the subject of that event is Jones, the exemplification of the property *runs* by Jones is an *event*, a running, not the event's subject. We can avoid this problem altogether if we distinguish instances from instancings (i.e., exemplifyings), since we can then maintain (1) that an instance of a property is the thing that has it (whether this is an object or an event), (2) that events just are (i.e., are identical with) exemplifyings of dynamic properties of objects in those objects, *and* (3) that an instance of a property of an event just is the event that has that property. Events, like any other entity, have properties by instantiating them, but their constitutive properties are not, according to PEA, properties that they possess. These distinctions are important to our solution to the problem of causal relevance, since only certain ways of developing the PEA will make that solution possible. For more on the distinction between static and dynamic properties, and the differences between Kim's and Lombard's versions of the PEA, see Macdonald (1989, 2005).

We now prefer to avoid the term 'instances' entirely, since it suggests a trope view of properties, which we reject. But, since many parties to the dispute concerning the problem of mental causation (e.g. Pettit [1993] and Kim himself [1993, 1998]), regularly talk of events as instances of properties—intending the universalist view of properties as multiply-exemplifiable entities that can be (wholly) present in many places at the same time—we will, for present purposes, speak in these terms too.

[7] An event's constitutive property can no more be viewed as a property *of* it than its constitutive object can be viewed as a property of it. The claim that *P* is a constitutive property of *e* entails, not that *P* is a property of *e*, but rather, that being an exemplifying of *P* is a property of *e*. Thus, for example, the claim that the property, *firing*, is a constitutive property of the event which is Joe's firing a gun at *t* entails, not that *firing* is a property of that event, but rather, that being an exemplifying of the property, *firing*, is a property of that event.

it is in the nature of any event to be an exemplifying of a property (of its subject) in a subject at a time. Two conditions on events are essential to the account, one an existence condition and one an identity condition. These are formulated for monadic events as follows:[8]

Existence Condition: Event [*x,P,t*] exists if and only if the object *x* has the property *P* at time *t*.

Identity Condition: Event [*x,P,t*] is identical with event [*y,Q,t'*] if and only if the object *x* is identical with the object *y*, the property *P* is identical with the property *Q*, and the time *t* is identical with the time *t'*,

where expressions of the form '[x,P,t]' are construed as singular terms referring to events.[9] Kim takes expressions of this form to be canonical descriptions of events because they pick such events out in terms of their constitutive objects, properties, and times. Given that events can be described in ways that do not pick them out in terms of their constitutive 'components', there is no easy way of telling, for any given description of an event, whether it is a canonical description of that event. The importance of this point will emerge in our discussion to follow.

Although Kim himself assumes that events have only one, unique, constitutive property, the view that events are property exemplifications does not require this commitment, nor do the existence and the identity conditions, as formulated above, since neither condition states that events whose constitutive

[8] The exposition of the PEA here is based on this work of Kim's (see especially Kim 1973, 1976). According to Kim, although the first condition is indispensable to the theory, the second, as formulated, is not. The theory could proceed, for example, by defining the predicate 'is an event' over ordered *n*-tuples of objects, properties, and times. In this case, the ordered triple, <*x,P,t*>, would be an event if and only if *x* has *P* at *t*; and the principles of set theory would guarantee the existence of the triple (assuming, of course, that *x*, *P*, and *t* exist). But Kim himself appears to favour the first method over the second, and it is certainly the preferable one from the point of view of the phenomenon of causal interaction between events, where this is assumed to entail their positionality.

[9] More precisely, objects that are *minimal subjects* of events, since, as stated in the text, we prefer the version of the account developed by Lawrence Lombard (1986). According to this, an object, *x*, is the minimal subject of an event *e* if it is the minimally involved subject of *e*, where the notions of an object's involvement and minimal involvement in an event are defined as follows:

If *x* is any object, *e* is any event, and *t* is a time, then *x* is involved in *e* at *t* if and only if it is the case that if *e* occurs (or is occurring) at *t*, then *x* changes (or is changing) at *t*, and a change in *x* at *t* is identical with *e* at *t*; and

If *x* is any object, *e* is any event, and *t* is a time, then *x* is the minimally involved subject of *e* at *t* if and only if (a) *x* is involved in *e* at *t*, and (b) *x* is the smallest object which is such that a change in *x* at *t* is identical with *e* at *t*. (Lombard 1986: 122–3)

For more on the details of Lombard's version of the PEA, and his reasons for distinguishing subjects from minimal subjects of events, see Lombard (1986). That mental and physical events might have different subjects—mental ones having, say, persons, and physical ones having, say, brains—does not preclude identity between mental and physical events on the PEA, since the distinction between subjects and minimal subjects allows for the possibility that persons are not minimal subjects of mental events. For more on this, see Macdonald (1989).

properties are specified by singular terms referring to them do not possess *other* constitutive properties in addition to the ones specified by those terms. If they do, then such events will have more than one canonical description, and the identity condition will require that such events have *all* the same canonical descriptions.[10] Indeed, the version of the account we favour, developed by Lawrence Lombard (1986), explicitly allows for an event's having more than one constitutive property. This version is predicated on the assumption that events are paradigmatically and fundamentally changes, where these are *not* to be understood as states or persisting conditions. This assumption is founded on the intuition that some properties are such that their possession by an object at a time implies change, whereas others are not. Lombard labels these two sorts of properties 'dynamic' and 'static' respectively, and argues that only exemplifyings of the dynamic ones imply the existence of events.

If a material substance has a dynamic property during an interval of time, then it will be true that that substance is changing during that interval from having one static property to having another. This will be true because a dynamic property just *is* the property of first having one, then another, static property. Thus Lombard's version of the PEA, unlike Kim's, not only countenances the possibility that an event may have more than one constitutive property, but actually requires it.

As the above discussion suggests, in addition to constitutive properties, events also have characterizing properties. These are properties that events themselves possess, at least some of which they possess in virtue of having the constitutive properties they have. Thus, for example, the event that is the exemplifying of the property, *runs*, in Jones at time *t*, has as its constitutive property a property of Jones. That event has the property of being a running.

The PEA construes properties as abstract, multiply-exemplifiable entities that can have, but are not identical with, their exemplifyings. According to it, to say that mental events are identical with physical events is to say that each event which is (= is identical with) an exemplifying of a mental property of a subject in that subject at a time is identical with an exemplifying of

[10] See Lombard (1986), who points out that the view that events may have more than one constitutive property is *not* inconsistent with the existence and identity conditions of events as stated by the PEA and formulated in the text here:

Suppose that an event, e_1, is x's exemplifying of F at t, and that an event, e_2, is x's exemplifying of G at t, where F and G are distinct properties. Despite the fact that Kim's criterion of identity for events says that events are identical only if they are exemplifyings of the same property, that condition does *not* imply that e_1 and e_2 are distinct events. Nothing in that condition or in Kim's existence condition for events says that e_1 could not, in addition to being an exemplifying of F, be an exemplifying of G, and that e_2 could not, in addition to being an exemplifying of G, be an exemplifying of F. And if those were the facts, then e_1 and e_2 would be exemplifyings of the same properties by the same objects at the same times, and hence would be, according to Kim's criterion, identical. (Lombard 1986: 54–5) a physical property of that subject in that subject at that time. Crucially, this amounts to the claim that there is just *one exemplifying* of two properties, one mental, and one physical, by an object at a time.[11] That this is possible is apparent from determinable/determinate examples, such as that of being coloured and being red. The most natural understanding of the relation between these properties is that for an object to instance the latter (being red) just is for it to instance the former (being coloured): nothing further is required, once the latter is instanced, for the former to be instanced. Unlike the determinable/determinate property relation, the relation between mental and physical properties is not both metaphysical and conceptual. However, if non-reductive monism is committed to the view that mental properties supervene on physical ones in the sense specified in section 2, the result is that mental properties of persons *are not themselves constitutive properties of the events that are (identical with) exemplifyings of them*, but rather, supervene on those events' physical, constitutive properties. That is to say, a description such as 'Jones's having pain at t' (i.e., '[*Jones, having pain, t*]') is *not* a canonical description of the event which is Jones's having pain at *t*. And our view is that although the supervenience relation is a weaker metaphysical relation than the determinable/determinate one, both are cases where there can be a single exemplifying of distinct properties.

Thus, appealing to the PEA in order to rescue causal *efficacy* for mental events requires simply recognizing that an event can be a single exemplifying of both a mental property and a physical property. In the case of mental and physical properties of events, we claim that this is just what happens, and hence, that the following 'Co-Instantiation Thesis' for events is true:

> (CI) Two or more properties of an event can be co-instantiated in a single instance, i.e., there can be just one instance of distinct properties. (Macdonald and Macdonald 1986, 1995)

By the extensionality of the causal relation, if the physical event is causally efficacious, the mental one is. This shows that the PEA has the resources with

[11] Kim claims on behalf of the PEA that both mental properties (of persons) and physical properties (of persons) are constitutive properties of events, and, in his early work (Kim 1972) he concludes that token identity theories of the mind-body relation are false, on the grounds that mental properties are not identical with physical ones, but the PEA is *not* committed to this conclusion, for two reasons. First, it requires that an event cannot have more than one constitutive property, and the PEA need not be committed to this (cf. our discussion of Lombard's version of the PEA, which rejects it). Second, even if one does suppose it, one might claim—as we do—that mental properties of persons supervene on the constitutive properties of physical events, and so are not constitutive properties of those events (see subsequent pages of the text).We prefer this way of reconciling the PEA with token event identity because we think that a proper physicalism must be committed not just to an ontology of physical events, but also to providing an explanation of the relation between mental and physical *properties* which shows them to be, if not physical, not worryingly non-physical. The first way of reconciling the PEA with token event identity leaves the question of the relation between mental and physical properties completely open.

which to rescue the causal efficacy of instances of mental properties, which on any plausible account is necessary for the causal relevance of the properties themselves.

4. DOWNWARD CAUSATION*

The argument against emergent properties was tied to the incoherence of downward causation. We have accepted that some account of downward causation must be given, and it must respect the causal closure of the physical. The argument for downward causation goes like this. Emergent properties must have distinctive causal powers; they must be capable of being causally effective in bringing about their own distinctive effects. Suppose that they only bring about effects of the same (higher) level. These effects will be higher-level effects (given that emergent properties themselves are higher-level). But, given that supervenience holds, this means that the higher-level effects will have lower-level realizations. So, it is claimed, it is by causing instances of the lower-level (base) realizing properties that an emergent property will cause a higher-level effect. So, higher-level causation presupposes 'downward' causation.

Why, according to Kim, is downwards causation incoherent? Consider emergent properties *M1* and *M2*, where *M1*'s instantiation causes *M2*'s instantiation, *M1* being realized by *P1* and *M2* realized by *P2*. Given that *M2* 'arises out of' (is realized by) *P2*, *M2* would be instantiated by *P2*'s instantiation, regardless of whether *M1* had caused *M2*. Simplicity dictates that *M1* causes *M2*'s instantiation by causing *P2* to be instantiated, and this is the 'Downward Causation' conclusion. This conclusion holds for all higher-level supposedly autonomous causal action: given supervenience, 'we can no longer isolate causal relations within levels; any causal relation at level *L* (higher than the bottom level) entails a cross-level, *L* to *L-1*, causal relation' (Kim 2005: 40).

But given that *M1* is realized by *P1*, and given irreducibility (i.e., that $M1 \neq P1$), we now have two sufficient causes of *P2*, and this breaches the spirit of the principle of closure, which allows only one sufficient cause. Given physicalism, we are driven to the conclusion endorsed by Kim: 'The putative mental cause, *M1*, is excluded by the physical cause, *P1*. That is, *P1*, not *M1*, is the cause of *P2*'[12] (Kim 2005: 43). It follows that the emergent property *M1* is not independently causing *P2*'s instantiation: what is doing the causal work is what realizes *M1*, namely, *P1*. So the so-called emergent property has no (distinctive) causal power, and *M1* has no independent causal relevance.

* This section draws on work done in Macdonald and Macdonald 2007.

12 We have re-labelled Kim's 'P*' as 'P2' for consistency with what has gone before.

It looks as though Kim has a sound argument for the causal irrelevance of emergent properties. For the emergentist, only triviality looms. But we claim that the argument is not sound. In this section we want to show that there is a sense in which it is true that downwards causation is incoherent. But the route to that conclusion is significantly different from Kim's, and leads to different consequences. In particular, it rescues the possibility of the causal relevance of (some) higher-order properties, mental ones included.

What is of particular interest in the way the argument is set up is that it shuttles between talk of the downward causal power of properties and that of their instances. It is not that Kim is unaware of the importance of the instance–property distinction. He recognizes that

> Properties as such don't enter into causal relations; when we say *M* causes *M**, that is short for 'An instance of *M* causes an instance of *M**' or 'An instantiation of *M* causes *M** to instantiate on that occasion'. (Kim 2003: 155)

With this distinction in mind, we can express what is happening in putative 'downward causation' as: an instance of the lower-level property *P2* is caused by an instance of the higher-level *M1*, and *M1* does this while being realized by *P1*. Kim argues that either *P1* does all the causal work, or *M1*=*P1*. Kim opts for the latter solution, rescuing the *M1*-*M2* 'causal' relation by ensuring, via reducibility, that it is the same relation as the *P1*-*P2* 'causal' relation.[13]

Diagramically, his picture of the situation is this (Kim 2005: 55):

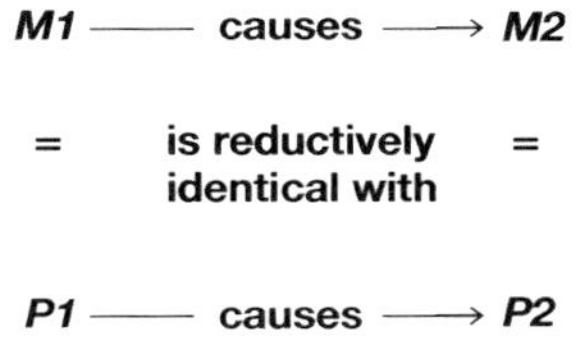

Figure 10.1.

But this picture plainly flouts the distinction Kim explicitly recognizes. The story should go: the putatively higher order *M1* has an instance, $M1_i$, that causes an instance of *M2*, $M2_i$, and does this while being identical with an instance of its realizing base, $P1_i$, thus causing an instance of *M2*'s realizing base, $P2_i$. Read this way, there is a sense in which we agree with Kim's conclusion: the causal relation between $M1_i$ and $M2_i$ *is the same as* the causal relation between $P1_i$ and $P2_i$. The picture looks like this:

[13] Our use of scare quotes around key terms here in this paragraph is intended to mark the equivocation we detect in the argument between talk of property-instances and causal efficacy, on the one hand, and talk of properties and causal relevance, on the other.

$M1_i$ —— causes ——→ $M2_i$

= is identical with =

$P1_i$ —— causes ——→ $P2_i$

Figure 10.2.

That is, causation between mental events just *is* causation between physical events, since, given physicalism, mental events are physical events. But the obvious question now is, why is the supervening property said to be either reducible or causally inert, when the natural assumption, one argued for earlier, is that the supervening and base properties are instantiated in a single instance? If there is just one instance of both the supervening and the base property, then it is true that there is no 'downwards causation', where this now means that there are *no higher-level instances* of properties that have as effects lower-level instances of properties. This, though, is because there is *no distinction of levels of instances*, only levels of properties; at the level of instances, the world is flat. But this is unremarkable, and does not have the consequences drawn by Kim. This 'fact' of no downwards causation does not by itself lead to the conclusion that the higher-level properties are causally inert, nor does it lead, without further argument, to the conclusion that they are reducible. The causal efficacy of the instance is as secure as the causal efficacy of the base instance, given there is here only one instance. All that is needed to secure the causal power of the supervening property is the plausible additional premise that if a property has instances that are causally efficacious then the property has causal powers. And if the higher-level property is irreducible, then it will have independent causal relevance; it will have a causal 'profile' different from that of its particular realizing properties.

It may be objected that the higher-level mental property *is* reducible, but this has only been asserted as a way of rescuing its causal efficacy and relevance. Since, however, we have defused that argument, there is nothing to stand in the way of the irreducibility claim, and familiar points about the multiple realization of mental properties support irreducibility. So the causal powers of supervening properties can have different profiles from those of the base properties on which they supervene. There can be emergent properties, properties that are causally relevant to the effects they produce, even though there is no 'pernicious' downwards causation.

What, then, drives Kim to his sceptical conclusion? There is an argument in Kim (1998) that looks as though it will still deliver the unwelcome conclusion. The critical move is made by the claim that where the realization relation holds between properties, the instance of the realized property has identical causal powers to that of the instance of the realizer property, so that, in a situation in which *P1* realizes *M1*, the causal powers of $M1_i$ and $P1_i$ will be the same. Now, Kim construes this as flowing from the *causal inheritance principle*, which says that, in cases of higher-order/lower-order causation, the instance of the higher-order property 'inherits' all its causal power from the instance of the lower-order property.[14] But this causal inheritance principle is not obviously derivable from the less controversial claim *that identical instances have identical causal powers*, and even this is controversial enough. Let's consider the identity claim first before returning to the inheritance claim.

The identity claim looks uncontroversial; indeed, it looks like it provides the ground for the conclusion that the supervening property is causally efficacious, and hence has causal power. It provides support for the efficacy claim because, as we have remarked before, 'is causally efficacious' is an extensional context. If this is all that is entailed by the causal inheritance principle, then there can be no objection to it. On our account, this reading of the causal inheritance principle is clearly unexceptionable, given that the most plausible view is that the realized and realizing properties have the same instance. But there is a way of reading the attribution of causal power to an instance that suggests that it is the property instanced, and not the instance itself, whose causal power is in question. What this ambiguity can do is camouflage an inference from the identity of what we will call instance causal power to a conclusion about the identity of causal powers of the property instanced. This inference would enable one to move from accepting the picture as presented in Figure 10.2 to accepting the picture as presented in Figure 10.1. And it is in fact this further inference that Kim needs in order to arrive at his sceptical conclusion concerning the incoherence of downward causation, and hence the rejection of emergent properties. But this inference is infirm, so the scepticism is unwarranted. Additional argument is required in order to be entitled to conclude, from a claim about the identity of the causal power of the instance of co-instanced properties, that the two properties thus instanced have the same causal power. The identity of instance causal power in our example ($CM1_i = CP1_i$) does not by itself license the inference to the conclusion that *M1* and *P1* have the same causal power, since this latter causal power, property causal power, is connected with instances of *M1* and *P1* other

[14] Where Kim and others would use 'higher-order/lower-order' terminology we use the terminology of 'higher-level/lower-level'. Higher-level properties should not be confused with higher-order ones. Higher-order properties are properties of properties, not properties of the things that have them in virtue of their possession of other properties. It is common, especially in functionalist treatments in the philosophy of mind, to use 'higher-order' rather than 'higher-level' when talking about mental properties such as being pain, or dispositional properties like being soluble. But this is quite different from the contemporary logician's usage (though similar to Russell's and Ramsey's). In contemporary terms, 'being soluble', like 'being a number', is a first-order predicate and so stands for a first-order property because its instances are particulars. However, both predicates might be classed as impredicative, i.e., specificiable by phrases that include second-order quantification over all properties, including those properties themselves. Thus, 'λx(x is soluble)' might be specified by something like, '$\exists F(Fx \,\&\, \forall y(Fy \,\&\, y$ is placed in a relevant liquid$\rightarrow y$ dissolves)', where we have a second-order quantifier, '$\exists F$', which ranges over all properties, including being soluble (just as the bound variable in '$x(\forall y)(xly \rightarrow x$ is taller than y)' impredicatively specifies the tallest person.

than $M1_i$ and $P1_i$. Further, given the possibility of multiple realization, it is clear that we are not entitled to conclude, from the fact that $M1_i = P1_i$, that every instance of *M1* is an instance of *P1*.

In the case being considered by Kim it is unlikely that an argument to this conclusion can be mounted that will not simply beg the question about the coherence of the notion that emergent properties have distinctive causal powers.

5. OBJECTIONS FROM THE OPPOSITION

We believe that our solution to the problem of downward causation is both simple and elegant, offers a clear and coherent account of how higher-level properties can be causally effective without overdetermining the effects they and their realizer properties have in the world, and makes much needed sense of what the relation between higher-level and lower-level properties is when these are related as realized to realizer. However, objections have come forward from sources that take a different emergentist view of the relation that mental and other higher-level properties bear to their physical realizers and that claim to make better sense of how emergent properties can exert a 'downward' causal influence on the workings of the physical domain. One very well known source of this view is to be found in the writings on 'program explanation' of Philip Pettit (1993, 2007; Jackson and Pettit 1990); another, strikingly similar view has been voiced in a number of writings by Carl Gillett (2002, 2006a, 2006b).[15] A key theme in both is that emergent properties exert their causal influence, not by being co-instantiated with their physical realizer properties, but by *non-causally ensuring* that the realizer properties will be instantiated, and, by so ensuring, themselves will be causally 'effective'. In Pettit's terminology, the realized properties, by ensuring that their realizers will be instantiated, 'program' for certain effects that instances of those realizer properties would not, had they not stood in this relation to the realized ones, have had. In Gillet's terminology, the realized property instance non-causally determines some of the causal powers that its realizer property instance contributes to individuals that have it. Such powers are 'conditioned' powers, ones conditioned by the presence, and instantiation, of the emergent, realized properties.[16]

[15] Pettit places three conditions on a property's being a 'programming property':

1. Any instantiation of the higher-order property non-causally involves the instantiation of certain properties—maybe these, maybe those—at a lower order.
2. The lower-order properties associated with instantiations of the higher-order, or at least most of them, are such as generally to produce an E-type event in the given circumstances.
3. The lower-order properties associated with the actual instantiation of the higher-order property do in fact produce E. (Pettit 1993: 37)

[16] Thus, he says, 'The key question is whether the emergent property instance H is realized and focusing upon the non-causal nature of the determination exerted by the instance of H shows that

Given, as we have seen, the importance of the property/instance distinction in discussions of the causal efficacy and 'downward causation' of mental and other higher-level properties, the question of prime importance here is how it is that the realized properties can non-causally ensure or determine that the realizer property will be instantiated, thereby bringing about an effect that otherwise it would not have brought about. Both Pettit and Gillett acknowledge the property/instance distinction (although Gillett in particular is not careful to observe it[17]), as well as the point that properties, being abstract entities, are not the sort of things that can be causes. Their view is that it is instances of properties that are causally efficacious. So, in the case of realizers of mental (and other higher-level) properties, it is their instances that are causally efficacious. The question is how it is that an instance of a higher-level property can 'non-causally ensure/determine' that an instance of the realizer property will have an effect that it would otherwise not have brought about without being epiphenomenal.

In his earlier work, Pettit does not tell us how it can do so, except to say that it does not do so by being co-instantiated with an instance of a realizer property, as our solution would have it. Indeed, he objects to our position precisely on the grounds of its commitment to the co-instantiation principle.[18] Fundamentally,

it is indeed realized in this case.The central point that we need to emphasize is that H is *not causing* P1 to contribute certain powers' (Gillett 2006a: 282).

17 So, for example, despite the emphasis placed in the quote in note 16 on the distinctness of the emergent property instance and the realizer property instance, and despite his claim that 'It should also be marked that under Alexander's concept it is plausibly property instances that are emergent' (2006a: 292), he says, 'an emergent property is identical to a combination of such properties which is itself realized.But the emergent property is not metaphysically "nothing but" the realizers. For no *merely* microphysical set of properties by themselves account for the causal powers contributed by this combination of properties to individuals. Only by taking the lower level properties to be "new" in realizing H, i.e., the emergent property instance, can we account for certain of the powers of the new "constellation" of microphysical properties. Thus, in such a case, a realized property like H, albeit one identical to some combination of lower level properties and relations, can be a necessary member of a set of properties that are only jointly sufficient for contributing a certain causal power, in Cx, to an individual such as *a*1' (Gillett 2006a: 282).

18 According to him, the program account 'would make sense of the causal relevance of intentional states, improving considerably on the most influential current alternative: the story that gives them relevance through construing them as identical with electronic or neural states. That story had the defect that it would make a state like the belief that p causally relevant but in virtue of a property other than that of being the belief that p: relevant in virtue of being such and such a neural or electronic state. . . . Consider the case where the belief that p happens to have the same realiser state as the belief that q. Both the belief that p and the belief that q will have the same neural or electronic character but the program account explains how the belief that p may be causally relevant to an action of A-ing in a way in which the belief that q is quite irrelevant (Macdonald and Macdonald 1986)' (Pettit 1993: 38). We take it that the first accusation, that 'the belief that p would be causally relevant in virtue of being a neural state' is the accusation that the co-instantiation model, by identifying instances of these two properties, makes the property, being a belief that p, causally *efficacious* by being co-instantiated with the property of being a certain neural state. So it does, but since our solution to the problem of mental causation distinguishes causal efficacy from causal relevance, the former being a necessary but not sufficient condition for the latter, the co-instantiation model does not thereby establish that mental properties are causally relevant only 'in virtue' of the causal relevance of their realizing properties. Further, given that both we and Pettit

the Program Explanation (or PE) strategy construes the notion of a mental property's 'determining an effect' as non-causal. PE thus bites the bullet: mental (and presumably all other special science properties) are *not* causally efficacious, in the sense that events that are instances of them do *not* bring about the effects they do in virtue of being instances of such properties. Mental properties are taken to be higher-level properties that supervene on physical properties of events. In any case where a mental property is thought to be causally efficacious in the production of an action (in the sense just specified), what really happens is that the instantiation of the higher-level (mental) property 'ensures that' a lower-level (physical) property is instantiated, this lower-order property doing the causal work (again, in the sense that the event that is an instance of that lower-level property brings about the action in virtue of being an instance of that property). As Pettit puts the point,

> The general idea in the program model . . . is that a higher-order property is causally relevant to something when its instantiation ensures or at least probabilifies, in a non-causal way, that there are lower-order properties present which produce it. (Pettit 1993: 37)[19]

So an instantiation of a mental property will 'program for' the instantiation of those physical properties required for the production of the physical effect. The 'ensuring that' and 'programming for' are non-causal relations so there is no causal competition between mental and physical properties, and so no overdetermination.

Like Pettit, Gillet doesn't really give us much of an idea as to how instances of emergent properties non-causally determine or ensure that their 'realizers'—instances of their realizer properties—will occur without being epiphenomenal. He does say that the relation between the emergent property instance and its realizer property instances is a 'part-whole' one, this being so because the former is identical with a combination, or sum, of instances of properties and relations between them that realize the emergent one (Gillett 2006b). Since the aggregate property instance just is a sum or collection of instances of the microphysical properties and relations between them that realize the emergent one, and the emergent property instance is identical with the aggregate property instance, the emergent property instance bears a part-whole relation to the instances of the properties that constitute the aggregate property instance. However, the mere fact that the instances of microphysical properties

accept that mental (and higher-level) properties supervene on their realizers, the possibility that mental properties could be realized by the same physical ones is ruled out. So the second part of the quote again must refer to instances, not properties, and causal efficacy, not causal relevance. But given the co-instantiation model, causal efficacy of the mental property is secured and cannot be eclipsed by that of its physical realizer.

[19] Again, where Pettit and others would speak of mental properties as higher-order properties, we would use the term 'higher-level'. See note 14.

that constitute the realizing property aggregate instance bear a part-whole relation to the emergent property instance does not by itself explain how the latter is, or can be, causally effective, since the part-whole relation is not a causal relation. Further, since the emergent property instance is identical with the combination of instances of microphysical properties that realize the emergent one, it is hard to see how any genuinely new causal powers are contributed by the emergent one to individuals that instance the microphysical property-realizers.

So neither version of the alternative emergentist doctrine gives a genuine alternative account of how 'downward' causation is possible, or occurs.

We think that these proposed alternatives to the co-instantiation model are fuelled by two thoughts. One is that, given that emergent properties are distinct from and irreducible to their realizer properties (where 'properties' means 'property types', not 'property instances'), their instances must also be distinct. The second is that, given the distinctness of the instances of such properties, one can only avoid troublesome overdetermination problems, and with it the threat of epiphenomenalism to the emergent ones, by denying that the 'influence' that they exert is genuinely causal. The consequence is a peculiar mixture: emergent property instances are indeed 'causally effective' but they are so, not by actually causing their realizer instances, but rather, by non-causally ensuring that those instances occur. We think that, in effect, this is to concede epiphenomenalism, since the emergent property instances are not themselves causally efficacious. This being so, it is hardly a solution to the problem of mental causation, nor, more generally, to the problem of 'downward' causation of emergent properties.

There is a suggestion in recent work by Pettit (2007) that the program model can accept the co-instantiation in some cases (the ones that we bring to bear on the discussion of the problem of mental causation) but has the virtue of being more general than our solution in allowing for programming to occur in cases of causation involving realized and realizer properties where, it is claimed, the co-instantiation model will not work, via distinct instances of the higher-level and realizer properties. Though we do not see how it can accommodate the co-instantiation without robbing it of the distinctive appeal over our solution claimed for it, what is important is the claim that the program solution handles other cases with ease that cannot be handled by the co-instantiation one. For it is here where we and Pettit disagree about the virtues of the program model and about whether it can do distinctive work in the area of higher-level causation.

We claim that our solution is by far the most plausible and does the work required of it in accounting for the causal efficacy of higher-level, emergent properties, in cases that involve properties in the domains of the special sciences, where those properties are best understood as standing as realized to physical realizers: biofunctional properties in biology, intentional properties in psychology, and so on. Pettit, however, thinks that there are cases of ordinary macrophysical causation, where macrophysical properties are also best understood as standing in a relation of realized to physical microphysical property realizers, but where the co-instantiation model does not get the metaphysics of the situation right. Now, we have not maintained that this model generalizes to all cases of higher-level/lower-level 'causation'; it may be that it will work only for cases of higher-level properties in the special sciences (and that might be good enough for us!). Nevertheless, in closing, we'd like to say something in response to the particular example Pettit brings to bear on the debate between us, since we don't think that it establishes what he thinks it does.

Like us, Pettit thinks that there are two 'problems' of mental causation; one concerns causal efficacy of property instances, and the other concerns the causal relevance of properties themselves (the *qua problem*). And, like us, he thinks causal relevance concerns in part causal efficacy, in part explanatory potential. Finally, like us, he thinks that causal relevance concerns the potential of the mental (or other higher-level property) to explain, not why a particular effect, under just any description, occurred, but why an effect of a certain *type* occurred. His claim is that the program model is designed to handle the problem of causal relevance, not causal efficacy.

Where he parts company with us is in his understanding of the sense in which causal relevance concerns in part causal efficacy. For us, the property that is causally relevant must itself have instances that are themselves causally efficacious; that is an absolute requirement on causal relevance. When Pettit supposes that he can accept this within the confines of the program model, then, he is mistaken, *for the program model does not require this*. Pettit makes this clear in both his recent and his earlier work. And it is this that is the source of dispute between us. Bearing it in mind, let's look at the example that Pettit thinks will not fit the co-instantiation model, see what our model can say about it, and compare it with what Pettit's own model can say about it.

Suppose that there is a closed glass flask containing water that is boiling, with a mean molecular motion of such and such. And suppose that one of the molecules in this, in the aggregate of molecules that has the property of boiling, is moving with a particular momentum and position in such a way that it breaks a molecular bond in the flask. As a result of this, the flask cracks.

Pettit says that, intuitively speaking, the property of being at boiling temperature programmes for the 'production of the breaking' and that this is a property of an aggregate of molecules, though the production of that effect—the cracking of the flask—is not brought about by an event involving the entire aggregate but only by an event involving one of the molecules in the aggregate, one of its components. This programming property might be realized in any number of ways, by different numbers of molecules having different momentum-position properties, but in each case there will be one molecule in any such aggregate whose instancing of a particular momentum-position is sufficient to break the bond in the glass and thus to crack the flask.[20] In this situation, Pettit says,

> A property of the component event—the momentum-position property—programs at the same time, but in a more specific manner, for the production of that very same effect. And the more general programmer programs for the effect via the programming of the more specific programmer. The program architecture still holds. . . . It is important to see that the program model may be extended to more complex cases like this . . . in many of these cases it would require procrustean efforts of reconstrual to be able to argue that all the relevant programming properties are co-instantiated. (2006: 223)

Pettit's basic complaint here seems to be that, because the relation between water and the constituent molecules of the aggregate with which (he seems also to hold) water is identical is a part-whole, or mereological, one, and because only one of these constituent molecules instantiates the property that is causally relevant to the cracking of the flask and whose instance is causally efficacious in bringing that effect about, although the entire aggregate has the property of being at boiling temperature, the instance of being at boiling temperature (instantiated by the aggregate as a whole) cannot be construed as identical with the instance of having a certain momentum-position, a property had by the constituent molecule. Instead, the instance of the property, being at boiling temperature (the programming property), non-causally ensures that the property of having a certain momentum-position will be instantiated, thereby bringing about the cracking of the flask.

This still leaves unresolved how it is that an instance of the programming property non-causally ensures that the realizing property will be instantiated. Given that Pettit rejects the co-instantiation solution, his explanation is that

> The event that realizes the more general programming property involves the mass of water molecules, and the event that realizes the more specific programming property involves a part of that whole: the particular molecule that does the damage. The instance of the property of the whole—the boiling of the mass of molecules—will be one event, the instance of the property of the part—the vibration of the efficacious molecule—will be another; they will be non-identical. But they will still not be distinct events. This appears in the fact that the change in the whole cannot cause the change in the part; the change in the whole is partly constituted by the change in the part: it is superveniently determined by the changes of motion in that and other parts. (2006: 224)

What do we have to say about this example? In contrast to Pettit, we think that it is not at all clear that there is no case of co-instantiation here, and in other cases like it. In order to establish this, Pettit would need to show that in general,

[20] In this Pettit's example differs from the kind of case that Gillett's view is concerned with, a case where the emergent property instance is mereologically related in a part-whole way, to all of the instances of the lower-level, microphysical properties that realize it. See Gillett 2006b.

when an object with parts changes in such a way as to bring about an effect of a certain type, and it changes solely because one of its parts changes, in order for the change in the whole to be identical with the change in the part, the change in the part must be of the same type as the change in the whole. But this seems false. Suppose, for example, that I stammer, and that it is because my stammering is a stammering, i.e., is an instance of the property of being a stammering, that my conversational partner blushes with embarrassment. But my stammering just is—is identical with—my tongue's catching, so that the property of being a stammering and that of being a tongue-catching are here co-instantiated.

It is true that being a tongue-catching will not causally explain why my partner blushed with embarrassment. In our terminology, that property is not causally relevant to that effect, as described. We think that it is this kind of consideration that leads Pettit to say that the property of the whole (here the property of being a stammering; in his own example, the property of being at boiling temperature) programs for but is not co-instantiated with the property of one of its parts (the tongue-catching, and the property of being a tongue-catching; in his example, the constituent molecule's vibrating, and the property of being a vibrating) that is effective in bringing about the effect. But this seems to us just to confuse causal efficacy with causal relevance. Note that our example, like his, concentrates on an ordinary case of macrocausation.

So our response is to issue a challenge: Pettit's claim that the co-instantiation model won't work in a case such as his trades on an assumption that seems to us to be false, and our stammering/tongue-catching example brings this out. More generally, however, we are not convinced that Pettit's example serves his own model of an emergent property exerting a 'downward' influence, and so we are not convinced that this is a case where the problem of emergent causation arises. The issue depends on what role the property of having a mean molecular motion is playing in the example and argument.

As we indicated, Pettit's assumption seems to be that water is identical with an aggregate of molecules, and, along with this, that the event that is water's boiling is identical with 'an aggregate event', the momentum-positions of the molecules. The instance of the property of being at boiling temperature, the property which he says is a property of the aggregate of molecules, seems to be either identical with, or realized by, the instance of the property of having a mean molecular motion. The most plausible understanding of the example, as he describes it, is that the instance of being at boiling temperature is identical with the instance of having a mean molecular motion, especially if the boiling is identical with the aggregate event.

Given this understanding of the relation between the two property instances, the claim is that being at boiling temperature/having mean molecular motion, programs for the production of the cracking of the flask by virtue of its instance (= the instance of the property of having mean molecular motion) non-causally ensuring that an instance of one of its realizer properties, the property (had by one of the molecules constituting the aggregate constituent molecules) of having a certain momentum-position, will break a particular bond in the surface of the glass, thereby cracking it. The non-causal ensuring is accounted for by the fact that the realizing instance bears a part-whole relation to the realizer, higher-level instance.

It seems to us that this explanation, like Gillett's, effectively concedes that the higher-level macrophysical property is epiphenomenal, since the part-whole relation is not a causal one. Further, since the higher-level macrophysical property instance is identical with the combination of instances of microphysical properties that realize it, it is hard to see how any genuinely new causal powers are contributed by the former. In contrast with the view that the higher-level macrophysical property is emergent, it seems rather to be a good example of a resultant property, one whose causal powers are a product of, or function of, the causal powers of its realizer properties, and so a case of 'upward' rather than 'downward' causation.

A resultant property of a complex entity, Kim tells us, is arrived at by 'mere' addition or subtraction of properties of its parts or components (Kim 2006).Examples of such properties are those of the shape of a table, its mass, or its weight; properties that are 'resultants' of the properties of the table's microstructure. This notion of being 'additive' covers different kinds of case. One is where the property of the whole is a function of properties of the *same* type had by the constituents. An example is Pettit's case of the property of having mean molecular energy and having a certain momentum position. Another kind of case is one where, although the 'resultant' property of the whole is a function of the properties of its microstructure, the properties of the microstructure in virtue of which the object has its resultant property are not properties of the same type. So, for example, this 50kg table's having the property of weighing 50kg is a result of the properties of its microstructure; the latter properties are not themselves weight properties.

In both kinds of case it seems that the nature of the resultant property is fixed entirely by the properties out of which it is constituted, and the entity's having it doesn't introduce a pattern of behaviour at the micro- (or lower-) level that differs from the sort of behaviour that would occur in its absence. This being so, the property is said not to be emergent in the sense that it contributes new causal powers that require new laws or forces not already in place at the lower, realizing level.

Pettit's example is one, not only of a resultant property, but of an 'averaging' property. This makes it unsuitable for the purposes of adjudicating between different emergentist versions of non-reductive monism, one based on the program model, the other based on the co-instantiation model. The reason is that an averaging property, even more obviously than other resultant properties, is one that a complex entity has solely in virtue of the properties of its constituents, and one whose causal powers are exhausted by those of its constituent properties.

It is therefore not the type of property that raises problems for emergentism generally or for non-reductive monism specifically. It could not raise the threat of overdetermination, the very threat that lies at the core of emergentist views. That threat is specifically one concerning the coherence of 'downward' causation.

REFERENCES

Alexander, S. 1920. *Space, Time, and Deity*, 2 vols. London: Macmillan.

Armstrong, D. 1978. *Universals and Scientific Realism*. Vol. 2: *A Theory of Universals*. Cambridge: Cambridge University Press.

Bigelow, J. and Pargetter, R. 1989. 'A Theory of Structural Universals'. *Australasian Journal of Philosophy* 67: 1–11.

Chalmers, D. 2006. 'Strong and Weak Emergence'. In P. Clayton and P. Davies (eds), *The Re-Emergence of Emergence: The Emergentist Hypothesis from Science to Religion*. Oxford: Oxford University Press, 244–54.

Clayton, P. and Davies, P. (eds) 2006. *The Re-Emergence of Emergence: The Emergentist Hypothesis from Science to Religion*. Oxford: Oxford University Press.

Di Francesco, M. 2005. 'Filling the Gap, or Jumping Over it? Emergentism and Naturalism'. *Epistemologia* 28: 95–122.

Ellis, G. 2006. 'On the Nature of Emergent Reality'. In P. Clayton and P. Davies (eds), *The Re-Emergence of Emergence: The Emergentist Hypothesis from Science to Religion*. Oxford: Oxford University Press, 79–107.

Gillett, C. 2002. 'Strong Emergence as a Defense of Non-Reductive Physicalism: A Physicalist Metaphysics for "Downward' Determination" '. *Principia* 6: 89–120.

—— 2006a. 'Samuel Alexander's Emergentism: Or, Higher Causation for Physicalists'. *Synthese* 153: 261–96.

—— 2006b. 'The Hidden Battles Over Emergence'. In P. Clayton and Z. Simpson (eds), *Oxford Handbook of Religion and Science*. Oxford: Oxford University Press, 261–96.

Harman, G. 1977. 'Ethics and Observation', in G. Harman, *The Nature of Morality*. New York: Oxford University Press, 1–10. Reprinted in G. Sayre-McCord (ed.), *Essays on Moral Realism*. Ithaca, New York: Cornell University Press, 1988, pp. 119–24.

—— 1985. 'Is There a Single True Morality?'. In D. Copp and D. Zimmerman (eds), *Morality, Reason and Truth*. Totowa, NJ: Rowan and Allanheld, 27–48.

—— 1986. 'Moral Explanations of Natural Facts: Can Moral Claims be Tested?'. In N. Gillespie (ed.). *Southern Journal of Philosophy* 24 (Supplement on Spindel Conference, 1986: *Moral Realism*): 57–68.

Jackson, F. and Pettit, P. 1990. 'Program Explanation: A General Perspective'. *Analysis* 50: 107–17.

Johnson, S. 2002. *Emergence: The Connected Lives of Ants, Brains, Cities, and Software*. New York: Touchstone.

Kim, J. 1972. 'Phenomenal Properties, Psychophysical Laws, and the Identity Theory'. *Monist* 56: 177–92.

—— 1973. 'Causation, Nomic Subsumption, and the Concept of Event'. *Journal of Philosophy* 70: 217–36.

—— 1976. 'Events as Property Exemplifications'. In M. Brand and D. Walton (eds), *Action Theory*. Dordrecht: D. Reidel, 159–77.
—— 'Supervenience and Nomological Incommensurables'. *American Philosophical Quarterly* 15: 149–56.
—— 1984. 'Concepts of Supervenience'. *Philosophy and Phenomenological Research* 45: 153–76.
—— 1993. *Supervenience and Mind*. Cambridge: Cambridge University Press.
—— 1998. *Mind in a Physical World*. Cambridge, MA: MIT Press.
—— 1999. 'Making Sense of Emergence'. *Philosophical Studies* 95: 3–36.
—— 2003. 'Blocking Causal Drainage and Other Maintenance Chores with Mental Causation'. *Philosophy and Phenomenological Research* 67: 151–76.
—— 2005. *Physicalism, or Something Near Enough*. Princeton, NJ: Princeton University Press.
—— 2006. 'Being Realistic About Emergence'. In P. Clayton and P. Davies (eds), *The Re-Emergence of Emergence: the Emergentist Hypothesis from Science to Religion*. Oxford: Oxford University Press, 189–202.
Lennon, K. and D. Charles. 1992. *Reduction, Explanation, and Realism*. New York: Oxford University Press.
Lombard, L. 1986. *Events: A Metaphysical Study*. London: Routledge and Kegan Paul.
Macdonald, C. 1989. *Mind-Body Identity Theories*. London: Routledge.
—— 2005. *Varieties of Things: Foundations of Contemporary Metaphysics*. Oxford: Basil Blackwell.
—— and Macdonald, G. 1986. 'Mental Causation and Explanation of Action'. *Philosophical Quarterly* 36: 145–58.
—— 1995. 'How to be Psychologically Relevant'. In C. Macdonald and G. Macdonald (eds), *Philosophy of Psychology: Debates on Psychological Explanation*. Oxford: Basil Blackwell, 60–77.
—— 2006. 'The Metaphysics of Mental Causation'. *Journal of Philosophy* 103: 539–76.
—— 2007. 'Beyond Program Explanation' in G. Brennan, R. Goodin, F. Jackson, and M. Smith (eds), *Common Minds: Essays in Honour of Philip Pettit*. Oxford: Oxford University Press, 1–27.
Macdonald, G. 1986. 'The Possibility of the Dis-unity of Science'. In G. Macdonald and C. Wright (eds), *Fact, Science, and Morality*. Oxford: Basil Blackwell, 219–46.
—— 1992. 'Reduction and Evolutionary Biology'. In K. Lennon and D. Charles (eds), *Reduction, Explanation, and Realism*. New York: Oxford University Press, 69–96.
Macdonald, G. and Wright, C. (eds). 1986: *Fact, Science, and Morality*. Oxford: Basil Blackwell.
Marx, K. 1972. 'Theses on Feuerbach'. In R. Tucker (ed.), *The Marx-Engels Reader*. New York: W.W.W. Norton & Co, 107–9.
McLaughlin, B. 1995. 'Varieties of Supervenience'. In E. Savellos and U. Yalcin (eds), *Supervenience: New Essays*. Cambridge: Cambridge University Press, 16–59.
Morgan, C. Lloyd. 1931. *Emergent Evolution*. New York: Henry Holt.
Morowitz, H. 2002. *The Emergence of Everything: How the World Became Complex*. New York: Oxford University Press.
O'Connor, T. 1994. 'Emergent Properties'. *American Philosophical Quarterly* 31: 91–104.
—— and Wong, Hong Yu. 2005. 'The Metaphysics of Emergence'. *Noûs*, 39: 658–78.
Pettit, P. 1993. *The Common Mind: An Essay on Psychology, Society, and Politics*. New York: Oxford University Press.

Pettit, P. 2007. 'Joining the Dots'. In G. Brennan, R. Goodin, and M. Smith (eds), *Common Minds: Essays in Honour of Philip Pettit*. Oxford: Oxford University Press, 215–338.

Simon, H. A. 1996. *The Sciences of the Artificial*. Cambridge, MA: MIT Press.

Smuts, J. C. 1929. 'Holism'. *Encyclopedia Britannica*, 14th edn, vol. 11: 640.

Sperry, R. 1980. 'Mind-Brain Interaction: Mentalism, Yes; Dualism, No'. *Neuroscience* 5: 195–206.

—— 1987. 'Consciousness and Causality'. In R. Gregory (ed.), *The Oxford Companion to the Mind*. Oxford: Oxford University Press, 164–66.

Yablo, S. 1992: 'Mental Causation'. *Philosophical Review* 101: 245–80.

11

Identity with a Difference: Comments on Macdonald and Macdonald*

Peter Wyss

1. BALANCING DISTINCTIVENESS AND DEPENDENCE

Two powerful ideas drive emergentism and non-reductive physicalism. The first is that mental properties are irreducibly distinct from physical properties, and the second is that mental properties depend for their presence on (physical) base-properties. Since these ideas pull in different directions, they create two challenges for balancing distinctiveness and dependence. The first challenge is to reconcile non-reductivism with physicalism or, in the case of emergentism, with naturalism. If non-reductivists accept mental properties that are too distinct, or too little dependent, they may cease to be physicalists or naturalists; and if they accept mental properties that are too dependent, or not distinct enough, they may cease to be non-reductivists. The second challenge, which arises if we flesh out the distinctiveness of mental properties in causal terms, is to balance the causal independence of mental properties with their ontological dependence on base-properties.

In order to achieve a stable balance, the Macdonalds (this volume, Chapter 10, see also Macdonald and Macdonald 2006, 2007) combine the thought that mental properties, and higher-level properties in general, are emergent properties characterized by distinctive causal powers, with a property-exemplification account (PEA), according to which events are identical with an 'exemplifying' of a property in a subject at a time (for details, see this volume: 149ff.). I will begin this comment by showing how classical emergentism supports certain aspects of the Macdonalds' position. Nonetheless, I will then challenge the stability of their balance between distinctiveness and dependence, and in particular, their assumption that physicalism is compatible with emergentism (this volume: 145).

* I would like to thank Alan Brown and Sarah Patterson for their valuable suggestions, and Cynthia Macdonald for her generous and encouraging help with the initial comment.

2. 'AN IDENTITY DOCTRINE OF MIND AND BODY'

Samuel Alexander, one of the leading proponents of early twentieth-century emergentism, argues in favour of an 'identity doctrine of mind and body' (1920 vol. 2: 9). Many philosophers are puzzled by this, not only because distinctiveness seems lost once identity is brought into play, but also because of the widespread assumption that reduction and emergence are mutually exclusive (see, e.g., Calkins 1923: 204; McLaughlin 1992: 66; Stephan 1992: 30ff., 1999: 65).[1]

In barest outline, Alexander regards entities as more or less permanent collocations, or groupings, of some basic stuff, which may become differently qualified in virtue of their increasingly complex configuration (1920 vol. 1: 341, 347; vol. 2: 48–50).[2] 'Emergence' is Alexander's technical term to describe the diachronic appearance of new qualities (i.e. properties in our sense) from, or in, this basic stuff (1920 vol. 2: 45). Since all particulars in this inherently dynamic universe are ultimately just more or less complex collocations of basic stuff, mental processes are (identical with) physical or neural processes (see, e.g., Alexander 1914: 290; 1920 vol. 2: 5, 7, 45, 63, 65; 1922: 615). Therefore, mental particulars and physical particulars are both 'expressible without residue' in terms of more basic spatio-temporal processes (Alexander 1920 vol. 2: 45).

In striking correspondence with the Macdonalds (2006: 542), Alexander employs this identity theory to secure the causal *efficacy* of mental properties. The problem with mental causation for non-reductivists rests on the assumption that a psychophysical event is both an instance of a physical base-property (B_i) and an instance of a mental (emergent) property (E_i), which then creates a causal competition between these two instances. Given the causal closure and the non-overdetermination theses of physicalism, the E-instances systematically lose out. Not so on Alexander's proposal and on the Macdonalds' PEA, because there is no competition between *one* exemplification of two properties (see Macdonald and Macdonald, this volume: 153–6). If an emergent property E is co-located, or co-instantiated, with its base-property B, an instance of E is causally effective because it is an instance of B (cf. Alexander 1920 vol. 2: 12–13).

The commitment to emergent properties apparently entails a commitment to downward causation. This consequence is problematic, for downward causation is inconsistent with causal closure.[3] Since causal closure is constitutive of their physicalism, the Macdonalds need to reconcile downward causation with

[1] This exclusivity is disputed; see, e.g., Van Gulick (2001) and Wimsatt (1997).

[2] See Alexander (1922) for a charming summary of his position; for detailed expositions of Alexander's work, see Stiernotte (1954) and Brettschneider (1964).

[3] The Macdonalds think that downward causation is incoherent (this volume: 155). But downward causation is incoherent only in its synchronic variation, according to which the higher-level property causes the instantiation of the base-property on which it depends for its very presence. In contrast, the diachronic version of downward causation is coherent (for details, see Kim 1999).

this principle (this volume: 148, 154). Again in conspicuous agreement with Alexander, they do so neatly by disposing of 'downward' causal efficacy altogether: on the PEA, no higher-level property causes a lower-level property, because the properties combined in an instance are not causally related, and because the causal *relata* are events, i.e. property-exemplifyings, rather than properties (Macdonald and Macdonald, this volume: 155 ff., 159).

Although all particulars in Alexander's world are ultimately complexes of basic stuff, they have a 'distinctive complexity of spatio-temporal structure which makes them the bearers of their distinctive empirical qualities' (1920 vol. 2: 369). For instance, a material complex that self-replicates and metabolizes has not 'merely' material properties, but 'also' vital properties. On the other hand, this entity is not 'merely' a living, but 'also' a material thing (Alexander 1920 vol. 2: 6–8, 46).[4] Congruent with this, the Macdonalds' PEA offers a natural solution to the interpretative puzzle mentioned above, as it combines the identity of the instances, or Alexander's processes, of two or more properties ($B_i = E_i$) with the diversity of properties involved ($B \neq E$).[5] Indeed, to strike a balance between dependence and distinctiveness consistent with non-reductivism, the properties must be distinct, for their instances are not.

3. THE DISTINCTIVENESS OF EMERGENT PROPERTIES

Both Alexander and the Macdonalds relate the distinctiveness of emergent properties to their causal relevance. In other words, emergent properties introduce independent and non-redundant causal powers that transcend those of their base-properties (see Macdonald and Macdonald, this volume: 146; Alexander 1920 vol. 2: 69–70). Alexander's adamant rejection of epiphenomenalism and the claim that the reality of mental properties amounts to their causal relevance is now celebrated as 'Alexander's dictum' (Kim 1992: 134; cf. Alexander 1920 vol. 2: 8). So, the thought is that even though E is co-instantiated with B, E is distinct because of its causal profile. Hence, an entity in which E arises has some of its causal effects in virtue of E, i.e. *qua* being an entity with this property (cf. Macdonald and Macdonald 2006: 542; this volume: 149).

According to the Macdonalds, emergent properties must meet two necessary conditions in order to have independent causal powers, or independent causal relevance, namely their instances must be causally effective, and they must be irreducible (this volume: 156, 159, note 18). The first condition, as discussed above,

[4] Just as life is the quality of chemical and biological processes, mind is the quality of living processes (Alexander 1920 vol. 2: 45ff.). If we follow Alexander and regard mind as a higher-level property relative to vitality, a mental process (or particular) is 'stratified', as he nicely puts it (1920 vol. 2: 68).

[5] Gillett offers an alternative (and divergent) interpretation of Alexander's identity theory based on part-whole considerations (2006: 266, 275).

is satisfied by identifying the instances of emergent properties with the instances of their base-properties. They defend the second condition, which makes their physicalism 'minimal' (2006: 541), by referring to the argument from multiple realizability (e.g., this volume: 156). In contrast, classical emergentists argue that emergent properties are irreducible because they cannot be derived, or deduced, from the base-properties. Broad, for instance, says that a (collective) property is reducible if it is 'logically entailed' by the properties that the system's parts have alone, in other systems, or in other configurations. In contrast, a property is emergent if it is *not* 'logically entailed' (see Broad 1933: 268–9, 1925: 61; see also Beckermann 2000). Emergent properties are distinct, then, because they are related neither logically nor conceptually to their base-properties (for a similar thought, see Alexander 1920 vol. 2: 46).

Even though the Macdonalds are suspicious of non-deducibility (this volume: 144), they express sympathy for the idea that emergent properties introduce a 'categorical difference' (this volume: 141) or a 'new type of property' (this volume: 142), and that emergent properties are not structural properties (this volume: 145).[6] A little-noticed thread in Alexander's work can boost the Macdonalds' hunch here, namely the insight that an emergent property individuates the thing that has it as a being of a new kind (see also Lovejoy 1927). As illustrated above, pieces of matter that metabolize and self-replicate are no longer mere pieces of matter, but also living organisms. Thus, vitality is the property that determines, or makes it the case, that the things exemplifying it are living things, which are 'kindred' in virtue of having this property. Alexander's general idea is that for any particular entity of kind K, an emergent property E is K-distinctive in the sense of being individuative of instances of this kind. Emergent properties, therefore, are distinct because they make both a causal and an ontological difference. These aspects are related, as I will now explain.

In Alexander's dynamical universe, all things are individuals, whose kind-identity is determined by the universal they instantiate (1920 vol. 1: 208ff.). A universal is the 'pattern of construction of the particular' (1920 vol. 2: 68), or the constitutive 'plan' of the things that instantiate it (1920 vol. 1: 214). So, particulars that share universal U are kindred nomologically, that is, they fall under laws associated with U that specify (and limit) how these entities behave in various situations. To fall under a kind K hence is to instantiate a universal that is individuative of K. Now, Alexander suggests that emergent properties are such that they individuate a new kind, and hence introduce distinctive new nomological realms, or levels, associated with a new basic universal (1920 vol. 2: 46–7, 70; 1922: 612).[7] Given Alexander's principle that properties without

[6] See also their critique of Gillett (this volume: 160–1).

[7] He writes: 'when mind emerges it is the distinctive quality of many finite individuals with minds' (1920 vol. 2: 361). This shows, *pace* Gillett (2006: 292, note 21), that Alexander's *emergenda* are universals (i.e. entities that can have instances), rather than property-instances.

underived causal powers lack reality, and the idea that properties are individuated by their causal profiles (see, e.g., Shoemaker 1980), there is but a short step to the idea that an emergent property individuates things of kind *K* in virtue of the causal powers it bestows on *K*-things.

This is also in line with the thought that kind-individuative emergent properties demarcate conceptually closed domains, which means that there is no conceptual relation, such as entailment, between the predicates denoting emergent properties and those denoting their base-properties. If we interpret irreducibility as non-deducibility, this lack of a conceptual relation could explain why emergent properties are not reducible.[8] Overall, these ideas mesh well with the Macdonalds' claim that the causal relevance of properties is associated with distinctive causal profiles, and systematic structures or patterns of covariation between causes and effects relative to specific causal-explanatory contexts (this volume: 149; for more details, see Macdonald and Macdonald 2006: 565ff.). Such an approach in terms of kind-individuation or clusters of causal relevance would depart refreshingly from the customary discussion of emergence in quasi-mereological terms.

However, if sound, these considerations about distinctiveness suggest that emergence is not compatible with the Macdonalds' physicalism. If a mental property *M* emerges, a physical event is now 'also' a mental event, i.e. falls under a new kind in virtue of *M*. Since *M* bestows a new identity on this event, it seems right to regard *M* as the constitutive property of it. If we accept this, we have an elegant and strong account of *M*'s causal powers, for it is in virtue of *M* that the event has the effects it has.[9] Yet, even though the PEA expressly tolerates multiple, and even non-physical, constitutive properties of events, the Macdonalds must reject non-physical constitutive properties in order to be physicalists.

4. THE DEPENDENCE OF EMERGENT PROPERTIES

Emergent properties are not only distinctive from, but also dependent on, their base-properties. The Macdonalds explicate ontological dependence in terms of realization (this volume: 158). When a base-property *B* realizes an emergent property *E*, an event or a thing has *E* in virtue of, or, as the Macdonalds also say, *by* having *B*, which is *E*'s realizer on this occasion (see also Macdonald

[8] This is reminiscent of Kim's (e.g., 2006) account of 'functional reduction' (see also Macdonald and Macdonald, this volume: 147). Here, the thought is this: if *E* is reducible to *B*, then *E* cannot be individuative of kind *K* (though *B* would be); in general, kind-individuative properties are irreducible. It makes no sense to say that *E* is reducible to *B*, which means that everything that can be said about *E* can be said in terms of *B*, yet *E* is individuative of a distinctive kind—*E* introduces a new domain that cannot be 'characterized' otherwise (cf. Alexander 1920 vol. 2: 46).

[9] In other words, emergentists might accept that emergent properties, rather than base-properties, are constitutive of the thing (or event) that instantiates them. We might even say that this event has a physical property *P* in virtue of *M*, which would be congruent with Lloyd Morgan's idea of (upwards) 'dependence' (1923: 15–17).

and Macdonald 2006: 563–4). The thought is that realization guarantees that emergent properties are not 'worringly non-physical' (this volume: 145).[10] However, it appears that on this view, realized properties are too dependent on, or too little distinct from, their base-properties. The idea that an emergent property is had 'by' having a base-property undermines its distinctiveness, which is reason to deny that emergent properties are realized properties. Here are four reservations about realization.[11]

The first reservation concerns causal relevance. As all co-instantiated properties on the PEA are causally relevant (since they are all causally effective), some of them need to be excluded in a principled manner. Furthermore, the physicalists' view that physical properties are ontologically prior to mental properties demands that only physical properties or properties realized by physical properties can be causally relevant. Thus, realization is supposed to endow non-constitutive properties with causal relevance, or 'property causal power' (Macdonald and Macdonald this volume: 157). But how can a dependent (emergent) property have independent causal powers, if it is had *by* having a base-property? If E is realized by B, then its causal powers are conditional, and therefore dependent, on its co-instantiation with B. This runs counter to the claim that independent causal powers are definitive, and required for the distinctiveness, of emergent properties (and so for 'minimal' physicalism).[12] Moreover, since B alone is constitutive of a causal event, it is not clear what it means for this event to cause an effect *qua* E, if it has E *by* having B, that is, in a *derivative* way.[13] In other words, the Macdonalds' by-relation is incompatible with the claim that events can be causes *qua* mental (or in virtue of instantiating an emergent property), and the claim that (emergent) mental properties have underived causal powers (this volume: 147).

The second reservation is about explanatory completeness. The Macdonalds reject the widespread view that mental properties are higher-order properties (2006: 550, note 23), i.e. the view that a realized property is the property of having some other property that meets a certain condition associated with the realized property, such as a causal role (see, e.g., Kim 2003: 578f.). Since realized properties are defined over their realizing properties, the higher-order approach entails a conceptual relation between realized and realizing properties.

[10] Although they are not always careful to distinguish between orders and levels of properties (see, e.g., this volume: 155), the Macdonalds also believe that realization best captures the relation between higher-level properties and lower-level properties (this volume: 158; see also the PD thesis in 2006: 564). It is doubtful, however, whether realization yields a substantive account of ontological levels.

[11] Similarly, it is realization, rather than supervenience, that gives rise to the worry that emergent properties are epiphenomenal (cf. Welshon 2002).

[12] In contrast, Gillett (2002) suggests that B bestows its causal powers conditional on, or in virtue of, realizing E. Since this implies the denial of causal closure, the Macdonalds reject this suggestion (this volume: 158ff.).

[13] I borrow the terminology, but not the content, from Baker (2000: 46ff.).

Notably, and consistent with classical emergentists, the Macdonalds deny such a conceptual relation (this volume: 153). This is important, for a conceptual relation between the emergent property *E* and its base-property *B* would suffice to reduce *E* to *B* (if a failure of reduction is a failure of deduction), which would jeopardize their non-reductivism.

However, realization should also be explanatory: when *B* realizes *E*, the presence of *B* not only entails, but also explains, the presence of *E* (see Kim 2003). Without a conceptual relation, it is hard to see how the realization-relation can comply with this explanatory requirement. To be sure, there must be some kind of nomological covariation between realized and realizing properties, for otherwise *E* would not even supervene on *B*. This would be unacceptable for emergentists as well as the Macdonalds, who claim that higher-level properties supervene on the constitutive properties of events that instantiate them (this volume: 152–3; see also 2006: 560ff.). For emergentists, however, the covariation between base-properties and emergent properties is a brute empirical fact (Alexander 1920 vol. 2: 46–7; 1922), or a fundamental *a posteriori* law (Broad 1925: 77–9). Hence, if the Macdonalds reject a conceptual relation between realizing and realized properties, their realization is as brute as the emergentists' emergence.[14] This is in conflict with the explanatory aspirations constitutive of their physicalism, namely the aim for explanatory completeness or exhaustiveness (cf. Macdonald and Macdonald 2006: 561, note 39; this volume: 144, 153, note 11). Since explanatory completeness is obviously not a feature of emergentism, it follows once again that physicalism and emergentism are inconsistent.[15]

The third worry concerns ontological redundancy. On the higher-order approach, realized properties are not additions to our ontology (see, e.g., Kim 1998: 103ff.). It seems that the Macdonalds' by-relation cannot avoid this consequence either. If an entity has a property by having another property, there is nothing more to this state of affairs than this entity's having the latter property, as the Macdonalds' allusion to the determinate/determinable relation suggests (2006: 563; this volume: 153). Suppose an apple is coloured by being red, and red is a way of being coloured, then arguably there is but one (determinate) property instantiated in this case.[16] If this is true, realized properties (like determinables) assume an air of irreality, and seem little more than multiply satisfiable predicates. Consequently, realized properties would not be new and distinctive. Therefore

[14] Similarly, neither the CI thesis (Macdonald and Macdonald 2006: 562; this volume: 153) nor supervenience *qua* modalized property-covariance explain why (or how) emergent properties relate to their base-properties.

[15] As Crane (2001) suggests, it is this divergence in epistemic attitude that separates emergentism from non-reductive physicalism.

[16] 'Arguably', because on the PEA, this instance of red is *also* an instance of colour. It is not clear, however, how to reconcile this with the thought that a realizing property is a way of being a realized property, and with the Macdonalds' explicit view that in this case there is just one property instantiated. Note also that the Macdonalds qualify the analogy between realization and the determinate/determinable relation (2006: 563, note 43; this volume: 153).

they could not be emergent either; or worse, they become ontologically redundant and 'might as well be abolished' (Alexander 1920 vol. 2: 8). If so, emergent properties cannot be realized properties.

The final reservation concerns the identity of properties. The idea that *E* is had by having *B* contravenes the claim that emergents make a categorical difference. As suggested above, without *E*, something could not be the kind of thing it is, even if it had *B*. If they are individuative of kinds, emergent properties are not identity-dependent on their base-properties. For if *E*'s identity is determined by specific causal powers, and if these causal powers are not individuative of *B*, then *E*'s identity is not dependent on *B*'s identity. We could say that something has *E* essentially, rather than derivatively 'from' *B*. Therefore, an emergent property is neither a 'way of being' a base-property, nor does it need to be made what it is by its base-property. If this is right, emergent properties are unrealized or basic properties.[17]

The Macdonalds' fundamental thought is that physical base-properties vouchsafe the causal relevance of emergent properties. In contrast, the basic emergentist thought is that emergent properties bestow their causal powers in virtue of themselves, which is the point about their distinctiveness. The initial similarity between the Macdonalds and Alexander thus gives way to a crucial difference: Alexander and the emergentists in his wake are not physicalists. This is not only because they reject realization, but also because they accept a 'democratic' universe with a plurality of real properties (see Alexander 1914: 280), and consequently do not grant exclusivity to a particular class of properties. On the other hand, the Macdonalds cannot be emergentists. For if the worries about realization are sound and emergent properties are not realized, yet realization is definitive of the Macdonalds' position, it follows that their combination of physicalism and emergentism is incoherent.

REFERENCES

Alexander, S. 1914. 'The Basis of Realism'. *Proceedings of the British Academy 1913–1914*. London: Oxford University Press, 279–314.

—— 1920. *Space, Time, and Deity: The Gifford Lectures at Glasgow 1916–1918* (2 vols). London: Macmillan. Available online at <http://www.giffordlectures.org> (accessed October 2007).

—— 1922. 'Natural Piety'. *The Hibbert Journal* 20: 609–21.

Baker, L. R. 2000. *Persons and Bodies: A Constitution View*. Cambridge: Cambridge University Press.

17 Identity-independence is compatible with ontological dependence, which merely says that things can have *E* only if they also have *B*, i.e. the base-property is necessary (but not sufficient) for the dependent property. Realization is stronger, because the realizer suffices for the realized.

Beckermann, A. 2000. 'The Perennial Problem of the Reductive Explainability of Phenomenal Consciousness: C. D. Broad on the Explanatory Gap'. In T. Metzinger (ed.), *Neural Correlates of Consciousness: Empirical and Conceptual Questions*. Cambridge, MA: MIT Press, 41–55

Brettschneider, B. D. 1964. *The Philosophy of Samuel Alexander: Idealism in 'Space, Time and Deity'*. New York: Humanities Press.

Broad, C. D. 1925. *The Mind and its Place in Nature*. London: Kegan Paul.

—— 1933. 'The "Nature" of a Continuant'. In *idem*, *Examination of McTaggart's Philosophy* (vol. 1). Cambridge: Cambridge University Press, 264–78.

Calkins, M. W. 1923. 'The Dual Rôle of the Mind in the Philosophy of S. Alexander'. *Mind* 32: 197–210.

Crane, T. 2001. 'The Significance of Emergence'. In C. Gillett and B. Loewer (eds), *Physicalism and its Discontents*. Cambridge: Cambridge University Press, 207–24.

Gillett, C. 2002. 'Strong Emergence as a Defense of Non-Reductive Physicalism: A Physicalist Metaphysics for "Downward" Determination'. *Principia: Revista Internacional de Epistemologia* 6: 89–120.

—— 2006. 'Samuel Alexander's Emergentism: or, Higher Causation for Physicalists'. *Synthese* 153: 261–96.

Kim, J. 1992. ' "Downward Causation" in Emergentism and Nonreductive Physicalism'. In A. Beckermann, H. Flohr, and J. Kim (eds), *Emergence or Reduction? Essays on the Prospect of Nonreductive Physicalism*. Berlin: W. de Gruyter, 119–38.

—— 1998. *Mind in a Physical World*. Cambridge, MA: MIT Press.

—— 1999. 'Making Sense of Emergence'. *Philosophical Studies* 95: 3–36.

—— 2003. 'Supervenience, Emergence, Realization, Reduction'. In D. W. Zimmerman (ed.), *The Oxford Handbook of Metaphysics*. New York: Oxford University Press, 556–84.

—— 2006. 'Emergence: Core Ideas and Issues'. *Synthese* 151: 547–59.

Lloyd Morgan, C. 1923. *Emergent Evolution: The Gifford Lectures at St Andrews 1922*. London: Williams and Norgate. Available online at <http://www.giffordlectures.org> (accessed October 2007).

Lovejoy, A. O. 1927. 'The Meanings of "Emergence" and its Modes'. *Journal of Philosophical Studies* 2: 167–81.

Macdonald, C. and Macdonald, G. 2006. 'The Metaphysics of Mental Causation'. *Journal of Philosophy* 103: 539–76.

—— 2007. 'Beyond Program Explanation'. In G. Brennan, R. Goodin, F. Jackson, and M. Smith (eds), *Common Minds: Themes from the Philosophy of Philip Pettit*. Oxford: Oxford University Press, 1–27.

McLaughlin, B. P. 1992. 'The Rise and Fall of British Emergentism'. In A. Beckermann, H. Flohr, and J. Kim (eds), *Emergence or Reduction? Essays on the Prospects of Nonreductive Physicalism*. Berlin: W. de Gruyter, 49–93.

Shoemaker, S. 1980. 'Causality and Properties', in P. Van Inwagen (ed.), *Time and Cause*. Dordrecht: D. Reidel Publishing Company, 109–35.

Stephan, A. 1992. 'Emergence: A View on its Historical Facets'. In A. Beckermann, H. Flohr, and J. Kim (eds), *Emergence or Reduction? Essays on the Prospects of Nonreductive Physicalism*. Berlin: W. de Gruyter, 25–48.

—— 1999. *Emergenz. Von der Unvorhersagbarkeit zur Selbstorganisation*. Dresden: Dresden University Press.

Stiernotte, A. P. 1954. *God and Space-Time: Deity in the Philosophy of Samuel Alexander*. New York: Philosophical Library.

Van Gulick, R. 2001. 'Reduction, Emergence, and Other Recent Options on the Mind/Body Problem'. *Journal of Consciousness Studies* 8: 1–34.

Welshon, R. 2002. 'Emergence, Supervenience, and Realization'. *Philosophical Studies* 108: 39–51.

Wimsatt, W. C. 1997. 'Aggregativity: Reductive Heuristics for Finding Emergence' *Philosophy of Science*, 64 (Supplement): S372–84.

12

Can Any Sciences Be Special?

David Papineau

INTRODUCTION

Non-reductive physicalism accepts the primacy of the physical while aiming to avoid the constraints of traditional reduction. It respects physicalism via the doctrine that all properties metaphysically supervene on physical properties. It avoids traditional reduction via the thesis that many properties cannot be type-identified with physical properties.

The viability of non-reductive physicalism has been extensively discussed over the half-century since it was first explored by Putnam (1960, 1967) and Davidson (1970). Most of the debate has focused on whether non-reductive physicalism can accommodate non-physical *causes* (cf Kim 1993; Robb and Heil 2003: sect 6.) However, there has been far less discussion of whether non-reductive physicalism can accommodate non-physical *laws* (though see Block 1997; Kim 1992; Macdonald 1992; Millikan 1999; Papineau 1985, 1992). In this chapter I wish to focus first on the issue of non-physical laws. This will turn out to cast some useful light on the question of non-physical causation.

Not all non-reductive physicalists think that there are non-physical laws. Davidson, for example, does not (1976). Even so, it is widely supposed that there can be laws in 'special sciences' like biology, psychology, and economics even though their categories do not reduce to physical types. The *locus classicus* for this position is Fodor's 'Special Sciences' (1974). Fodor made his analysis graphic in what must be the most-reproduced diagram in philosophy.

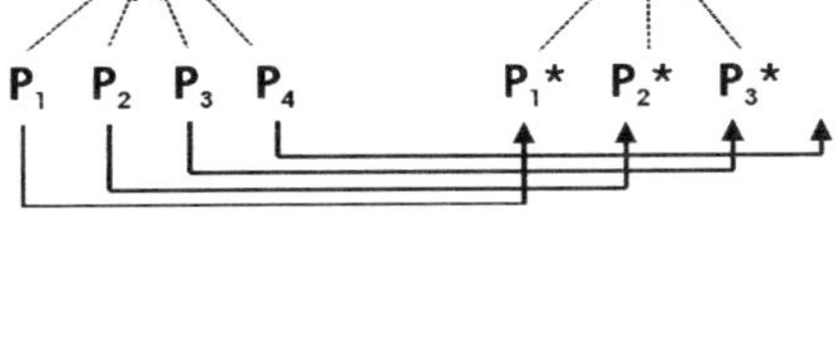

The idea is that S_1 and S_2 are special kinds. S_1 -> S_2 is a special law. Thus S_1 might be an increase in demand for some good, and S_2 an increase in price. P_1, P_2, . . . are the different physical ways that S_1 might be realized, and P^*_1, P^*_2, . . . different physical ways in which S_2 might be realized. (Thus think of all the different physical systems that can underpin economic exchanges—all the different kinds of monetary and non-monetary forms of exchange.)

Realization should here be understood in terms of metaphysically necessary supervenience: P_1 realizes S_1 in the sense that it is metaphysically necessary that any system that has P_1 will have S_1. At the same time, not every system that has S_1 will have P_1, or any other physical kind, since there are always other physical ways (P_2, . . .) in which S_1 can be realized. This is why S_1 is not type-reducible to any physical kind, even though it metaphysically supervenes on the physical facts.

At the physical level, the different realizers of S_1 generally give rise to realizers of S_2. Thus, when S_1 is realized by some P_i, this will instigate physical processes that give rise to a P^*_i, which in turn then determines S_2. These physical processes are thus consonant with the special law S_1 -> S_2.

A P_i -> P^*_i link need not hold in every single case. Some of the P_is that realize S_1 will fail to give rise to a P^*_i that determines S_2. This is why, according to Fodor, the laws of the special sciences only hold *ceteris paribus*. The physical shadowing of the S_1 -> S_2 law will not be perfect, and so the law will have exceptions.

SPECIAL LAWS IN QUESTION

I have always been puzzled by Fodor's picture of the special sciences (Papineau 1985, 1992). Here is the obvious worry. If the realizations of S_1 are all so physically different, then how come they all give rise to a similar result, namely, some physical state that determines S_2? Will it not be an unexplained coincidence that they should all display this common result? Unless more can be said about what ties the P_is together at the physical level—as would be provided by a traditional reduction—will the variability of the P_is not undermine the idea that S_1 is regularly followed by S_2?

Here is an example that will illustrate the point (cf Papineau 1993: ch. 2.) Suppose we find some initial evidence that people who eat reheated Brussels sprouts (S_1) come to suffer from inflamed knees (S_2). However, when we investigate this phenomenon, we find that there is no common feature that accounts for this syndrome. Rather, in one case the sprouts harbour a virus (P_1) that infects the knees (P_1*). In another the sprouts contain a high level of uric acid (P_2) that leads to gouty attacks (P_2*). In a third the sprouts involve some toxin (P_3) that depletes the cartilage that protects the knee joints (P_3*). And so on.

This story doesn't hang together. It beggars belief that reheated Brussels sprouts should always give rise to inflamed knees, yet the physical process that mediates this should be different in every case. Surely either there is some further feature of the sprouts that can explain why they all yield the same result, or we were mistaken in thinking that there was a genuine pattern in the first place, as opposed to a curious coincidence in our initial sample of cases.

Yet this looks just like the picture that Fodor is inviting us to accept for special scientific laws. So I am inclined to say just the same about Fodor's picture. Either there is something more to say about why S_1 should always give rise to S_2, or it can't be a genuine pattern to start with.

Does it help that Fodor's special science laws are only supposed to be *ceteribus paribus* and not strict? Not really. Note that the puzzle about the reheated Brussels sprouts leading to inflamed knees doesn't depend on this being an invariable pattern. In the absence of a uniform explanation, it would be just as puzzling if *most* people who eat reheated Brussels sprouts get inflamed knees—or even if reheated Brussels sprouts merely *raises the probability* of inflamed knees. Any such correlation would seem to call for a uniform explanation. It would be mysterious that reheated Brussels sprouts should so much as increase the probability of inflamed knees, if the mechanism were different each time it did so.

Some readers may wonder whether an analytic functionalist account of special science concepts can resolve the puzzle. Analytic functionalism defines concepts in terms of causal structures. Thus it might be definitionally required that something only counts as an 'S_1' if it gives rise to an S_2. For example: something might only count as a 'pain' if it leads to efforts to avoid the source of the pain; something might only count as 'inflationary pressure' if it generates a fall in the value of money, and so on. Given this kind of definition, it will scarcely be a surprise if many different physical kinds P_i realize S_1 and yet all give rise to a $P_i{}^*$ that determines S_2. After all, if they did not do this, then they would not count as realizations of S_1 in the first place. Something that doesn't generate avoidance behaviour just isn't a 'pain'; something that doesn't lead to a fall in the value of money isn't an 'inflationary pressure'; . . . So, given this, it will be inevitable that all S_1s will lead to S_2s, notwithstanding their variable realization, for that's what it takes to count as an 'S_1'.

Unfortunately, nothing in this line of thought helps explain variably realized special science *laws*. It may explain how definitional truths can be variably realized, but that is a different matter. Genuine laws can be expressed by synthetic statements with the antecedent definitionally independent of the consequent, as opposed to the analytic truths that result when 'S_1' is defined as a precursor of S_2. And that is precisely why there is a puzzle about their variable realization. Given that the antecedent circumstance S_1 in a genuine law can be identified independently of whether it produces the consequent S_2, we expect there to be some further account of why such S_1s are always (or at least unusually often) followed by S_2s—and that is what the variable realization seems to preclude.

KINDS OF KINDS

Despite the points made so far, it may seem that there cannot really be a problem about variable realized laws as such. After all, surely there are plenty of familiar examples of such laws. What about the law that a temperature of 100°C will make water boil? Are there not many different molecular movements that can realize a water temperature of 100°C? Yet clearly there isn't any puzzle about why water boils in all these cases.

But this is a different kind of set-up. To see why, we need to be a bit more explicit about the idea of 'variable realization'. For a category S to be variably physically realized, it isn't enough that the instances of S display *some* differences at the physical level. We wouldn't want to say that being square, say, is variably physically realized just because different square things have different masses. Nor should we say that being in pain is variably physically realized just because different people have different-sized C-fibres. For a category S to be genuinely variably realized, the requirement is not the weak demand that there be some physical differences between the Ss, but rather that there should be *no* physical property that is peculiar to them. The members of a genuinely variably realized kind will share no physical property that is not also shared with non-members.

With temperatures, there is, of course, just such a common physical property. All samples of water at a given temperature have the same mean molecular kinetic energy, notwithstanding any further differences between the specific motions of their constituent molecules. And that, of course, is why there is no puzzle about why water boils at 100°C. Despite the different molecular motions involved, all water at 100°C shares the same mean molecular kinetic energy, and this allows a uniform physical explanation of the boiling. By contrast, if there is no common physical feature to some category, then there is no room for such a traditional type-type reduction of any patterns it enters into.

Might Fodor just be saying that special science categories are like temperature? That is, might he simply be pointing out that there can be physical differences between different instances of some special type, like an increase in demand for some good, just as there are differences between different samples of water at 100°C, and that this is consistent with their having some physical commonality that will explain why they fit into some uniform pattern?

But this suggestion is not consistent with other claims Fodor makes. Thus consider his original response to the obvious query raised by his diagram: why isn't the disjunction P_1 v P_2 v P_3 . . . a physical property with which S_1 can be type-identified, thereby yielding a traditional physical reduction of S_1? Fodor's response is that even if we can formulate this disjunction, it will not represent a genuine physical *kind*, as opposed to a heterogeneous collection of different physical kinds. Correspondingly, even if we can write down the generalization P_1 v P_2 v P_3 . . . -> P_1* v P_2* v P_3*. . . , this won't constitute a genuine physical *law*, as opposed to a representation of a bunch of different physical processes. There is, of course, an element of circularity here, in that the standard explications of kinds is that they are categories that figure in genuine laws, while the standard explications of laws are that they are patterns that involve genuine kinds. But any such circularity does not affect the point currently at issue, which is that Fodor is explicit that there is no single physical kind that characterizes all instances of his special Ss.

A DILEMMA FOR FODOR

Given the points just made, the challenge facing Fodor can be put in the form of a simple dilemma. If the realizations of special S_1 and S_2 are genuinely variable and don't form kinds, then doesn't this immediately imply that the empirical generalization S_1 -> S_2 will not be a law, but rather a collection of heterogeneous processes? Alternatively, if the realizations of S_1 and S_2 do form kinds, doesn't this mean that P_1 v P_2 v P_3 . . . -> P_1* v P_2* v P_3*. . . will be a genuine law that constitutes a traditional reduction of S_1 -> S_2? (Cf Kim 1992.)

Fodor responds to this putative dilemma in his splendidly named 'Special Sciences: Still Autonomous After All These Years' (1997). He argues that the dilemma begs the question. True, he allows, special categories are not identical to *physical* kinds, and so any generalizations involving them will not be *physical* laws. But that's not decisive, he insists. For it is still possible that these categories constitute *special* kinds, in virtue of entering into *sui generis special* laws. Fodor takes it to be a datum that psychology, economics, and the other special sciences contain genuine laws covering categories that can't be type-reduced to physics. Given this, he concludes that the categories of such sciences are *kinds* all right, in virtue of entering into these special laws. From this perspective, the Brussels sprouts example is misleading: it appeals to our intuitive knowledge that there is no real law in the case and that *reheated Brussels sprouts* is thus not a medical kind. By contrast, Fodor suggests, in areas where there are real laws covering physically heterogeneous categories, like psychology and economics, we have every reason to ascribe kindhood to those variably realized categories.

At first pass, this response may seem reasonable enough. There is no immediate reason why the only laws of nature should be physical laws. After all, it is clearly consistent with supervenience physicalism that there should be a finite few cases in which, say, eating reheated Brussels sprouts leads to inflamed knees via disparate physical processes. So there can scarcely be any outright contradiction in supposing that such a variably realized pattern should be repeated indefinitely.

However, note that special categories do not just enter into laws connecting them with other special categories. They are also systematically related to *physical* categories. For example, a drought in cocoa-producing areas will raise the price of chocolate. Economic growth without environmental regulation will lead to an increase in atmospheric CO_2. And so on. (Indeed, such interaction is surely part of the underlying rationale for physicalism: if special categories did not interact causally with physical ones, there would be no reason for supposing that they must supervene on the physical realm to start with (Papineau 2002: ch 1).)

But this now reinstates the dilemma once more. If special categories are going to feature in physical laws, then does this not mean that the disjunction of their physical realizations itself will need to be a physical kind? As before, there is clearly something wrong with the idea of a *physical* law that is variably realized at the *physical* level. If kinds are categories that feature in laws, then the special categories that feature in physical laws will need to be type-identical with physical kinds.

We can make the point graphic by considering situations where a variably realized special category has some uniform physical cause and physical effect. For example, if a human's arm is immersed in ice-cold water, this will engender pain, and this will lead the human to remove their arm from the water. But now suppose that the category of human pain is not physically reducible. Then there will be quite different physical processes mediating between the initial physical cause and the final physical effect. What then ensures that all these different intermediary processes converge on the same final effect? It is not as if the pain exerts some independent causal influence to bring this about—that would require interactive dualism and 'causal gaps' in the physical realm. Rather, the causal influence of the pain in each instance is exhausted by the causal influence of its physical realization. But then we seem to be left with a mystery. The initial cause, the freezing water, generates a divergent range of intermediary neurological effects, but then these inexplicably converge on the same physical result, removal of the arm from the water.

METHODOLOGICAL ISSUES

So far my argument has proceeded on an abstract metaphysical level. But if it has any substance, some definite methodological implications must follow.

Fodor's terminology of 'autonomy' suggests that the special sciences will be threatened as independent academic disciplines if their categories are reducible to those of physics. The thought is that type-reduction would mean that any special laws would simply be special cases of the physical laws that reduce them, and the special sciences therefore little more than sub-departments of physics.

But is this a serious worry? There is, of course, a sense in which the reducibility of some special science means that it is not independent of physics—in principle its laws will follow from physical laws. But this in-principle possibility need have no practical implications. For the in-principle derivability may be practically unfeasible, in which case the reducibility of the special science will make no methodological difference to its practitioners. They will still proceed to investigate the relevant special laws using direct empirical evidence. This is surely how it goes in many science departments. Nobody doubts, I take it, that chemical, meteorological or geological laws have uniform physical explanations. But at the same time nobody tries to derive these laws from basic physics, at least once we are dealing with systems more complex than the hydrogen atom. Instead, special scientists investigate the relevant complex systems directly, using observation and experiment to ascertain the laws they obey—which is why we have separate chemistry, meteorology and geology departments in universities.

My methodological concern is the opposite of Fodor's. I am not worried that the special sciences will be undermined if they are reducible to physics. My concern is rather that they will be undermined if they are *not* reducible to physics. The argument so far suggests that only physically reducible categories can enter into genuine laws. If special sciences need laws, physical reducibility will therefore be a precondition of special sciences. This reducibility need make no methodological difference to the practice of the science, for the reasons just given. But there had better *be* a type-reduction, at the metaphysical if not the methodological level, otherwise there will be no laws to investigate empirically.

Many philosophers take it to be obvious that special categories cannot be type-reduced to physical categories. If this is right, the argument so far suggests that there will be no special laws. The non-reducibility will ensure all the autonomy Fodor could wish for. But it looks as if the special sciences will have nothing left to study.

SELECTIONAL PATTERNS

There is a gap in my argument so far. As a number of writers have observed (Block 1997; Macdonald 1992; Papineau 1985, 1992), one possible explanation for variably realized laws involving physical kinds is that they are the result of *selection processes*. Consider this example. In all electrical hot water heaters, the current is switched off at some temperature below boiling point. But when we look at the physical process that mediates between the high temperature and the switching off, we find that it is different in each case. Each heater contains a thermostat, but there are many different kinds of thermostat, each using different physical components in different combinations (including bimetallic strips, expansion gases, mercury bulbs, and thermocouples).

Given this, we can imagine someone asking why so many different physical processes should all lead to the same effect—namely breaking the circuit. If there is no uniform physical explanation for this commonality, is it not a mystery that all the divergent effects of temperature increases should converge on this single effect?

But of course in this case there is an obvious answer. All these different physical processes were *designed* to produce the same effect. The people who construct heating systems make sure they contain a thermostat. They want a device that will shut off the current when the temperature gets too high, and any of the different thermostats on the market will serve for this purpose. That's why we can have a genuine law with physical antecedent and consequent even though the intermediate process is variably realized. Designers want the antecedent to produce the consequent and there are different ways of achieving this.

I have illustrated the point with an example of human design, but the point generalizes. There are other selection processes in nature apart from conscious design by intelligent agents, such as the intergenerational selection of genes, or the selection of cognitive and behavioural elements in the course of individual learning. These selection processes can also give rise to variably realized laws.

Take the paradigm of a putatively variably realized special scientific category—*pain.* It is widely supposed that pain is variably physically realized across different life forms, yet nevertheless enters into laws mediating between physical causes and effects, such as the law that bodily damage gives rise to pain and the law that pain in turn leads to avoidance of the source of the damage. Here, too, there would be an obvious answer if someone asked why all the disparate physical processes caused by bodily damage have the same effect. Natural selection favours organisms that have *some* mechanism that mediates between bodily damage and the avoidance thereof. It doesn't care too much about how this is done. Or, to speak less metaphorically, natural selection will foster any mechanism that plays the pain role within a given species. This is why pain mechanisms can be different across different species, yet all underpin the same damage-avoidance law.

Here is another example. Animals who maintain individual territories will respond to the presence of conspecifics with some territorial display that makes the invaders retreat. Here there is a regular antecedent–consequent pattern—invasion followed by retreat—but the displays that play the intermediary role on this pattern will vary widely from species to species. But once more the explanation is clear enough—natural selection will encourage any display that plays this role, even if it is different from species to species.

We can expect something similar at the level of individual psychology. Grown-up human beings in the West respond to untied shoelaces by tying them. Yet they have different ways of doing this, whose only common feature is that they get the shoelaces tied. How do all these different responses to untied shoelaces produce the same effect? Again the answer is obvious enough. Humans learn in large part by trial and error. If by chance they light on some behaviour that produces a successful result, then they will persist in this behaviour. That's why different humans end up with different ways of tying shoelaces. Learning ensures that they will find some way of doing the job, but doesn't mind exactly how they do it.

Many other examples offer themselves. Most mature humans will have some way of recognizing and thinking about common objects (cats, dogs, telephones, bicycles) but there is no reason to suppose that they use the same brain states to achieve this. Most mature humans will have some technique for solving common intellectual problems (numerical addition, planning tomorrow's activities, balancing their budgets) but these will vary across individuals. Most mature humans will have some way of putting others at ease, but they won't all do this in the same way. And, in general, people with shared ends will generally work out some way of achieving their common aim, but will light on different means of doing this (cf Millikan 1999).

In all these cases, the variability of the means that lead to some given result can be explained by selection processes operating during individual development. Humans and other complex animals are learning machines. They embody a hierarchy of processes that operate at many different levels to preserve items that produce such-and-such effects. These items may well be physically different in different individuals, but this won't matter to the selection mechanisms, provided they produce the reinforcing effects. So the means by which the effects are produced will be variably realized at the physical level across different individuals.

SPECIAL SCIENCES

Do these kinds of selection-based patterns vindicate the possibility of 'special sciences' in the sense of sciences whose categories are variably realized at the physical level?

One possible worry is that the kind of selection-based patterns described in the last section are not precise enough to count as laws. After all, pains don't always lead to avoidance of the source of damage, territorial displays don't always succeed in repelling invaders, and untied shoelaces don't always get tied. These regularities look more like rules of thumb than anything worth dignifying with the name of 'law'.

I don't think this is a decisive consideration against the possibility of 'special sciences'. It may be some reason for withholding the terminology of 'laws', but there are surely plenty of sciences in good standing whose laws need to be understood probabilistically or as *ceteris paribus* claims. This was why the problem I originally posed for Fodor's picture was not how there can be strict exceptionless special laws, but rather how there can be so much as projectible correlations involving variably realized kinds. And the selection-based patterns from the last section certainly amount to projectible correlations. They carry information about as-yet unobserved cases, and they support counterfactuals. (Any damaged animal will respond by avoiding the source of the damage; if some animal were damaged, it would avoid the source of the damage . . .) These projectible patterns may be a lot less precise than the fundamental laws of physics, but they still display the characteristic properties that distinguish genuinely projectible patterns from merely accidental regularities.

However, selection-based patterns arguably fall short of the requirements for a genuine 'science' in a different respect. Paradigm examples of natural kinds enter into *lots* of laws, not just single ones. For paradigm natural kinds, we can project a wide range of properties. Thus, chemists can study many properties of gold: its density, colour, melting point, electrical conductivity, and so on. And this hinges on the fact that all samples of gold have a uniform physical realization. It is precisely because all gold has the same atomic structure that there are many different further features that all samples of gold have in common.

The point is not restricted to basic chemical kinds, but applies to any kind with a uniform physical realization. For example, there are many general truths about chickenpox: its gestation period, characteristic symptoms, ease of transmission, susceptibility to various drug treatments, and so on. Again, it is because of a common structure at the physical level that we are able to assume that all these different features will hold good across different instances of chickenpox.

This kind of multiple projectibility will not apply to the variably realized kinds that enter into selection-based patterns. Take pain, considered as a category that is variably realized in different species. This enters into the law that pain leads to damage-avoidance, as this is part of the role for which pain mechanisms are selected. But there is no reason to expect that the category of pain will enter into any further laws. Thus there won't be any cross-species laws about the sensitivity of pain mechanisms to stimuli, their susceptibility to analgesics, or the time it takes pains to abate. Precisely because the physical basis is different, such things will vary across different species.

The same point applies to other variably realized categories. There is no cross-species science of territorial behaviour, nor any cross-person science of shoelace-tying or bicycle-recognition. And this is precisely because these categories are variably realized. We can say that, in general, territorial behaviour will tend to repel invaders, but the fact that different species repel invaders in different ways blocks any other generalizations about territorial behaviour as such. The same goes for shoelace-tying and bicycle-recognition. We know that all normal people can do these things, but there are no further general facts about the means they adopt, precisely because the means vary across individuals.

We can emphasize the point by comparing variably realized categories with some of their more specific instantiations. Take human pain, as opposed to cross-species pain. Given that it seems highly likely that this has a uniform realization across humans,[1] it makes perfect sense to investigate the many properties of human pain as such (sensitivity to stimuli, effective analgesics, and so on). Again, there would seem to be no barrier to a complex of laws about the territorial displays of some particular bird species—goldfinches, say—covering triggers to aggressive behaviour, song patterns, seasonal variation, and many other things. Here, too, there are many laws because the physiological basis of the behaviour is presumably constant across robins. There could even be a range of general truths about a particular individual's shoelace-tyings or bicycle-recognitions, given that there is likely to be a uniform physical basis for these abilities within any given individual.

Biologists distinguish between *analogous* and *homologous* traits. Analogues are independently derived products of convergent evolution that serve a common purpose, like the wings of insects and birds. Homologues are traits that share a common descent, even if they now serve divergent functions, like the flippers of seals and the hands of humans. The last few paragraphs explain why homologous categorizations are standardly taken more seriously by biologists than analogous ones (cf. Brigandt and Griffiths 2007). Analogues do enter into common patterns, but they are once-off selection-based patterns. Both insect and bird wings lead to flight, but beyond that there is not much they have in common, because they have no common underlying physical basis. Homologues, by contrast, will be physically similar, even if they serve divergent functions, and because of that they will share a wide range of further developmental, structural and other similarities.

HUMAN SCIENCES

Where does this leave human sciences like psychology, economics, and political science? Does the fact that variably realized categories fail to underpin multiple laws undermine these disciplines' claims to science?

A first point to make here is that we should not take it for granted that the human sciences are *special* sciences, if this is understood as meaning that their kinds are variably realized at the physical level. For it seems highly likely that many of the categories that matter to these sciences are uniformly realized at

[1] Remember that this doesn't mean that there are *no* physical differences between individuals' pain mechanisms—just that there is enough physical commonality to yield uniform physical explanations of patterns involving pain.

the physical level *within* humans, even if they are variably realized across other species.

I have already made the point in connection with human pain. There is every reason to suppose that the pain mechanism is uniformly realized across humans, and that as a result there will be a rich nexus of laws about human pain. The same applies to many other cognitive abilities. Sensory mechanisms in general are uniformly realized across humans, which is why there is a substantial set of laws about human perception. The basic mechanisms that underpin human learning are physically similar across humans, which is why we have a wide range of generalizations about human learning as such. Again, it seems plausible that the basic mechanisms of reasoning—the processes that govern interactions between learned and other cognitive states—will be uniformly realized in all humans, and that here again we can expect a serious collection of generalizations about human reasoning.

It should not be supposed that the only attributes that are uniformly realized in humans are those that are genetically determined. Many of the physically uniform processes that occur in human ontogeny will hinge on interaction with environments as well as on common genetic endowment. (This may well include interaction with other humans as well as with the physical environment.) The question at issue is whether the overall developmental process produces a uniform physical structure, not whether this structure is determined by the human genome on its own.[2]

To the extent that human categories are uniformly physically realized, then, they will function as scientific kinds in the fullest sense. There will be a wide range of projectible general truths about various facets of human pain, human vision, human learning, and human reasoning, etc. (Indeed, to the extent that the physical basis for these mechanisms is shared with other mammals, as with many sensory abilities and some basic forms of learning, much of this range of projectible generalizations will carry over to these cases too.)

Still, many human sciences go beyond matters that are uniformly realized within humans. Maybe certain branches of psychology restrict themselves to processes underpinned by physically uniform mechanisms. But many other human sciences aim beyond this. Such subjects as economics, sociology and even social psychology do not just study sensory and other basic cognitive mechanisms. They also aim to generalize about the varied products of these mechanisms, including many of the different things that people learn about and subsequently reason over.

[2] Some philosophers explicate 'innate' as 'a product of normal development that is not due to learning' (Samuels 2002). If we assume that the products of learning are generally not uniformly physically realized, for reasons indicated in previous sections, then anything that is physically uniform across humans will need to be 'innate' in the suggested sense, since not due to learning. However, it is highly controversial whether 'due to normal development but not learning' is a legitimate reading of 'innate' (Mameli and Papineau 2006).

For instance, economists will generalize about the way people buy more when the price goes down, sociologists about the way that dispersed empires keep bureaucratic records, social psychologists about the way that people recognize and defer to authority. And here things will work differently. The patterns observed in such cases will not be the manifestation of common physical structures, but of similar selective pressures operating in different contexts. The humans involved will have been shaped to achieve the same results, but they will often have different ways of doing so. There are different ways of buying more of a product, of keeping bureaucratic records, of identifying people who wield authority, and so on. And this will limit the range of general truths we can expect to find in such cases. We might be confident that certain categories of people will all have some way of achieving some end, but characteristically there will be little to say about the many idiosyncratic ways in which they achieve this.

Does all this mean that the human sciences are not really *sciences* in the full sense? I don't think that this is a particularly fruitful question to press. As we have seen, the subject matter of the human sciences contains both physically uniform cognitive mechanisms and variably realized selectional categories. Correspondingly, some human kinds will enter into a thick nexus of projectible laws and others into a few thin selection-based laws. Once we are aware of this, there seems little point in continuing to ask whether economics as a whole, say, is a 'science'. The answer is that it resembles a paradigm science like chemistry in some respects, but not others.

The more interesting issue is to discover how much of the human sciences is grounded in uniform physical mechanisms and how much depends on common selectional pressures. I have been writing as if the dividing line is reasonably clear-cut, but on reflection it is by no means obvious where it lies. This is because the subject matter of the human sciences is largely constituted by human cognition, and the role of learning and other selective processes in the ontogeny of human cognition is a highly disputed matter. I would say that this should be a central issue for those thinking about the methodology of the human sciences. If we want to know about the kind of general truths we can hope to find in the human sciences, it is crucial that we work out which might rest on uniform physical mechanisms and which are the products of selection.

MORE AND LESS PRECISE CAUSES

I turn now to the question of whether special properties—that is, properties that are not identical to physical properties—can be causally efficacious. This issue has received a lot more attention in the literature than the possibility of non-reduced special laws. The discussion of laws so far in this chapter will cast some new light on the issue.

I shall assume that causes are in some sense property-involving. This will be true if causes are facts, or 'Kim-events', or even if they are Davidsonian-events, that enter into causal relations in virtue of some of their properties (see Papineau 2007: section 1.4.) The differences between these views will not matter for the arguments that follow. I shall talk henceforth as if causes are facts.

The problem facing special causes is that their physical realizations threaten to pre-empt them as causes. According to the causal completeness of the physical realm, every physical effect has a full physical cause (insofar as it has a cause at all). But if special properties are not type-identical to physical properties, then it is difficult to see how facts involving them can be identified with those physical causes,[3] and this argues that they are not themselves causes of those physical effects. And if this is so, then they will also be disqualified as causes of any facts that so much as supervene on the physical facts, for a cause of any such supervenient fact must surely proceed by causing the physical realization which determines that supervenient fact.

Sometimes this worry is raised unnecessarily. For example, it is sometimes suggested that the temperature of 100°C cannot be the cause of the boiling, because it is out-competed as such by the specific molecular movements. (The property of being at 100°C cannot be type-identified with the molecular movements, since other volumes of water will share the temperature property but have different molecular movements.) However, the natural answer here is to insist that the 100°C temperature is a perfectly good cause in its own right, given that it is a uniform physical kind that enters into the paradigm physical law that water at 100°C (at standard pressure) commences to boil.

We should not assume that, whenever some category is variably realized at some more precise physical level, as here with temperature and molecular movements, that the more precise physical facts will always outcompete the general property as the cause of any physical effects. There seems to be no good basis for this assumption. The metaphysical constitution of the causal relation is not well understood, but there is good reason to suppose that it is constituted at the level of thermodynamic phenomena, rather than at the level of the basic dynamics of fundamental particles. After all, causation has a preferred direction in time, which is true of thermodynamic processes, but not of basic dynamical ones. If this is right, then there will be a level of physical precision—the level of temporally symmetric basic dynamic processes—where causation disappears, so to speak. Clearly, precise physical facts at this basic level will not eclipse more general supervenient physical facts as causes. So we cannot, in general, assume that the more precise physical facts will always causally outcompete more general ones.

We might wonder, given the point just made, whether the vindication of a more general cause as the cause of some physical effect—such as the 100°C and

[3] But see Macdonald and Macdonald 1986, 1995.

the boiling—will always eclipse the more precise fact—the specific movements of the water molecules—as a cause of that effect. I have no view on this matter. Maybe there is a good argument that will establish this point. In the example at hand, it is certainly not out of the question to hold that the particular molecular movements do not cause the boiling—after all, the water would have boiled just as well even if the molecular movements hadn't been the same, provided the temperature was still 100°C. But in what follows I shall not assume that there must only be one cause in such cases. I shall leave it open that both the temperature and the molecular movements can happily qualify as causing the boiling.

VARIABLY REALIZED CAUSES

I say that the temperature counts as a cause because there is a physically uniform law connecting it to the boiling. In this case, then, the causal efficacy of a variably realized category derives from the presence of a uniform physical law connecting it to the effect. But what of those special categories that are not uniform physical kinds, such as cross-species pain, or deference to authority, or increases in supply? Can they be causally efficacious, even though they do not enter into uniform physical laws?

Much of the recent literature has been distracted from this issue by worries about overdetermination. Kim has insisted that it is unacceptable to have a physical effect caused by both a variably realized kind and its realizer (see Kim 1993). Orthodox non-reductive physicalists have retorted that this kind of 'overdetermination' is perfectly benign, due to the intimate connection between the kind and the realizer, and not to be conflated with real overdetermination by two genuinely distinct causes, as when someone is simultaneously shot and struck by lightning (Bennet 2003).

But this orthodox answer does not yet address the prior question of what qualifies the special fact as a cause of the physical effect in the first place. Let us allow that there would be nothing wrong with the 'benign overdetermination' of physical effects by both special causes and the physical realizations. Still, why count the special fact as a cause at all? In the case of the variably realized 100°C, we had a uniform physical law connecting the temperature with the boiling. But with variably realized special facts, there will be no such uniform physical law, precisely because they are variably realized at the physical level.

If pressed on this question, most non-reductive physicalists would probably respond that the special fact qualifies as a cause in virtue of relevant counterfactual truths—if I hadn't felt a pain, I wouldn't have pulled my hand out of the fire. But, notwithstanding all the recent enthusiasm for counterfactual theories of causation, it is by no means clear that the mere truth of such a counterfactual is sufficient to vindicate a special fact as a cause.

Consider this case. Let us define 'ricketiness', in a car, as present if some of the parts that are supposed to be joined together become disconnected. Now suppose that my car is rickety because the wire that joins the ignition to the starter motor is broken. The general property of ricketiness, that is present in any car with any disconnected parts, is here realized by the broken ignition wire. As a result my car does not start. Now it is true that if my car were not rickety, it would start. If it were not rickety, the ignition would still be connected to the starter motor. But does the ricketiness per se cause the non-starting?

It is easy to misread this question. On one understanding of 'ricketiness', it refers to the specific realizer property that is present in this case, of having a broken ignition wire. And it is certainly true that this realizer property causes my car not to start. But that is not the issue. The question is whether the variably realized role property caused the non-starting—that is, the property that is shared, not just by cars with broken ignition wires, but also those with loose door handles, faulty boot locks, and so on. And, once we have this question clearly in focus, a positive answer seems implausible. Surely it's not the ricketiness per se that stopped my car from starting. Plenty of cars are rickety, yet start perfectly well. What stops my car from starting is not that it has *some* part disconnected, but the more specific fact that the ignition wire is disconnected.

What seems to be needed, then, is some kind of general connection between the special fact and the relevant physical effect, over and above the special fact's supervening on a physical cause of the effect. If the (cross-species) pain stands to my arm movement merely as my car's (role) ricketiness stands to the non-starting, then there does not seem to be a good case for counting it as a cause of my arm movement.

THE CAUSAL IRRELEVANCE OF SPECIAL LAWS

I suspect that many philosophers are persuaded implicitly to think of the (cross-species) pain as a cause of the arm movement because they know that there is a law connecting pains with arm movements. It's not just that my pain is realized by a physiological process that causes my arm movement. It is also a general truth, holding across species, that pain leads to removal of the relevant body part from the source of the damage. This marks a contrast with the ricketiness example. It is not generally true, across rickety cars, that ricketiness leads to non-starting. So this makes it plausible to think that the pain is more seriously connected with the arm movement than the ricketiness is with the non-starting.

However, I don't think that this line of thought will serve to vindicate the pain as a *cause* of the arm movement. True, there is a serious empirical law connecting the pain with the arm movement. But the trouble is that it is a selection-based law. And on reflection it seems clear that this kind of variably realized selection-based law is the wrong kind of connection to ground a causal relation between the pain and the movement. Think about the aetiology of the law. Biological natural selection favoured different pain mechanisms in different life forms *because* these different mechanisms all had the right causal profile—they were activated by damage and gave rise to avoidance. The selection-based law was thus an upshot of the causal powers of the different pain mechanisms. Given this, it would seem odd to regard it as grounding some further causal powers. It's not as if the cross-species category of pain is constituted as a cause of avoidance movements in virtue of its role in the selection-based law. Rather, the law was cobbled together by natural selection, so to speak, because all the different realizations of pain already had just the right causal qualifications.

The point generalizes to the many variably realized special categories that enter into selection-based laws. These laws will mean that they are generally followed by specific effects, and to this extent they will be distinguished from categories like ricketiness, which isn't per se generally followed by non-starting. But this by itself doesn't seem to warrant counting these special categories as *causes* of the relevant effects, any more than we should count the ricketiness as the cause of the non-starting. The selection-based laws are based on pre-existing causal powers, and don't add to them.

All in all, I'm inclined to conclude that non-reduced special kinds are not causes. Even if they are connected to their putative effects by selection-based laws, they are not really different from ricketiness. They range over cases with quite different causal structures. The selection-based laws are a red herring. They are not the kind of laws that can constitute anything as causally efficacious.

Of course, this point can be obscured, as with ricketiness, if we read terms like 'pain' as used in a specific context as referring to the physical property that realizes the pain role in that context—that is, as referring to the physical property that uncontentiously causes the arm movement. In fact, I am quite open to the thesis that this is the most natural way to understand 'pain' talk. (After all, given the argument of this section, this is the only way to have 'pains' causing behaviour, so to speak.) But the point remains that the role property per se does not cause the behaviour.

A related point is that a special property can well be causally explanatory even if it is not causally efficacious (see Jackson and Pettit 1990). I would say that all explanations of particular facts need to mention the cause of those facts. But you can mention a cause without explicitly citing the property that makes it causally efficacious. Now, not all such indirect mentions of causes will be explanatory. It is not explanatory to say that X was due to the cause of X. But some indirect mentions of causes certainly are explanatory. Thus I might explain the high temperature of the room by reference to the setting on the thermostat, the improved performance of my car by its new carburettor.

It is plausible that explanation is related to prediction and causal control, and that therefore explanations need to cite properties that fit into laws—even if those cited properties themselves are not causally efficacious. Variably realized properties that enter into selection-based laws would seem to fit this bill perfectly well. These laws may not display the substance of the relevant causes, but they serve well enough to indicate what consequences to expect and how such things might be brought about. This is why, in addition to explanations citing artefacts, we find explanations citing all the other kinds of variably realized categories that enter into selection-based laws. Her shoes aren't loose because she has learned how to tie her laces. He pulled his arm away because he felt a pain. Falcons detect prey with their excellent eyes. And so on.

CONCLUSION

Let me sum up briefly. Non-reduced special kinds cannot play a role in full-fledged sciences involving a rich network of laws. Still, selective processes mean that they can enter into once-off laws. However, this is not enough to constitute them as causally efficacious as opposed to explanatory properties.

REFERENCES

Bennet, K. 2003. 'Why the Exclusion Problem Seems Intractable, and How, Just Maybe, to Tract It'. *Noûs* 37: 471–97.

Block, N. 1997 'Anti-Reductionism Strikes Back'. *Philosophical Perspectives* 11: 107–32.

Brigandt, I. and Griffiths, P. 2007. 'The Importance of Homology for Biology and Philosophy'. *Biology and Philosophy* 22: 633–41.

Charles, D. and Lennon, K. 1992. *Reduction, Explanation and Realism.* Oxford: Oxford University Press.

Davidson, D. 1970. 'Mental Events'. In L. Foster and J. Swanson (eds), *Experience and Theory.* London: Duckworth.

—— 'Hempel on Explaining Action'. *Erkenntnis* 10 (1976): 239–53.

Fodor, J. 1974. 'Special Sciences: Or the Disunity of Science as a Working Hypothesis'. *Synthese* 28: 77–115.

—— 1997. 'Special Sciences: Still Autonomous After All These Years'. *Philosophical Perspectives* 11: 149–64.

Jackson, F. and Pettit, P. 1990. 'Program Explanation: A General Perspective'. *Analysis,* 50: 107–17.

Kim, J. 1992. 'Multiple Realizability and the Metaphysics of Reduction'. *Philosophy and Phenomenological Research* 52: 1–26.

—— 1993. *Supervenience and Mind: Selected Philosophical Essays.* Cambridge: Cambridge University Press.

Macdonald, C. and Macdonald, G. 1986. 'Mental Causes and Explanation of Action'. In L. Stevenson, R. Squires, and J. Haldane (eds), *Mind, Causation, and Action.* Oxford: Blackwell.

—— 1995. 'How to be Psychologically Relevant'. In C. Macdonald and G. Macdonald (eds), *Philosophy of Psychology: Debates on Psychological Explanation*, vol. 1. Oxford: Blackwell.

Macdonald, G. 1992. 'Reduction and Evolutionary Biology'. In D. Charles and K. Lennon, *Reduction, Explanation and Realism*. Oxford: Oxford University Press.

Mameli, M. and Papineau, D. 2006. 'The New Nativism: A Commentary on Gary Marcus's *The Birth of the Mind*'. *Biology and Philosophy* 21: 559–73.

Millikan, R. 1999. 'Historical Kinds and the "Special Sciences" ' *Philosophical Studies* 95: 45–65.

Papineau, D. 1985. 'Social Facts and Psychological Facts'. In G. Currie and A. Musgrave (eds), *Popper and the Human Sciences*. Dordrecht: Nijhoff.

—— 1992. 'Irreducibility and Teleology'. In D. Charles and K. Lennon, *Reduction, Explanation and Realism*. Oxford: Oxford University Press.

—— 1993. *Philosophical Naturalism*. Oxford: Blackwell.

—— 2002. *Thinking about Consciousness*. Oxford: Oxford University Press.

—— 2007. 'Naturalism'. In E. Zalta, ed., *Stanford Encyclopedia of Philosophy*. Available at: <http://plato.stanford.edu/entries/naturalism/>.

Putnam, H. 1960. 'Minds and Machines'. In S. Hook (ed.), *Dimensions of Mind*. New York: New York University Press.

—— 1967. 'The Nature of Mental States'. In W. Capitan and D. Merrill (eds), *Art, Mind, and Religion*. Pittsburgh: Pittsburgh University Press.

Robb, D. and Heil, J. 2003. 'Mental Causation', in *Stanford Encyclopedia of Philosophy*. Available at <http://plato.stanford.edu/entries/mental-causation/>.

Samuels, R. 2002. 'Nativism in Cognitive Science'. *Mind and Language* 17: 233–65.

13

Can Any Sciences be Special? Comments on Papineau

*Michael Esfeld**

David Papineau, Jerry Fodor, and many others wonder how the conjunction of the following three positions can be true:

(1) *Special science laws*: There are lawlike generalizations in the special sciences. These sciences trade in kinds that are such that statements about salient, reliable correlations that are projectible and that support counterfactuals apply to the tokens coming under these kinds.
(2) *Non-reductionism*: The laws of some of the special sciences cannot be reduced to physical laws.
(3) *Physicalism*: Everything there is in the world supervenes on the physical, that is, is fixed by the distribution of the physical properties in the world.

The obvious problem is that (3) implies that the similarities among tokens in the world, accounting for the kinds in which the special sciences trade, and the correlations among such tokens, accounting for the laws of the special sciences, are fixed by the distribution of the physical properties. By contrast, (2) implies that some of the laws seizing such correlations are not reducible to physical laws. By using the term 'token', I mean a particular instantiating a property.

Papineau's proposal to reconcile these three positions is to account for (2) in terms of selection (this volume: 185–7): there can be laws in the special sciences that are not reducible to physical laws if and only if these laws focus on effects that are selected for in a given context independently of the mechanisms by which they are brought about. Thus, the fact of there being such laws and their non-reducibility to physics do not contradict physicalism (3). The drawback is that the kinds that figure in such laws cannot enter into a rich network of laws

* I would like to thank Christian Sachse for discussions about the content of this comment and for allowing me to integrate material resulting from common work; the example from genetics is his.

and that nothing can be causally efficacious insofar as it is a member of such a kind.

In these comments, I shall try to push Papineau further in the direction of a reductive physicalism, thus solving the problem by simply abandoning (2). The obvious gain of such a move is that the scientific quality of the special sciences is vindicated by being systematically linked to physics and the spectre of epiphenomenalism thus banned. The prima facie argument for reductionism goes like this: for every single case of a correlation among tokens that comes under a law of a special science, there is a physical explanation available why that correlation obtains. If tokens coming under different physical kinds all give rise to correlations covered by the same law of a special science, then there is in principle a physical explanation available why all these tokens are similar in that respect, on pain of violating global supervenience. So when it comes to laws of the special sciences that can be accounted for in terms of selection for specific effects, why should such laws in principle be irreducible to physics?

Biology is a paradigmatic case of a special science focusing on selection. Take classical genetics as a theory describing lawlike correlations between genes and phenotypes, whereby these correlations are brought about in different physical manners. For instance, there are certain kinds of genes of *Escherichia coli*, a well-known bacterium in genetic research, coding for the production of proteins that play a predominant role in the flagella, the capsule, or the cell wall. Certain sequences of DNA bring about the proteins in question provided that there are enough resources for the protein synthesis. These sequences can be of different molecular kinds. Consequently, there are different physical mechanisms to produce the proteins in question. However, these different ways to produce the proteins for which the genes of *E. coli* code are systematically linked with side effects such as the speed or the accuracy of the protein production (see Bulmer 1991).

For any such difference in side effects, there is a physical environment possible in which that difference is pertinent to selection. Oversimplifying, under certain conditions, an accurate and fast production of cell-protecting proteins may be important for the survival of the bacterium in question. Classical genetics has the means to consider these fitness differences. For any kind S of a gene of *E. coli*, it is possible to conceive within classical genetics sub-kinds S_1, S_2, S_3, etc., taking into account these side effects of the speed and accuracy of the protein production by means of considering the resulting measurable fitness differences. Thus, S_1 may be about the gene tokens defined by the effect to produce the protein in question (like all gene tokens coming under S) *and* the consideration of a certain time index of the protein production or the corresponding fitness value (in order to distinguish the gene tokens in question from the gene tokens coming under the other sub-kinds S_2, S_3, etc.). This more precise sub-kind S_1 may be something like 'S and production of the protein PR in t_1' or 'S and fitness contribution c_1', while the sub-kind S_2 may be something like 'S and production of the protein PR in t_2' or 'S and fitness contribution c_2'.

The point just made generalizes. Let S stand for any kind of the special sciences that can be multiply realized by tokens coming under different physical kinds P_1, P_2, P_3, etc. Since the issue is about selection, we can assume that S is defined in terms of certain salient effects that can be brought about in physically different manners, these manners being expressed in terms of P_1, P_2, P_3, etc. However, if the relevant difference between tokens coming under P_1 and tokens coming under P_2 consists in different manners to bring about the effects that define S, then although all these manners coincide in producing a salient effect of the same kind, they are distinguished by certain side effects linked with the production of the main effect. These different manners can be conceptualized in terms of the special science in question, thus leading to the conception of sub-kinds S_1, S_2, S_3, etc. of S. These sub-kinds are no longer multiply realizable. They are nomologically coextensive with the physical kinds P_1, P_2, P_3, etc., although the meaning (primary intension) of the concepts 'S_1', 'S_2', 'S_3', etc. is different from the meaning of the concepts 'P_1', 'P_2', 'P_3', etc. (as the meaning of 'water' is different from the meaning of 'H_2O').

Given this nomological coextension, it is possible to reduce the special science theory trading in S to physics: (1) Within an encompassing physical theory, one conceives the kinds P_1, P_2, P_3, etc. for the different configurations of physical tokens that all come under the special science kind S. (2) One makes the conceptualization of S more precise by constructing sub-kinds S_1, S_2, S_3 of S, seizing the systematic side effects linked to the different manners of producing the effects that define S. These sub-kinds are nomologically coextensive with the physical kinds P_1, P_2, P_3, etc. (3) One can reduce the theory of S to physics via S_1, S_2, S_3 and P_1, P_2, P_3. Starting from the encompassing physical theory, one constructs the concepts 'P_1', 'P_2', 'P_3' and then deduces the concepts 'S_1', 'S_2', 'S_3'from 'P_1', 'P_2', 'P_3' given the nomological coextension. One gains 'S' by abstracting from the conceptualization of the side effects contained in 'S_1', 'S_2', 'S_3'(see Esfeld and Sachse 2007).

Any law couched in terms of S can be made more precise by being conceived as a law about S_1 without thereby losing its lawlike character. If there are laws about S, these are also laws about S and fitness contribution c_1, etc. Indeed, as Papineau stresses (this volume: 189), it is likely that there are many more laws about S_1 than there are about S as such, since S_1 is coextensive with the physical kind P_1. From the laws about S_1, etc. one can gain the laws about S by abstracting from the conceptualization of the side effects, that is, the particular manner in which the effects characterizing S are produced. In other words, what remains when one abstracts from the conceptualization of the production mechanism contained in the laws about S_1 is projectible to all tokens coming under S, thus yielding the laws about S as such. Hence, the fact that some theories of the special sciences focus on effects that are selected for in a given context without

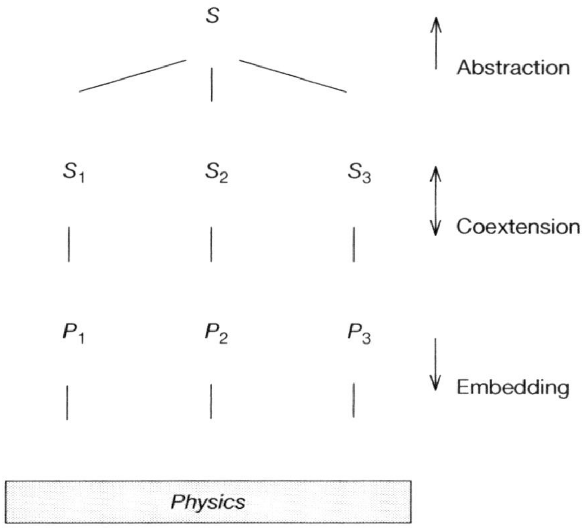

paying attention to the manner in which these effects are brought about does not prevent these theories from being reducible to physics, because the different manners to produce these effects can be relevant to selection and consequently be taken into account in terms of the special sciences.

Note that the sub-kinds S_1, S_2, S_3, etc. are not restricted to particular species. There is no question here of focusing on species-specific realizers and thereby carrying out what is known as a local or species-specific reduction (for instance, the concept of pain reduces in one species, say humans, to one physical concept—e.g. 'firing of C-fibres'—it reduces in another species, say octopuses, to another physical concept, etc.). The sub-kinds S_1, S_2, S_3, etc. are not construed in a species-specific manner, but in terms of purely functional differences only. They are distinct only by conceptualizing the different manners in which the effects are produced that define the kind S: they are heterogeneous as regards these manners of production, but homogeneous insofar as the effects are concerned that define S. The concept 'S' always has the same substantial meaning in 'S_1', 'S_2', 'S_3', etc.—a meaning that is only further specified by paying heed to the manner in which the effects in question are produced. Therefore, 'S_1', 'S_2', 'S_3', etc. clearly express for the scientist of the special science in question what the referents of these concepts *functionally* have in common and what their *functional* differences are, which may in certain circumstances be of interest for the special science in question—instead of splitting up the homogeneous kind S into species-relative kinds, thereby losing the homogeneity of S and thus paving the way for its elimination (cf. the so-called new wave reductionism advocated by Bickle 1998).

Nonetheless, Papineau is of course right in pointing out that there are laws of the special sciences that are non-physical in the sense that there is no single physical law having the same extension as any of the laws about *S*. But this is not a deep metaphysical or epistemological fact preventing reduction. It is simply a matter of the division of scientific labour. When talking about complex objects such as, for example, genes, or whole organisms, the physical concepts focus on the composition of these objects. Due to selection there are salient causal similarities among effects that such complex objects produce as a whole although they differ in composition. The concepts seizing these similarities are therefore not considered to be physical concepts, but classified as concepts of the special sciences. However, these concepts can be gained from physical concepts in the way sketched out above by the deduction of concepts for sub-kinds and reaching the kinds in which the special sciences trade by abstracting from the conceptualization of the side effects that distinguish these sub-kinds. Consequently, there is a systematic, deductive way of how to get from the physical laws to the laws of the special sciences.

It is obvious why there has to be such a way: the physical laws are about causal correlations among physical tokens, insofar as they explain why there are certain tokens in the world. They cover all the causal correlations among tokens in the world. If the laws of the special sciences offer explanations of why there are certain complex objects having certain specific effects, there has to be a systematic way to get from the physical laws to these laws. It is only that we need specific concepts of the special sciences when it comes to the effects that certain complex objects produce as a whole, since in that case the physical concepts focus on the composition of these objects rather than on the effects they have as a whole.

In sum, the scientific quality of the special sciences consists in offering lawlike generalizations concerning correlations among tokens that are projectible and that support counterfactuals. Although these laws are non-physical in that there are no physical laws that are coextensive with them, they do not come into conflict with the completeness of physics and the supervenience of everything on the physical if and only if there is a systematic, reductive way of how to reach them on the basis of the physical laws. Providing for such a way thus vindicates their scientific quality and shows that an eliminativist attitude towards them on the basis of the completeness of physics and global supervenience is unjustified.

The mentioned scheme applies all the way down to fundamental physics. It therefore also paves the way for accounting for the causal efficacy of objects insofar as they come under kinds of the special sciences. The debate about physics being causally complete and, consequently, the special sciences, insofar as they are not reducible to physics, being haunted by the spectre of epiphenomenalism, tacitly assumes that causation is a fundamental physical feature. Papineau, however, voices reservations about that assumption (this volume: 192). He locates causation at the level of thermodynamical phenomena because these phenomena have a preferred direction of time, which is not found among basic dynamical phenomena. This view is objectionable: consider quantum mechanics. If one accepts a version of quantum mechanics that includes the reduction of quantum superpositions to classical physical states, then the best concrete physical proposal for such a version is the one of Ghirardi, Rimini and Weber (GRW). The GRW-equation is a candidate for a fundamental physical law and it includes a preferred direction in time, since state reduction is irreversible (see Albert 2000: chapter 7). If one does not endorse a version of quantum mechanics that includes state reductions, then nevertheless there are processes of decoherence that lead to the appearance of a classical world to local observers. Decoherence is for all practical purposes an irreversible process in the same way as are thermodynamical processes; the latter ones are not strictly irreversible either if one takes thermodynamics to be reducible to statistical mechanics. Thus, if there is causation on the level of thermodynamical phenomena, then it follows there is causation on the level of basic dynamical phenomena, and nothing therefore speaks against taking causation to be a basic feature of the world.

Since Papineau does not regard causation as a fundamental physical feature, he does not consider it necessary to take a stance on the issue of whether or not there are determinable as well as determinate causes—such as the object's being *S* is a determinable and the object's being *P* a determinate cause of a given effect, without these two causes being identical (this volume: 193). However, as Gillet and Rives (2005: section 3) point out, the determinate properties include by definition all the causal powers of the respective determinables. Consequently, the determinate properties are sufficient to bring about all the effects that the determinables could cause. Unless one acknowledges token identity between the determinable and the determinate properties of objects, one thus faces again the epiphenomenalism objection, and it is doubtful to say the least whether admitting some sort of systematic overdetermination is able to counter that objection (not to speak of endorsing an interactionism that contradicts the completeness of physics).

Any token coming under a kind in which a special science trades can cause the effects that characterize the special science kind in question only by bringing about the effects that a certain configuration of physical tokens produces qua configuration. For instance, any gene token can produce certain proteins only by having all the molecular effects that a certain configuration of nucleic acids has qua configuration, for it is through those effects that the protein comes into being. To take another example, any pain token can cause pain behaviour only by producing the neuronal effects that a certain configuration of neurons has qua configuration because it is through those effects that the pain behaviour comes about. Thus there is a good argument for taking the way (mode) insofar an object comes under a certain special science kind *S* to be identical with the way (mode) insofar that object comes under a certain physical kind *P*, although these kinds are not coextensive; but that sort of multiple realization is no obstacle to reduction as sketched out above. If there is token identity, it does not make sense to ask whether an object brings about a certain effect in virtue of coming under *P* or in virtue of coming under *S*, since both are one and the same way (mode) an object is. Consequently, an object is not epiphenomenal insofar as it comes under *S* in the same way as it is not epiphenomenal insofar as it comes under *P*.

To sum up, coming back to the problem sketched out at the beginning of this comment, there are good reasons for abandoning the attitude of a principled non-reductionism with respect even to those special sciences that focus on selection, since there is a way open for a conservative reductionism, vindicating the scientific quality of the special sciences and the causal efficacy of objects insofar as they come under kinds in which the special sciences trade.

REFERENCES

Albert, D. Z. 2000. *Time and Chance*. Cambridge, MA: Harvard University Press.

Bickle, J. 1998. *Psychoneural Reduction: The New Wave*. Cambridge, MA: MIT Press.

Bulmer, M. 1991. 'The Selection-Mutation-Drift Theory of Synonymous Codon Usage'. *Genetics* 129: 897–907.

Esfeld, M. and Sachse, C. 2007. 'Theory Reduction by Means of Functional Sub-types'. *International Studies in the Philosophy of Science* 21: 1–17.

Gillet, C.and Rives, B. 2005. 'The Non-existence of Determinables: Or, a World of Absolute Determinates as Default Hypothesis'. *Noûs* 39: 483–504.

14

Emergence vs. Reduction in Chemistry

Robin Findlay Hendry

1. INTRODUCTION

In contemporary philosophy of mind, leading physicalists have come to agree that: (1) any physicalism worth the name ought to exclude the existence of causation 'downward' from the entities, properties and laws studied by higher sciences to their microphysical constituents; (2) successful quantum-mechanical explanations of chemical bonding render unlikely the existence of downward causation from the chemical to the physical; (3) the very idea of downward causation is murky, with some formulations being trivial and less interesting than emergentists think, while stronger formulations are implausible or even incoherent.

In this chapter I argue against claims (2) and (3). Claim (2) does not withstand investigation of either the early history or the mathematical structure of quantum chemistry. In section 2 I distinguish the issues of intertheoretic reduction and ontological reducibility as they apply to chemistry, and attempt to disentangle the intertwined debates about them. I distinguish two levels of reduction in chemistry: substances to molecules; molecules to quantum-mechanical systems of electrons and nuclei. To answer claim (3), I propose in section 2 a counternomic criterion for downward causation which attempts to capture the emergentist views of C. D. Broad, and in section 3 argue that on this criterion, the evidence supports the ontological emergence of molecular structure at least as well as it supports its reducibility. Some physicalist worries about emergence can be rejected, for they are motivated by the assumption that microphysical causation is closed from above, and so presume the non-existence of downward causation.

2. REDUCTION, EMERGENCE, AND CAUSATION

If chemical entities, properties and laws are real in some ontologically robust sense that precludes their being merely the worldly shadows of chemical concepts and modes of explanation, they are either reducible to more fundamental physical items, or they are emergent from them. Physical entities, properties and laws are more fundamental for two kinds of reason. One is mereological: during the nineteenth century, chemists began to theorize about the most fundamental items of classical chemistry, atoms and molecules. From the beginning of the twentieth century, physicists and chemists investigated the internal structure of atoms, identified their physical parts, and began to craft explanations of chemical structure and bonding in the light of these discoveries. In short, physics studies the parts of the most fundamental entities of chemistry. The second kind of reason concerns generality: physics studies the most general relationships between the electrical and mechanical properties that individuate the fundamental entities of chemistry. Hence physical laws must be fixed and prior constraints on any physical explanation of chemical bonding. However, the reducibility of the chemical does not follow, even if one accepts both the above lines of thought, that physical laws are more fundamental, and also that the intimate interaction between physics and chemistry from the late nineteenth century onwards genuinely deepened chemical explanations. That *y* is ontologically prior to *x* does not entail that *x* is reducible to *y*. To explore that point further requires a discussion of reduction itself.

Before we go on, chemical and physical properties need to be distinguished. In the mind–body problem, the parallel distinction between mental and physical properties is easily made, because they appear to be instantiated by radically different kinds of thing, and at least some mental properties appear to have phenomenal aspects which resist physicalist explanation. The contrast between chemical and physical properties is slightly more difficult to make clear, because in the sense in which philosophers of mind often use the word, chemical properties just *are* physical, although some use the broader term 'physico-chemical' instead. That sense of 'physical' achieves a trivial reduction of the chemical by set-theoretic inclusion, but fortunately it is easy to delineate chemical properties extensionally. From the seventeenth century onwards, the discipline of chemistry formed around the systematic study of its core phenomena, transitions between various chemical substances wrought by fire. To explain these phenomena, chemists identified, during the eighteenth and nineteenth centuries, a class of substances—the elements—which are the components of compound substances, but themselves have no chemical substances as components (Hendry 2010: ch. 2). The chemical behaviour of compound substances was explained at first by their composition from the various elements, and, later on in the nineteenth century, by the attribution of hypothetical structures to their molecules. The attribution of such molecular structures was not yet a subsumption by physics of chemical law, for in the nineteenth century they were proposed on the basis of chemical evidence alone: a successful explanatory engagement of physical theory with molecular structure and bonding began to be achieved only in the twentieth century. With this historical information to hand, we can identify the targets in the reduction of chemistry as chemical substances, and the molecular structures by which, since the nineteenth century, they have been known to be individuated. The putative reduction base in each case is also easily identified. The reduction of chemistry therefore has two stages: first comes the reduction of chemical substances to their molecular structures; then comes the reduction of molecules to systems of electrons and nuclei.

Discussion of the reduction of special sciences has centred traditionally either on the definability of special-science predicates in terms of those of more fundamental sciences, or on the derivability or non-derivability of special-science laws. If one takes the existence or non-existence of these relationships as exhausting the question of reduction, then that issue begins and ends with an investigation of logical or explanatory relationships between extant theories. This seems unsatisfactory: even if reductive explanations do not at present exist, any scientific theory may be overthrown at any time by new evidence or theoretical arguments, and any such change will require a reassessment of the relevant explanatory relationships. Hence reductionists can always see a failure of reduction as grounding either a theoretical argument for the elimination of special-science theory or a duty for physicists to adjust their theories. Non-reductionists will see either of these diagnoses as adopting too easily the hubris of physics: either in the airy dismissal of special-science discoveries, or in the continuing myth that physical theories have some special duty of full coverage, when the historical evidence is that physics often develops in blessed innocence of how its theories square with the results of other sciences. All this leads to an impasse: even if it is agreed that classical intertheoretic reductions are not currently available, temperamental reductionists and non-reductionists will differ in how optimistic they are for their future achievement, and may even be able to justify their optimism or pessimism given the historical track record of interdisciplinary scientific explanation. As long as reduction is seen as a dated intertheoretic achievement, however, the issue is essentially future-directed: both sides must wait and see, even if they would bet different ways. But the impasse is unstable, because there are reasons why the intertheoretic reduction of a special science may fail, which do not satisfy reductionists that the issue of reduction is thereby settled, and ought not to satisfy *non*-reductionists who are scientific realists about the special science in question. The reasons are twofold, and can be illustrated by brief reflections on (1) complexity and the mathematical structure of physical theories and (2) the nature of scientific disciplines.

First consider the kind of physical theory, like quantum mechanics, whose explanatory successes ground reductionist arguments. The basic principles of such theories are expressed mathematically, at very abstract levels, and applying them to particular kinds of system requires a complex process of what Nancy Cartwright has called 'theory entry' (1983: ch. 7). Special-science systems are typically complex, and their mathematical descriptions often give rise to insoluble equations. This is a familiar situation, faced by scientists ever since

Newton grappled with the moon's orbit, and can be finessed by introducing approximate and idealized models. Reductionists see the situation as follows: the exact equations are insoluble, but the approximate or idealized models can serve in their place in explanations of special-science phenomena. In principle there are explanations that have the exact equations as explanantia. Hence the explanatory link between physics and the special science has in some sense been forged: call this the 'proxy defence' (see Hendry 2010: ch. 7, section 2), but it is only the beginning of a defence of reduction. What has to be shown is that explanations using the proxies cite only features of them that are shared by the exact solutions. In any case the phrase 'in principle' needs spelling out, because the intractability of some equations, like those for the three-body problem in classical mechanics, is a matter of mathematical principle, not just of analytical methods or computing power.

Second, consider the nature of disciplines. Given some acquaintance with the history of science, even the most robust scientific realist must accept that scientific theories are human creations, the contingent products of disciplines whose existence and development are themselves contingent. The development of a discipline is the work of a community of scientists who may be relatively isolated, deliberately or accidentally, from the work of neighbouring disciplines. Each discipline may have its *own* theoretical concepts, styles of explanation and judgements of theoretical plausibility, so there can be no guarantee that physics and chemistry will mesh even if, ultimately, their subject matter is the same. Chemical concepts may have no obvious counterpart in physical theory. Even if physical laws undergird chemical explanation to the extent that, for every chemical explanation there is a complex physical reason why it is correct, the physical counterparts to chemical explanations may not seem explanatory to a chemist. Once again, reductionists and non-reductionists will interpret this situation differently. Reductionists will say that, even if a special-science law cannot be eliminated from chemical explanation, this is either unfinished business for physics (see above), or reflects chemistry's human component—its historically contingent concepts or styles of explanation—rather than how chemical reality ultimately is. That last option amounts to a concession, an acceptance of failed reduction with the intimation that it doesn't matter. It invites a non-realist interpretation of the relevant parts of chemistry, because it locates chemistry's independence from physics in the human sphere, but the situation is unstable for both the reductionist and the non-reductionist. On the one hand, the reductionist should be pressed: why doesn't the failure of this reduction matter? Answering that question will require a principled distinction between failed intertheoretic reductions that don't matter because they involve only failures of fit between the explanatory and conceptual schemes associated with different disciplines, and those that do constitute a problem for reductionism, because they arise from a disunity in nature that is deeper than the failure of intertheoretic reduction. On the other hand, the non-reductionist who cleaves to a robust scientific realism must regard the reductionist's concession of the failure of intertheoretic reduction on the grounds of conceptual mismatch as a pyrrhic victory, because it motivates a non-realist interpretation of chemistry. Hence the non-reductionist will feel pressure to develop a more robustly ontological non-reductionism.

If the reduction debate is to develop beyond this impasse, then it must go beyond intertheoretic reduction, addressing instead the ontological relationships between the entities, processes and laws studied by different sciences, which are fallibly and provisionally described by their theories. One obvious requirement on ontological reduction is that it must be a substantive metaphysical issue that transcends the question of what explanatory relationships exist between theories now, or might exist in the future, although intertheoretic relationships are obviously relevant evidence. Ontological reducibility is regarded plausibly as one end of a spectrum of ontological dependence relations that includes weak existential dependence close to the other end. Weak existential dependence I take to be close to one end because a version of physicalism according to which everything (concrete) is weakly existentially dependent on the physical can be accepted even by those who think that the ontology of science is far from unified (see for instance Crane and Mellor 1995: 85; Dupré 1993: 91–2). Reducibility is at the strong end of the spectrum because it is the limiting case in which we deny the distinct existence of what is dependent: the reductionist slogan is that *x* is reducible to *y* just in case *x* is 'nothing but' its reduction base, *y*. One can imagine many ways to cash this slogan out, depending on the aspect under which the reduced is held to be 'nothing but' its reduction base, but a consensus has emerged in recent philosophy of mind that the relevant aspect should be causal.

One reason for this is the familiar dialectical situation of supervenience in Kim's work, which was introduced into the mind–body problem by way of formulating non-reductive physicalism (see Kim 1998: ch. 1). However, as Kim succinctly puts it, 'mind-body supervenience in itself does not give us a theory of the mind-body relation' (1998: 12). Here's why not: supervenience is merely covariance between groups of properties. It may come in varying modal strengths, but even the strongest of these fails to be an ontological dependence relation. If a special-science property or group of properties supervenes on the physical, there is a (possibly messy) correlation between the special-science properties and the physical properties. Supervenience is to ontological dependence as statistical correlation is to causation, and so can be explained in one of two ways: either the special-science properties sometimes push around their physical supervenience bases (in which case there is downward causation), or the physical base properties are immune to such intervention because they are causally closed. This last feature of the reductive physicalist explanation reflects a central principle of reductive physicalism, the causal closure, or completeness of the physical, according to which physical effects are brought about by physical causes via physical laws (see Papineau 2002: 233–4). Hence supervenience is inadequate to articulate a mind–body theory because it is compatible with, and explainable by, either emergentism or reductive physicalism. Emergentism allows that supervenience can be explained either by upward determination (plus physical causation) and downward causation, while reductive physicalism appeals only to upward determination (plus physical causation).

This calls for some explication of downward causation. C. D. Broad's book *The Mind and its Place in Nature* (1925) provides an account of emergence from which a model of downward causation is readily extracted (for a fuller discussion see McLaughlin 1992: 75–89; Hendry 2006a: section 3). First consider the position that Broad calls 'pure mechanism'. According to this position, every material object is made from the particles of 'one fundamental kind of stuff' (1925: 44). One physical law governs the interactions between the particles, and if pure mechanism is right, this law determines the behaviour of every material object. According to Broad, the existence of irreducibly macroscopic qualities like colours and temperatures (1925: 50–1) shows that pure mechanism *must* fail. Such qualities are associated lawfully with microscopic states through 'trans-physical' laws, and cannot be accounted for by pure mechanism. Aside from the necessary emergence of properties involved in trans-physical laws, Broad countenanced emergent 'intra-physical' laws between physical properties. Breathing, for instance, is a type of movement, and if it is not determined to occur by interactions between the particles from which breathing systems are formed as governed by the one basic law, then it is emergent (1925: 81). Broad saw clear epistemic differences between trans-physical and emergent intra-physical laws, because a failure to account mechanistically for an intra-physical law may arise either from incomplete knowledge or from its being genuinely emergent (1925: 80–1). That an intra-physical law is emergent must always remain a hypothesis.

Within quantum mechanics, motions are described by Hamiltonian operators which are determined by the force laws which apply to a system. It is easy to formulate the difference between emergentism and Broad's pure mechanism. Pure mechanism expects only one force term to appear in the Hamiltonians that govern the behaviour of complex systems. By analogy with the well-known terminology for forces, call the appropriate one-force-term Hamiltonian for a system its 'resultant' Hamiltonian. The emergentist posits that non-resultant, or 'configurational' Hamiltonians govern the behaviour of at least some complex systems. Broad's pure mechanism is an extreme, of course, because it countenances only one basic kind of interaction, and physics countenances more. But the point is that the disagreement between emergentism and pure mechanism (and, more generally, reductive physicalism) is relative to some set of fundamental physical properties and laws.

Where does downward causation fit into this? For the emergentist, every complex system is composed of the same basic stuff, but some complex systems are covered by non-resultant or configurational Hamiltonians. In an emergent complex system, the behaviour of the basic stuff of which it is made is governed by a configurational Hamiltonian, which is different from what it would be were its behaviour governed by a resultant Hamiltonian. Since the Hamiltonian of a system determines the precise nature of the physical law that governs its behaviour, to say that some system exhibits downward causation is to make a counternomic claim about it: that its behaviour would be different were it determined by the more basic laws governing the stuff of which it is made. The emergentist and the reductionist can agree that a unified framework of physical law (quantum mechanics) governs how forces act, but disagree on the extent to which physical law is unified from a dynamical point of view, that is, on how many independent kinds of Hamiltonian operate in the world. Note that the truth of emergentism would be no barrier to genuine quantum-mechanical explanations of chemical structure and bonding. The difference between the emergentist and the reductive physicalist concerns only the form of those explanations: the emergentist expects that they will involve configurational Hamiltonians, the reductive physicalist that they will involve only resultant Hamiltonians (see Hendry 2006a: sections 3 and 4).

A direct connection to ontology comes via Alexander's dictum, the principle, often cited by Kim (see for instance Kim 1998: 119), according to which being real requires having causal powers. The precise formulation of this principle is in need of clarification (see Hendry 2010: chs 9 and 10), but in the present context, we seek only to focus the disagreement between the emergentist and the reductionist as involving a worldly fact of the matter, one that can transcend relationships between current and future scientific theories. If it is construed as a biconditional principle, Alexander's dictum does this nicely: it can be cited by both emergentists and reductionists, and is a neutral point of agreement. Bearing causal powers is both necessary and sufficient for a property's being counted as real, hence the criterion bites both ways. The reductive physicalist thinks special-science properties are no more than their physical bases because the causal powers they confer are just those conferred by their physical bases; the emergentist sees them as distinct and non-reducible just because the causal powers they confer are not exhausted by those conferred by their physical bases. This formulation fits the earlier requirement of a conception of ontological reduction: the reductionist and the emergentist disagree over a substantive issue, a complex matter of fact concerning the structure of the laws of nature. The counternomic requirement sharply expresses what is involved in downward causation. Although the issue cannot be directly operationalized, both sides can seek evidence in the sciences themselves, and it seems they must, unless a priori arguments can settle the matter, which seems unlikely when chemistry is the special science whose reducibility is at issue.

3. EMERGENCE VS. REDUCTION: THE SYMMETRY PROBLEM

Physics and chemistry meet in quantum chemistry, the interdisciplinary field which applies quantum mechanics to explain the structure and bonding of atoms and molecules. For any isolated atom or molecule, there is a Schrödinger equation which is determined by enumerating the electrons and nuclei in the system, and the forces by which they interact. Classical intertheoretic reduction would require the derivation of the properties of atoms and molecules from their Schrödinger equations. Quantum chemistry does not meet these strict demands, because its models bear only a loose relationship to exact atomic and molecular Schrödinger equations (see for instance Woolley 1976, 1991, 1998; Bogaard 1981; Primas 1983; Hofmann 1990; Scerri 1991, 1994; Hendry 2010; ch. 7). There is an exact analytical solution to the non-relativistic Schrödinger equation for the hydrogen atom and other one-electron systems, but these are special cases on account of their simplicity and symmetry properties. Carl Hoefer (2003: 1404) makes eloquent use of the elegance and precision of this solution. While he is quite right that it is impressive, caution is required in drawing any consequences for how quantum mechanics applies to real-worldly systems more generally, because it is such a special case. The simplicity and symmetry properties of the problem are central to the elegance of the solution, and molecules in particular cannot share those symmetry properties.

The Schrödinger equation for the next simplest atom, helium, is not soluble analytically, although accurate numerical methods are available. To solve the Schrödinger equations for more complex atoms, or for any molecule, quantum chemists apply a battery of approximate methods and models. Whether they address the electronic structure of atoms or the structure and bonding of molecules, these approximate models are calibrated by an array of theoretical assumptions many of which are drawn from chemistry itself. This failure of classical reduction may not impress reductive physicalists. As we saw in the last section, they have three main lines of defence: (1) to claim that the approximate models are merely 'proxies' for more rigorous treatments; (2) to wait for new developments (new models for quantum chemistry or new mathematical methods for deriving the old ones rigorously from exact Schrödinger equations); or (3) to accept the failure of intertheoretic reduction and retreat to ontological reducibility. In what follows I will argue that there are objections to any of these strategies with regard to molecular Schrödinger equations. In contrast I will make no claim about the non-reducibility of multi-electron atoms, because the application of models which are motivated by empirical information from the higher science is not *in itself* enough to ground an argument against ontological (as opposed to intertheoretic) reduction. There must also be issues of principle that *require* the application of such models. There are such issues in the molecular case, but not (I think) in the atomic case.

Although molecular structures cannot be derived directly from exact molecular Schrödinger equations, quantum-mechanical models do assume that molecules have them, for example in the explanation of microwave spectroscopy. Molecular structures are justified as approximate solutions to the exact Schrödinger equation in a way that does sound as if it will allow a proxy defence, via the Born–Oppenheimer approximation. The justification is as follows. The nuclei within a molecule are thousands of times more massive than the electrons, and so they can be regarded as approximately at rest when the electronic motions are considered. The trick is to solve a Schrödinger equation just for the electrons, in which a fixed nuclear geometry appears as a parameter. In principle, the electronic Schrödinger equation could be solved for many different arrangements of the nuclei to see how the electronic energy depends on nuclear geometry. This generates a potential energy surface whose local minima (and sometimes maxima) can be interpreted as corresponding to the geometries of chemically significant structures. In practice, it may be enough to consider only an equilibrium geometry that is known empirically, and small oscillations around it. The justification for this approximation is that using the Born–Oppenheimer solution instead of the exact solution makes only a small difference to the energy.

The justification, though mathematically correct, fails to meet the conditions required for the applicability of the proxy defence. The Born–Oppenheimer approximation makes only a small difference to the calculated energy of the molecule, but it makes a big difference to its symmetry properties (Woolley and Sutcliffe 1977). Although Schrödinger equations for complex polyatomic molecules cannot be solved analytically, much can be known about their solutions by considering the nature of the forces that appear in them. Of the four fundamental forces, three (gravitational, weak, and strong nuclear) can be neglected in calculating the quantum-mechanical states governing molecular structure. Hence physics itself tells us that the Coulomb (electrostatic) force is the overwhelming determinant of molecular structure, which should arise from the quantum mechanics of systems of charged particles moving under electrostatic forces. Now arbitrary solutions to exact Coulombic Schrödinger equations should be spherically symmetrical, but polyatomic molecules cannot be spherically symmetrical, for their lower symmetries are important in explaining their behaviour. Consider, for example, the hydrogen chloride molecule, which has an asymmetrical charge distribution which explains its acidic behaviour and its boiling point. In the Born–Oppenheimer approximation, the spherical symmetry that is expected of exact solutions to the full Schrödinger equation is simply replaced by a less symmetrical structure that is compatible with the asymmetrical charge distribution. Molecular structures cannot be recovered from the Coulomb Schrödinger equations, but not because of any mathematical intractability. The problem is that they are not there to begin with. The

Coulomb Schrödinger equations describe mere assemblages of electrons and nuclei rather than molecules, which are structured entities (Woolley 1991: 26). This is illustrated by the fact that isomers, which are distinct molecules sharing the same molecular formula, share the same Coulomb Schrödinger equation (Woolley 1998: 11). For instance ethanol (CH_3CH_2OH) is the active ingredient of whisky, and boils at 78.4°C. Methoxymethane (dimethyl ether, CH_3OCH_3) is sometimes used as an aerosol propellant, and boils at −24.9°C. These are distinct substances, though each contains carbon, oxygen, and hydrogen in the molar ratios 2:1:6. Clearly the distinctness of ethanol and methoxymethane as chemical substances must lie in their different molecular structures, that is, the arrangement of atoms in space, yet both are represented by the same Schrödinger equation. Since the proxy defence relies on the exact treatments sharing explanatorily relevant features of the models, it must fail in the case of the Born–Oppenheimer approximation because they have explanatorily relevant features which are not shared by the exact solutions. Without a quantum-mechanical justification for the attributions of structure within the Born–Oppenheimer models, they simply assume the facts about molecular structure that ought to be explained.

The spherically symmetrical states could perhaps be regarded as superpositions of asymmetrical states with opposite orientations, just as the spin states of a silver atom may be regarded as superpositions of spin-up and spin-down, or the quantum state of Schrödinger's cat can be regarded as a superposition of 'cat-alive' and 'cat-dead' states. If anything, this makes the symmetry problem look more intractable, and a proxy defence based on the Born–Oppenheimer approximation more obviously inadequate. In the quantum-mechanical measurement problem, the difficulty is in explaining why, when we have suitably entangled the cat's quantum state with a radioactive source which itself is in a superposed state with respect to a radioactive decay event, we find on measurement of the cat's state one of the determinate states (cat-alive or cat-dead), rather than their superposition. Imagine the following solution to the measurement problem: although the cat ought to be found in a superposition once its quantum state interacts with the radioactive source, I provide a proof that which of the three states the cat exhibits (cat-alive, cat-dead, superposition) makes little difference to the overall energy of the system. Whether the cat is found to be dead or alive, I explain the measurement result, and thereby 'solve' the measurement problem, by invoking the 'superposition approximation' in which the actual measurement result is modelled by the superposition, on the grounds that the substitution makes little difference to the overall energy. It is just as egregious a mistake to invoke the spherically symmetrical exact solutions to Coulombic molecular Schrödinger equations as explaining the lower symmetry of the real molecular structures found in nature. So much for the proxy defence.

What of the second response? To wait for new developments seems unpromising. The wait is for new kinds of molecular model meeting two conditions: they should have the right symmetry properties to explain the structure and bonding of molecules, and also be defensible as approximations to exact quantum mechanics in a way that allows a new version of the proxy defence. However, Woolley and Sutcliffe's symmetry problem arises from foundational features of how exact quantum mechanics is generated using the Coulomb force, and shows that the two demands cannot be met simultaneously. Hence the wait must either be for an entirely new framework for exact quantum mechanics, or the replacement of quantum mechanics itself, but I take it that that involves a retreat to ontological reducibility (see below). The symmetry problem arises in the first instance by considering the Schrödinger equation for an isolated molecule, and the only obvious way out is to appeal to the molecule's interaction with its environment, which would be represented by a symmetry-breaking non-Coulomb term in the molecule's Schrödinger equation. The particular form of the symmetry-breaking addition must be justified however, and it is quite mysterious how that could work if all one has in the environment are more molecules described by Coulombic Hamiltonians. The Coulomb Schrödinger equation for an *n*-molecule ensemble of hydrogen chloride molecules has precisely the same symmetry properties as a Coulomb Schrödinger equation for a 1-molecule system. If the particular form of the symmetry-breaking addition is not justified, then it is just ad hoc: a *deus ex machina*.

The final option is to wait for a replacement for quantum mechanics, and (for the moment) retreat to the ontological reducibility of chemistry, hoping that some future physical theory will achieve an intertheoretic reduction to reflect this metaphysical fact. It might be thought that this is unproblematic: the above issues bear only on intertheoretic reduction and not ontological reduction, because they concern the explanatory relationship between a physical theory, quantum mechanics, and chemical theories of molecular structure. But that would be too quick. We saw in the last section that if molecules are ontologically reducible to their physical bases, then they ought to have no causal powers beyond those that are conferred by those physical bases. That much follows if ontological reduction is committed to the causal completeness of physics, as physicalists generally require it to be (McLaughlin 1992; Horgan 1993; Kim 1998; Papineau 2002).

The symmetry problem impacts evidentially on ontological reduction via its commitment to the completeness of physics in two ways. The first is direct: if the acidic behaviour of hydrogen chloride is conferred by its asymmetry, and the asymmetry is not conferred by the molecule's physical basis according to physical laws, then ontological reduction fails because the acidity is a causal power which is not conferred by the physical interactions among its parts. Of course future physics and chemistry may be more amenable to ontological reduction, just as it may solve the related quantum-mechanical measurement problem, but proponents of ontological reduction are not entitled to presume that it will. On any conservative amendment to quantum mechanics, the explanation of why molecules exhibit the lower symmetries they do would appear to be holistic, explaining the molecule's broken symmetry on the basis of its being a subsystem of a supersystem (molecule plus environment). This supersystem has the power to break the symmetry of the states of *its* subsystems without acquiring that power from its subsystems in any obvious way. That looks like downwards causation. As for a non-conservative amendment to quantum mechanics, the bets are off, but there is no particular reason to think the successor to quantum mechanics will exclude downward causation. In fact the inductive evidence is that it will not, because its immediate predecessor, quantum mechanics, does not. This is not just a contentious appeal to the results of the foregoing symmetry argument: it may well be that quantum-mechanical entanglement should be interpreted more generally as indicating a failure of ontological reducibility (Humphreys 1997a: section 6).

The second way that the symmetry problem impacts on the completeness of physics is indirect. If the ontological reducibility of chemistry is not a default position that can be established by appeal to intuition, then its most important element, the causal completeness of physics, must be a substantive thesis requiring empirical support. The symmetry problem removes much of the empirical support that is claimed for the principle. Here's why. The completeness of physics involves the claim that the general framework of mechanics is able to unify the motion of any physical system by seeing it as arising from just a few forces which apply very generally. Is there any evidence *for* this principle? David Papineau (2002: appendix) sets out to explain what he sees as the consensus in twentieth-century science that physics is complete. Papineau explains that consensus as arising from the acceptance of two interlocking arguments, the 'argument from fundamental forces' (2002: 250), and the 'argument from physiology' (2002: 254). The argument from physiology may be relevant to Papineau's intended application of the completeness of physics to the mind–body problem, but only the argument from fundamental forces is relevant to the reduction of chemistry. In any case I am highly sceptical about how closely physics (as opposed to chemistry) has unified, or even been involved in physiological explanation at the cellular level. Hence I will concentrate on the argument from fundamental forces. The conclusion of this argument is that 'all apparently special forces characteristically reduce to a small stock of basic physical forces which conserve energy' (2002: 250). This conclusion, Papineau argues, was available to Hermann von Helmholtz because in order for the principle of the conservation of energy to hold, non-conservative, dissipative forces like friction must be reducible to conservative forces, which is ensured if dissipative forces all arise from a few basic forces. The reducibility of *all* special forces is, he argues, a natural generalization of this reduction.

Whether or not this is the right psychological explanation for the views of Helmholtz or any other nineteenth-century scientist, to have held that all special-science forces are reducible at that time would have been grossly premature, since there were precisely no detailed physical explanations of the 'special' forces invoked by physics' neighbouring science, chemistry. The more general reduction is, in any case, not required for conservation of energy but merely suggests itself as an explanation of it, as Papineau himself points out. In the nineteenth century, there were only speculative (and unsuccessful) attempts to explain the microstructure of chemical substances in terms of physical theory, partly because the microphysical structure of substances and molecules at that time was simply unclear. Only in the twentieth century was there any detailed and successful application of physics to the explanation of chemical structure and bonding that could ground an argument for the reduction of chemistry. That application is appealed to by Brian McLaughlin in his account (1992) of why the currency of emergentism declined in the twentieth century. McLaughlin sympathetically sets out the doctrines of emergentism, including the existence of configurational forces and downwards causation, and argues that the failure of any configurational forces to turn up in quantum-mechanical explanations of chemical structure and bonding undercuts emergentism. Now Papineau points out (2002: 254) that the reduction of chemistry hardly enforces the reduction of life or mind, but if McLaughlin is right then his argument is at least relevant to chemical emergence. However, we have already found that actual quantum-mechanical explanations of chemical structure and bonding presuppose unexplained symmetry breaking. In effect they appear to employ configurational forces. This undercuts any empirical support they could offer to the completeness of physics with regard to chemical systems, and with it, ontological reduction (see Hendry 2006a for a more detailed discussion of this point). I am also dubious of the suggestion that in twentieth-century physics and chemistry there was any consensus in favour of the completeness of physics. To invoke one important constituency, among the chemists who founded quantum chemistry, there were many temperamental non-reductionists like Linus Pauling, who saw the quantum-mechanical explanation of chemical structure and bonding as a process which drew equally on physical principles and chemical knowledge, adapting quantum mechanics significantly in the process (see Hendry 2003).

Though it is undeniable that (some) physical principles have a special role in chemical explanation, the physicalist arguments misread that role. Reductive physicalists assume that physical principles apply to the many systems studied by science by completely determining their motions. There is one obvious and weaker alternative principle of universality: the ubiquity of physics. This principle is more obviously consonant with the founding quantum chemists' view of quantum-mechanical principles. Under the ubiquity of physics, physical principles constrain the motions of particular systems though they may not fully determine them. Some physical principles are naturally understood this way, even by physicists: the second law of thermodynamics, and (ironically) the conservation of energy are obvious examples. Taken individually, the various force laws, which dictate the form of the potential terms in quantum-mechanical Hamiltonians, seem also to be understood this way, even within physics, since they operate together to produce the potential term which governs the overall motion of a system. The Coulomb law, for instance, is not regarded as being violated in systems in which (say) gravitational forces also act. Robert Bishop (2006: section 2) makes a closely related point, arguing that 'Physics itself does not imply its own causal closure' (2006: 45): the completeness of physics is a metaphysical principle, an important point to note since completeness is a necessary premise in any physicalist argument for the causal exclusion of the non-physical. In short, one may accept that physical principles apply universally without accepting that they fully determine the behaviour of the systems they govern. To accept the universal applicability of physical principles does of course imply that *ceteris paribus* (that is, absent *non*-physical influences) they fully determine the motions of any system they govern, but this leaves open what happens when *ceteris* (as the saying goes) isn't *paribus*. The difference between the two principles is subtle: in defending 'fundamentalism' against Cartwright's arguments against it (Cartwright 1999), Carl Hoefer rightly distinguishes fundamentalism, the claim that there are 'universal fundamental laws with which all phenomena are in accord' (2003: 1403) from the stronger 'thesis of the reducibility of biology, chemistry, or meteorology to physics' (2003: 1408). As formulated, Hoefer's fundamentalism seems to express ubiquity very nicely: he doesn't explicitly distinguish intertheoretic and ontological reduction, though he clearly intends to deny only intertheoretic reducibility. Therein lies the problem. He commits fundamentalism to the causal completeness of physics (and therefore ontological reducibility), although he defends only ubiquity against Cartwright's arguments. He endorses a quote from Richard Feynman clearly expressing ontological reductionism, glossing it only as ubiquity:

> What the fundamentalist believes in is a sort of no-conflicts thesis, between physical laws and higher-level phenomena. Feynman . . . expresses it nicely (Hoefer 2003: 1408–9 note 3).

The Feynman passage he quotes reads:

> For example, life itself is supposedly understandable in principle from the movements of atoms, and those atoms are made out of neutrons, protons, and electrons. I must immediately say that when we state that we understand it in principle, we only mean that we think that, if we could figure everything out, we would find that there is nothing new in physics which needs to be discovered in order to understand the phenomena of life. (Feynman 1965: 151)

Now there are two versions of fundamentalism in play in the argument between Cartwright and Hoefer's fundamentalist: one is committed only to ubiquity, the other to completeness. Hoefer's arguments defend only the former against Cartwright. There are two corresponding versions of Cartwright's alternative metaphysical picture of a 'patchwork of laws', one denying only completeness, the other denying ubiquity too. The former version, I think, is close to the nomological formulation of emergentism I have set out here, although Cartwright's intentions are primarily sceptical, and so she eschews the metaphysical terms required to formulate emergentism.

Emergentism is compatible with ubiquity, though not with completeness, understood so as to apply to (a specified list of) force laws. Completeness implies reducibility, as we have seen, while the ubiquity of physics does not: this is not surprising, as completeness is a logically stronger principle of the priority of physics than ubiquity. The mere applicability of physical principles to chemical bonding requires only ubiquity, and does not rule out downward causation. Non-reductionists can accept that physical principles are uniquely universally applicable, and that they have a unique priority and universality with respect to the special sciences. Both emergentism and reductive ontological physicalism are consistent with the successful application of quantum mechanics to explain chemical structure and bonding. The situation is not, however, symmetrical between the two. Emergentism is at least as well supported as reductive physicalism by the data of the explanatory interface between physics and chemistry, for two reasons. If emergentism were true, and configurational Hamiltonians really did govern the behaviour of molecules, then the disunified structure of quantum-mechanical models explaining molecular structure and bonding, including the unexplained symmetry-breaking through the imposition of determinate molecular structure by hand, is just what one would expect. Although it can be made consistent with this situation, reductive physicalism has to appeal to independent factors to explain the disunified structure of quantum chemistry, and must posit a mechanism for symmetry-breaking for which there is no independent evidence. In short, the overall philosophical package of emergentism gives the more unified explanation of how both physical and chemical theories, and physical and chemical properties, interact. On any evidential principle under which the hypothesis that gives the more unified explanation of the phenomena is better supported, emergentism wins out. Second, reductive physicalism embodies a logically stronger version of the priority of physics, and there are no good arguments for the excess content of the stronger version. Since the probability of a logically weaker principle is an upper bound on that of a logically stronger principle, evidence for completeness supports ubiquity at least as well.

4. CONCLUSION

In chemistry there are two main issues of reduction: (1) whether macroscopic substances reduce to their characteristic molecules, and (2) whether molecules as structured entities reduce to quantum-mechanical systems of nuclei and electrons. A distinction was also drawn between intertheoretic and ontological reduction, although there are tight evidential connections between the two issues because scientific evidence for ontological reduction must come chiefly in the form of intertheoretic relationships. The difference between (strict) ontological reductionism and emergentism comes down to a disagreement over the causal completeness of microphysics. Since the evidence for the causal completeness of microphysics is weak, and better supports a logically weaker principle of the *ubiquity* of physical laws that is compatible with emergentism, I have argued that the ontological emergence of molecular structure with respect to quantum-mechanical systems of nuclei and electrons interacting via Coulomb forces is at least as well supported by the available scientific evidence as its ontological reducibility.

Moving up a level, I am not committed to a position on whether or not chemical substances are reducible to their molecular structures. I have not presented an argument *against* that reduction, but even if the identities of particular substances are determined by their molecular structures, reducibility does not follow, but then neither does emergence (see Hendry 2010: chapter 2). How does my argument affect the relationship between the physical and the mental? Reductive physicalists whose primary interest is the mind–body problem and the causal efficacy of the mental may, with perfect consistency, accept that the chemical is emergent with respect to microphysics. On the model of emergence proposed in this chapter, this is only to accept that the 'physical', construed broadly so as to encompass the chemical, exhibits more nomic disunity than they might hitherto have supposed. But perfect consistency is not plausibility. A new thesis would appear as a premise in arguments for the causal exclusion of the mental: the completeness of the broadly physical. This thesis is also in want of justification, and, as before, I would bet that there would be no evidence for it over the logically weaker thesis of the ubiquity of the (broadly) physical, which would not be sufficient for causal exclusion arguments. This move would, in any case, remove the last shred of a connection between 'the completeness of physics' and any particular physical theories, which would render the undoubted explanatory successes of real physical theories within chemistry simply irrelevant to reductive physicalism in the philosophy of mind.

REFERENCES

Bishop, R. 2006. 'The Hidden Premiss in the Causal Argument for Physicalism'. *Analysis* 66: 44–52.

Bogaard, P. 1981. 'The Limitations of Physics as a Chemical Reducing Agent'. *PSA 1978: Proceedings of the Biennial Meeting of the Philosophy of Science Association*, vol. 2. East Lansing, MI: Philosophy of Science Association, 345–56.

Broad, C. D. 1925. *The Mind and its Place in Nature*. London: Kegan Paul, Trench and Trubner.

Cartwright, N. 1983. *How the Laws of Physics Lie*. Oxford: Clarendon Press.

—— 1999. *The Dappled World: A Study of the Boundaries of Science*. Cambridge: Cambridge University Press.

Crane, T. and Mellor, D. H. 1995. 'There is no Question of Physicalism'. In P. K. Moser and J. D. Trout (eds), *Contemporary Materialism*. London: Routledge, 65–89.

Dupré, J. 1993. *The Disorder of Things: Metaphysical Foundations of the Disunity of Science*. Cambridge MA: Harvard University Press.

Feynman, R. 1965. *The Character of Physical Law*. Cambridge MA: MIT Press.

Hendry, R. F. 2003. 'Autonomy, Explanation and Theoretical Values: Physicists and Chemists on Molecular Quantum Mechanics'. In J. Earley (ed) *Chemical Explanation: Characteristics, Development, Autonomy: Annals of the New York Academy of Sciences* 988: 44–58.

—— 2006a. 'Is there Downward Causation in Chemistry?' In D. Baird, L. McIntyre, and E. R.Scerri (eds), *Philosophy of Chemistry: Synthesis of a New Discipline*. Dordrecht: Springer, 173–89.

—— 2006b. 'Elements, Compounds and other Chemical Kinds'. *Philosophy of Science* 73: 864–75.

—— 2010. *The Metaphysics of Chemistry*. New York: Oxford University Press, forthcoming.

Hoefer, C. 2003. 'For Fundamentalism'. *Philosophy of Science* 70: 1401–12.

Hofmann, J. R. 1990. 'How the Models of Chemistry Vie'. *PSA 1990: Proceedings of the Biennial Meeting of the Philosophy of Science Association*, vol. 1. East Lansing, MI: Philosophy of Science Association, 405–19.

Horgan, T. 1993. 'From Supervenience to Superdupervenience: Meeting the Demands of a Material World'. *Mind* 102: 555–86.

Humphreys, P. 1997a. 'How Properties Emerge'. *Philosophy of Science* 64: 1–17.

—— 1997b. 'Emergence, not Supervenience'. *Philosophy of Science* 64: (Proceedings), S337–S345.

Kim, J. 1998. *Mind in a Physical World*. Cambridge MA: MIT Press.

McLaughlin, B. 1992. 'The Rise and Fall of British Emergentism'. In A. Beckermann, H. Flohr, and J. Kim (eds) *Emergence or Reduction? Essays on the Prospects for Non-Reductive Physicalism*. Berlin: Walter de Gruyter, 49–93.

Papineau, D. 2002. *Thinking about Consciousness*. Oxford: Clarendon Press.

Primas, H. 1983. *Chemistry, Quantum Mechanics and Reductionism*. Berlin: Springer.

Scerri, E. 1991. 'The Electronic Configuration Model, Quantum Mechanics and Reduction'. *British Journal for the Philosophy of Science* 42: 309–25.

—— 1994. 'Has Chemistry at least Approximately been Reduced to Quantum Mechanics?' *PSA 1994: Proceedings of the Biennial Meeting of the Philosophy of Science Association*, vol. 1. East Lansing MI: Philosophy of Science Association, 160–70.

Woolley, R. G. 1976. 'Quantum Theory and Molecular Structure'. *Advances in Physics* 25: 27–52.

—— 1991. 'Quantum Chemistry Beyond the Born–Oppenheimer Approximation'. *Journal of Molecular Structure* (Theochem), 230: 17–46.

—— 1998. 'Is there a Quantum Definition of a Molecule?' *Journal of Mathematical Chemistry* 23: 3–12.

—— and Sutcliffe, B. 1977. 'Molecular Structure and the Born–Oppenheimer Approximation'. *Chemical Physics Letters* 45: 393–8.

15

An Emergentist's Perspective on the Problem of Free Will*

Achim Stephan

INTRODUCTION

Over billions of years, many changes occurred in a universe, which was *lifeless* and *mindless* at the beginning. These changes include the development of organisms, that is, the formation of entities, which are alive and which have a variety of cognitive capacities. Depending on their species, organisms are capable of perceiving, learning, communicating, or action planning, to name just some of these capacities. Furthermore, some of the evolved organisms, namely we human beings, treat themselves as being responsible for what they do, at least sometimes. We do this on the basis of thinking of ourselves as being *free agents* who are capable of performing free actions and decisions.

Within the philosophy of mind, one major task is to give an account of how the cognitive features we usually describe in psychological terms on the personal (or macro) level relate to bodily features we usually describe in neuroscientific (or physiological, or even molecular) terms on a subpersonal (or micro) level. Within this debate, particular emphasis is given to states and processes known as phenomenal experiences (such as smells, pains, or feelings of thirst) and propositional attitudes (such as beliefs, hopes, and desires), recently also to emotional reactions (such as anger, sadness, or jealousy). Furthermore, the question of mental causation is one of the strongly disputed issues within this debate.

However, the feature of having the capacity to bring about free decisions or to perform free actions has not been discussed within the debate on how the personal and the subpersonal level relate to each other. The metaphysical debate on whether or not free will is compatible with determinism did not stretch

* I would like to thank Brian McLaughlin, Wolfgang Lenzen, and Sven Walter for helpful comments on an earlier version of this chapter.

out to the debate on the mind–body problem (except, may be, in the writings of Timothy O'Connor (see 2000: ch. 6, and 2002). This issue has entered the debate only recently through the impact from the neurosciences and their claim that the subpersonal processes of the brain, and not 'we' as persons, are 'responsible' for who we are and what we do.

So, let me outline how I will proceed. Since I cannot discuss here all possible positions with regard to the problem of free will, I have to narrow down what shall be considered. Emergentism, as I understand it, is committed, first of all, to a world that does not contain supernatural entities: all behaviours, states, processes, properties we know of are realized by systems, which are composed of physical entities only. Within the metaphysical debate on the free will problem, I therefore concentrate on positions that are compatible with physicalism in one way or another. These comprise versions of hard determinism, compatibilism, and libertarianism.[1] After having introduced these positions by referring to prominent authors of the recent German debate on free will, I characterize two types of emergentism, namely a weak and a strong version of synchronic emergentism, in order to comment on the free will debate from their perspective, respectively.

Since it is hotly debated whether or not free decisions exist at all—some philosophers and neuroscientists claim that it is an illusion when we think of ourselves as free agents—I prefer to start with differentiating between 'candidates' and 'non-candidates' of free actions and decisions. They have in common that on the *personal* level they are human behaviours, which makes it possible to relate them to processes on the *subpersonal* level. Even if it would turn out that we were completely wrong in assuming that some of our decisions are free decisions, the candidates for free actions and decisions would remain human behaviours.

CANDIDATES FOR FREE ACTIONS AND DECISIONS

Within the free will debate we should distinguish various topics of dispute. First, we have to consider actual behaviours (decisions as well as actions) of persons; among these are the *candidates* for free actions and decisions. Second, we have to consider the various *criteria* that are proposed for judging candidates for free actions and decisions as being genuine free actions and decisions. And third, we have to consider whether the candidates for free actions and decisions do in fact *fulfil* the proposed criteria.

[1] There are, of course, libertarians who take sides with dualism. However, some more recent discussions show that libertarians are neither committed to dualistic positions nor to agent causality (see, e.g., Kane 2002; Keil 2007).

So, let us turn first to the candidates for free actions and decisions. Among the actual behaviours of human beings are *candidates* for free actions and decisions, and *non-candidates*. Non-candidates are, for example, reflex actions such as the lid-reflex or yawning, dreaming, movements during sleep, feeling pain, or blushing. All these behaviours, although ours, have in common that we have not decided to perform them. Reflex actions (including yawning) are automatic and involuntary neuromuscular actions elicited by certain stimuli; unconscious processes behind dreaming, apart from creating non-volitional fantasies, can also produce movements we have not consciously decided upon. Also, feeling pain is nothing we decide to do, nor is blushing. It is not only that we do not decide to perform these behaviours; they also seem to be beyond our control in that we are unable to decide not to perform them. In fact, these behaviours befall us. It's not up to us whether or not we do these things. And, of course, this is reflected by the fact that nobody would regard him- or herself or somebody else to be responsible for any of these behaviours. To summarize, behaviours, which are not done volitionally, and which escape our control in that we cannot refrain from performing them, are definitely non-candidates for free actions.

Clear candidates for free actions and decisions seem to be, on the other hand, for example, booking the summer-holiday residence, applying for a senior position at Harvard, careful and minute planning of a bank robbery, or working on a philosophical paper about the free will problem (further examples are investing money at the stock exchange market, marrying one's beloved girlfriend, buying a house). These actions have in common that they are voluntary, that we have the impression that we could do or could have done something else instead, that we have thought them through, and that we have consciously decided to perform them, maybe after weighing pros and cons. Also, our decisions to perform these various actions seem to be free, we could have decided differently, they are under our control, or so it seems. It seems to be up to us whether we do these things.[2]

Between these rather clear-cut cases there are others where common sense has no clear intuitions; among them are, for example, laughing while being tickled—sometimes we are, sometimes we are not able to resist or to decide not to laugh; being jealous of someone—sometimes we can control these feelings, sometimes not; smoking one cigarette after the other, committing crimes like sexually abusing children or slamming another person to death, or the betrayal of secrets under torture. Here, it is not clear how much control we could gain

[2] There is an established distinction between free *actions* and free *decisions*. Accordingly, an agent's action might be called free, if he or she could have acted differently, and he or she often could have acted differently if he or she would have chosen differently. A decision, however, is called free if the agent *could* have chosen differently (cf., e.g., Moore 1912: ch. 6; see also Schopenhauer 1839: 536). When I speak of free actions here, I refer to actions that are free in the strong sense, namely that we *could* have chosen differently. I take it for granted that we often act freely in the weak sense, namely that we could have acted differently if we had chosen to act differently.

over these actions. But often people are held responsible, at least in some sense, for acting in this way.

CRITERIA FOR FREE ACTIONS AND DECISIONS

In the following section I will argue with a neuroscientist and two philosophers who have been influential in the current debate in Germany, which was triggered mainly by provocative claims of well-known neuroscientists such as Wolf Singer and Gerhard Roth. What is crucial for all three of them—Ansgar Beckermann, Geert Keil, and Wolf Singer—is that they really have taken sides: Wolf Singer, neuroscientist and the director of the Max Planck Institute for Brain Research at Frankfurt, opts for a determinist picture that leaves no room for truly free decisions; he is a *hard determinist*. Geert Keil, philosopher at Aachen University, opts for an indeterminist perspective that leaves room for free decisions; he is a *libertarian*, without being committed to substantival dualism. Ansgar Beckermann, philosopher at Bielefeld University, is a *compatibilist* who develops a notion of free will that allows accepting both determinism and the capacity of free decisions.[3]

Interestingly enough, there is considerable prima facie agreement among the three concerning the conditions that would need to be fulfilled by candidate-behaviours to be counted as genuine free actions or decisions. First and foremost, there is the *principle of alternative possibilities*,[4] which is referred to by Geert Keil (2007: 282) under exactly this label, whereas Ansgar Beckermann (2005: 111) calls it the 'could-have-done-or-chosen-otherwise condition'. Wolf Singer does not explicitly refer to the principle of alternative possibilities; however, it is implicitly contained in his considerations, particularly when he *denies* that in given situations we could decide differently than in fact we do (2006).

Widely accepted as a further criterion for free actions and decisions is, second, the *principle of origination* (or *authorship*): the choice that is made in a certain situation must depend on the agent, according to Beckermann—he calls this the 'authorship condition' (2005: 111). The principle of origination also lurks in what Robert Kane has called 'ultimate responsibility', which is the idea that for an agent to be ultimately responsible for an action, he or she must be responsible for anything that is a sufficient reason (condition, cause, or motive) for the occurrence of the action (see Kane 2002: 407).

A third principle often referred to within the context of free will is the *principle of intelligibility* (or *rationality*). Beckermann states that our decisions are free, 'if

[3] A similar position is taken by Henrik Walter who is both a psychiatrist and neuroscientist at Bonn University and a philosopher. He calls the position he opts for *revisionist compatibilism* (Walter 2004).

[4] See, e.g., Harry Frankfurt (1969: 829).

and only if, they rest on processes that can be influenced by rational arguments and considerations' (2005: 120). For libertarians it is evident, anyway, that free agents must be able to decide between possible alternatives, and that they do so consciously on the basis of reasons. Even Singer acknowledges this principle when citing the common sense view: 'We judge decisions as being free, which are based on conscious considerations of variables, i.e., on the rational deliberation [or negotiation] of contents accessible to consciousness' (2005: 156; my translation).

HOW TO RELATE PERSONAL- AND BRAIN-LEVEL TALK

Let us now consider how the three authors comment on the free will problem given some candidates for free actions and decisions. I start with Ansgar Beckermann who argues for a compatibilist version. According to him, our decisions are free, if and only if, *they rest on processes that can be influenced* by rational arguments and considerations. To illustrate what he means, Beckermann provides us with an example, in which he narrates how he might have been convinced by a colleague to jump out of his warm bed rather early in the morning to join an important faculty meeting. Without the colleague's call he would have preferred to stay in bed, but this tiny change in the world, his colleague's call, might lead to another decision, if weight is given to the arguments of the colleague. Beckermann does not present an argument for genuine alternative possibilities in the very same situation. He is not claiming that he could have decided differently under the very same conditions, that is, after having received the colleague's call; he only claims that he was open to think through his preferences on the basis of (new) arguments, which might come from somebody else.

Beckermann develops his position from the personal perspective; in his focus are the agent's deliberations and decisions. If those are sensitive to arguments and rational considerations they are classified as free. Note that Beckermann has found a tricky formulation to circumvent typical challenges such as: How could our conscious considerations influence what we will do or decide qua being *conscious considerations*? Being a naturalist and willing to accept thorough determination on the neural level, Beckermann sees only two alternatives: '*either* it is the case that not all decisions rest on natural processes *or* there are natural processes that can be influenced by considerations and arguments' (2005: 121). He clearly opts for the second alternative. Thus, Beckermann combines our two-level talk of both arguments and considerations on the one hand, and neuronal processes on the other, in suggesting that our decisions rest on processes that can be influenced by arguments. His position might be captured by the following graphic, see Fig. 15.1.

It is assumed that the processes underlying our decisions are neurophysiological ones; therefore each act of deliberation (some m_i) that moves into a decision (some m_i) rests on some corresponding physical process or state (the p_i). However,

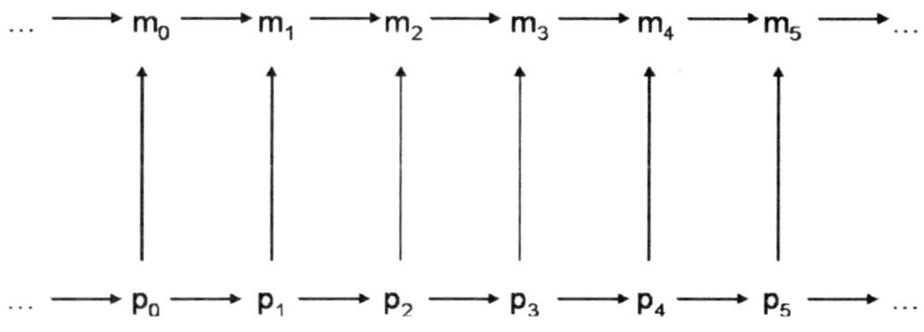

Figure 15.1. Beckermann's compatibilism.

resting on physiological processes does not make decisions less sensitive to arguments, according to Beckermann; it only shows that the brain mechanisms seem to be such that they can take care of semantics, critical thinking, and rational considerations. What is left open, however, is the problem of mental causation. Beckermann remains silent about how to explain the mechanisms that underlie what we perceive phenomenologically as conscious considerations of arguments.

In opposition to Beckermann's compatibilism are the incompatibilist positions as held by Wolf Singer and Geert Keil, which, of course, are also in opposition to each other. Singer claims that our distinction between free and non-free decisions, which seems to rest essentially on the idea that we have conscious access to the motives that seemingly lead to free decisions and actions, has no basis if looked at from the perspective of the brain. There are, of course, behaviours we perform without consciously having thought of them, and there are other behaviours where we consciously weigh arguments and deliberate upon them to arrive at certain decisions. But according to Singer there is no real difference, from the perspective of the brain, that allows us fundamentally to distinguish between these two types of behaviours with all the implied (social and evaluative) consequences. From the perspective of the brain, Singer cannot see a decisive difference between processes that are accompanied with conscious access to the motives we rely upon in supposed free decisions, and those which are not.

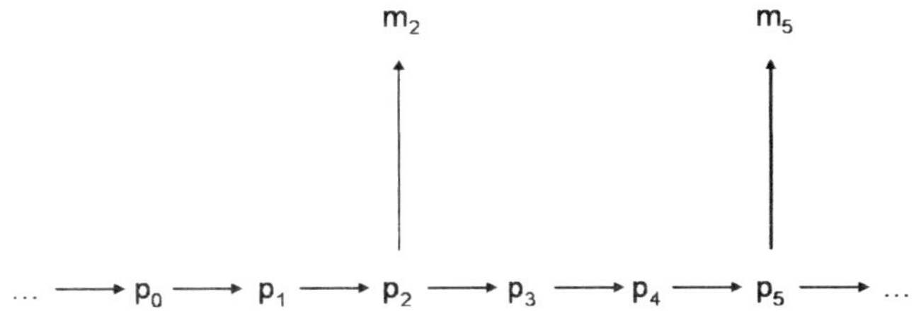

Figure 15.2. Singer's hard determinism.

According to Singer, one total brain state (some p_i) determines the next one (some other p_j); some of these processes reach the conscious level (some m_i), most don't (see Fig. 15.2). Therefore, from the perspective of the brain there would not be a real difference between the deliberating processes and the more automatic, non-deliberating processes. Since the 'solutions' which are found on the brain level do not differ, in principle, in cases where conscious deliberations are involved, from cases where no conscious consideration is involved, Singer concludes that the distinction we draw on the basis of our own experiences (from a 'participant's perspective') has no resilient factual basis. Hence, Singer opts for free will illusionism or free will eliminativism.

It is, however, not at all settled how much weight we should give to bottom-up-type arguments delivered by Singer. In order to support the claim that there is no difference in the brain processes in question, Singer only refers to 'general principles' that govern the (underlying) brain processes. At issue are, however, remarkably different macro-behaviours. Let me illustrate the situation by an example taken from mineralogy. If we refer to 'general principles' only, the same general quantum principles are at work in diamond and graphite, and still there could not be a greater difference between diamond and graphite on their macro-level, namely the difference in their hardness. In that case we would not claim that the well-established difference in their macro-properties might be illusionary, but ask for an explanation that *accounts for* the different dispositions of graphite and diamond with respect to their microstructure. Similarly, the task for the neuroscientist might be better to provide us with an account of the different kinds of human behaviours. Here, too, the issue should not be to claim that the established distinction between candidates and non-candidates for free will is spurious. To claim this, Singer would need arguments that show that there is *no counterpart at all* on the neurophysiological side that could account for the difference between so-called free and non-free decisions and actions we regularly notice on the personal level.

The arguments Singer really provides show that *conscious* processes play a less significant role in human behaviour than the common sense view and also many philosophers have assumed. There is, in fact, neuroscientific evidence that our decisions rest heavily on processes that occur in brain areas that are not open to conscious access, e.g. the influences from the limbic system or from glands that release neurotransmitters. Thus, it depends on whom Singer really wants to argue against. Is it someone who thinks that our conscious decisions rely *only* on what we can consciously deliberate about? Or, is it someone who thinks that besides the factors that, unbeknown to us, play an important role in shaping our considerations and behaviour, conscious deliberations have also an influence on what we will decide? It seems that Singer mainly argues against the first option.

In contrast to Wolf Singer, Geert Keil turns tables; he takes the manifest image we have of ourselves as a starting point, and this image depicts us as free agents who sometimes 'really' decide between different alternatives. This does not mean that Keil tries to take an unscientific path, not at all. Rather, he points

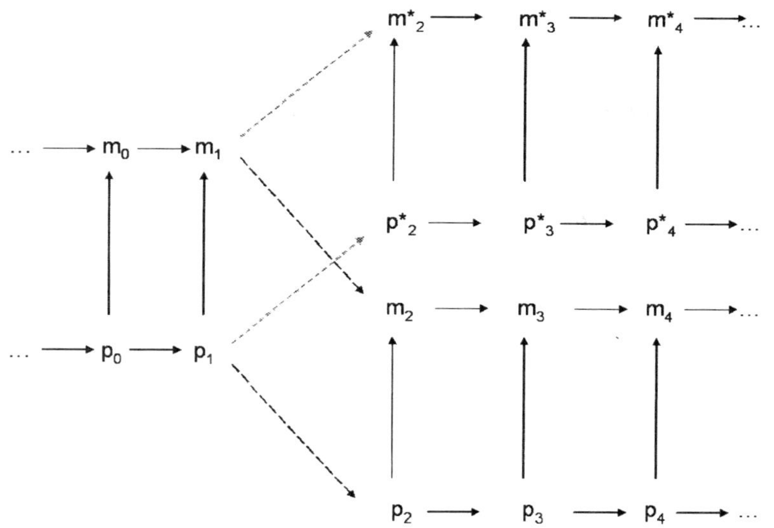

Figure 15.3. Keil's libertarianism.

to consequences we would have to accept on the neurophysiological level: if there are non-deterministic processes on the personal level, then there should exist corresponding non-deterministic processes on the subpersonal level. We can illustrate his ideas with the above graphic (see Fig. 15.3).

What we see here is a branching in the possible courses of subsequent events right after the one referred to by the pair (m_1, p_1). The agent is depicted, according to the libertarian, in the situation of decision, in which he can choose at least among some (here two) ways to carry on. Keil stresses that from a diachronic perspective the free agent has alternatives to decide and act: he can proceed either with what is called here (m_2, p_2) or with (m^*_2, p^*_2). However, Keil also stresses that even free agents do not have alternatives from a synchronic perspective. No agent can proceed with, say, what might be referred to by (m_2, p^*_2) or by (m^*_2, p_2), i.e., no agent is free to decide or act differently from what he does, given the neurophysiological processes (the neuronal correlates of these decisions and actions). Keil illustrates his point by means of an example from chess. A special movement of the king and the rook determines, for example, that this movement is a castling. But this movement, which cannot be not a castling, does not determine one of the next moves (Keil 2007: 287). Accordingly, Keil seems to acknowledge that mental events are physically realized, while stressing that this realization relationship has nothing to do with diachronic determinism, and as such is not limiting our freedom:

In fact, the can-do-otherwise of the libertarian is not a can-do-otherwise vis-à-vis an actual physical event, this would be absurd, but rather it is a can-do-otherwise by a given previous history. . . . Why should the fact that mental processes are physically realized, i.e. that something is happening in my brain while I am deliberating or wanting, endanger my freedom? Those who see a contradiction herein base their freedom indeed on dualism (Keil 2007: 287–8; my translation).

Interestingly enough, although the authors discussed thus far differ considerably with respect to their answers to the free will problem, they all agree that there is no variance in synchronous respects: neither Geert Keil, nor Ansgar Beckermann, nor Wolf Singer thinks that the very same brain event could go along with different decisions.

To explore the connection between synchronic and diachronic versions of determination a bit further, let us pause here and take a look at the taxonomies provided by emergentism. I will distinguish between *weak* and *strong* forms of *synchronic* emergence. Let me start with the weak notion.[5]

WEAK EMERGENCE

Weak emergentism specifies the minimal criteria for emergent properties. Its basic features—the thesis of *physical monism*, the thesis of *systemic* (or *collective*) *properties*, and the thesis of *synchronic determination*—are compatible with reductionist approaches. More ambitious theories of emergence have a common base in weak emergentism; they can be developed from it by adding further theses.

The first thesis of current theories of emergence concerns the *nature of systems* that have emergent properties. It says that the bearers of emergent features consist of physical entities only. According to it, all possible candidates for emergent properties such as, for example, being alive, feeling lonely, acting or deciding freely, are instantiated only by material systems with a sufficiently complex microstructure. The thesis of *physical monism* denies that there are any supernatural components such as an *entelechy* or a *res cogitans* responsible for a system's having emergent properties. Particularly, this means that living or cognizing systems consist of the same basic parts as lifeless or mindless objects of nature.

Physical Monism. Entities existing or coming into being in the universe consist solely of physical constituents. Properties, dispositions, behaviours, or structures classified as emergent are instantiated by systems consisting exclusively of physical entities.

Note that all authors discussed above would subscribe to the thesis of physical monism. Only some of the positions Wolf Singer likes to attack would not

[5] More elaborated developments and explications of the variants of emergentism can be found in Stephan (2007[1999]: part I, 1998), or more recently in (2006).

subscribe to physical monism, but this is also neither true of compatibilists such as Beckermann (and Walter) nor of libertarians such as Keil (or Kane).

While the first thesis places emergent properties and structures within the framework of a physicalistic naturalism, the second thesis—the thesis of *systemic properties*—delimits the types of properties that are possible candidates for emergents. It is based on the idea that the general properties of a complex system fall into two classes:[6] those that some of the systems' parts also have and those that none of the systems' parts has. Examples of the first kind are properties such as being extended and having a velocity. Examples of properties of the second kind are breathing, reproducing, having a sensation of an itch, or deciding to retire at fifty-nine. These properties are called systemic or collective properties.

Systemic properties. Emergent properties are systemic properties. A property of a system is systemic if and only if the system possesses it but no part of the system possesses it.

Both artificial and natural systems with systemic properties exist. Those who would deny their existence would have to claim that all of a system's properties are instantiated already by some of the system's parts. Countless examples refute such a claim. It is evident that human behaviours that are candidates for free actions and decisions are systemic properties (to behave in such a way)—no proper part of a human being decides or acts. When neuroscientists say, and they like saying it, that the 'brain decides' or the 'brain finds a solution' it is 'as if'-talk: brains are no agents, they do not decide, reason or look for and eventually find a solution. They are highly complex systems, in which thousands of neurophysiological and neurochemical processes are running simultaneously.

While the first thesis restricts the type of parts out of which systems having emergent properties may be built, and the second thesis characterizes in more detail the type of properties that might be emergent, the third thesis specifies the type of relationship that holds between a system's microstructure and its emergent properties as a relationship of synchronic determination:[7]

Synchronic determination. A system's properties and its dispositions (to behave in a certain way) nomologically depend on its microstructure. There can be no difference in a system's systemic properties without some difference in the properties of its parts or in the arrangement of its parts.

Anyone who denies the thesis of synchronic determination either has to admit properties of a system that are not bound to the properties and arrangement of its parts, or to suppose that some other, in this case non-natural, factor is responsible for the different dispositions of systems that are identical in their microstructure. Both options seem implausible.

[6] General properties are determinables, i.e., properties of a general type, such as having a weight, or being liquid; they are not determinates, i.e., specific properties, such as having a weight of 154.5 pounds or being liquid at a temperature of 1200 °C.

[7] Some colleagues use the label 'micro-determination' to refer to the idea I have dubbed the thesis of synchronic determination (cf., e.g., Bedau 2002: 14).

Within the last section we could see that all authors discussed thus far probably would subscribe to the thesis of synchronic determination (or micro-determination), although I am not certain whether Geert Keil would like to sign this principle explicitly. His example from chess is, if at all, a different case of determination than the one that is claimed to exist between neuronal mechanisms and an agent's deliberations and decisions: the chess example seems to refer to nothing but the fixing of the name 'castling'. In contrast, the synchronous determination that is thought to relate certain brain processes on the one hand to certain considerations and decisions on the other, does not seem to be just a type of name-giving: the language games of both considerations and decisions on a personal level and brain processes on a subpersonal level are already well established. The principle of synchronic determination is a claim of how the processes referred to by these different language games relate to each other. But nevertheless, Keil conceded that the can-do-otherwise-principle of the libertarian is not a can-do-otherwise vis-à-vis actually given brain processes or events (Keil 2007: 287). Thus, being able to do something different from what one really does (or having been able to do so) does not amount, according to the libertarian, to doing something else while keeping the neurophysiological correlates constant.

But isn't there a certain tension for the free will friend when learning that even one's free decisions are synchronically determined by corresponding brain processes? How could an agent be treated as the author of his or her decisions and doings if what he or she does is micro-determined by brain processes? Ansgar Beckermann explicitly takes up these worries when he says: 'If in biological creatures all decisions are based on neuronal processes—and it is exactly this that neuroscience seems to show—how then should these processes be influenced by rational arguments and considerations? It seems to me that this objection does not go through. [. . .] There is only the following alternative: Either it is the case that not all decisions are based upon neuronal processes, or it is the case that neuronal processes exist that can be influenced by considerations and arguments. [. . .] The fact that something is a neuronal process, does not exclude that the *very same* process is a process of consideration' (2006: 301–2; my translation).

What we encounter in the apparent tension is an old problem for the philosophy of mind, although not often discussed within the debate about the problem of free will: the so-called 'qua'-problem of mental causation: How could we cause what we do by a conscious decision qua it's being a conscious decision? In Kane's comprehensive 638-page *Handbook of Free Will* (2002) we find, for example, exactly two sentences and one footnote on 'mental causation'.[8] Moreover, those philosophers who have contributed to the debate on free will

[8] In his own contribution Kane says: 'The preceding account of libertarian free will [. . .] does appeal to a notion of *mental causation*. It assumes that choices and actions can be caused or produced by efforts, deliberations, beliefs, desires, intentions, and other reasons or motives of the agent' (2002: 426).

have also been silent about the synchronous relationship of the personal level to the subpersonal level. But it is exactly this relationship that has come into focus through the contributions by the neuroscientists. Within the philosophy of mind, this relationship has been discussed particularly by reductionists and emergentists in terms of reductive explanation.

Let us now see how the free will problem looks if we move towards more ambitious emergentist positions.

STRONG (SYNCHRONIC) EMERGENCE

Synchronic theories of emergence are of great importance for the discussion of the psychophysical problem, particularly to the formulation and analysis of non-reductive, but nevertheless substantival monist positions in the philosophy of mind. Key questions concern the relation between mental and physical properties, for example whether mental properties such as the having of intentional or phenomenal states can be explanatorily reduced to a physical basis. If we answer 'no' we hold a strong emergentist position; we claim mental properties to be irreducible, and thus to be synchronically (i.e., in a strong sense) emergent.[9] But what does it mean for a property not to be reductively explainable?

Generally, we ask for reductive explanations when we want to understand *why* and *how* a certain entity instantiates a certain property, in fact a property that is only attributed to the system as a whole. The aim of each reductive explanation is to explain (or predict) a system's having its dispositions and properties solely by reference to its components, their properties, arrangement, and interactions. For a reductive explanation to be successful several conditions must be met:

(1) the property to be reduced must be re-construed (or construed) in terms of its causal (or functional) role;
(2) the specified functional role must result from the properties and behaviours of the system's parts and their mutual interactions;
(3) the behaviour of the system's parts must result from the behaviour they show in simpler systems than the system in question.[10]

[9] In the heyday of emergentism during the 1920s C. D. Broad was the first who explicitly characterized a synchronic theory of emergence; he did so within his discussion of mechanism as a metaphysical theory. There he claimed for various chemical, biological, and psychical properties that they are 'emergent properties' due to not being mechanically explainable (Broad 1925: ch. 2). Given the successful developments of the natural sciences many chemical and biological properties initially thought to be emergent were discarded from being good candidates for synchronic emergence (see McLaughlin 1992).

[10] Among others, Kim and Levine do without the last condition. Since mental properties are considered as remarkably intractable, those interested in giving reductive explanations of mental phenomena, better provide *everything* available that could count as a reduction base: the complete physical system including its components, their arrangement, properties, and interactions, as well

What we are looking for, then, are functional characterizations of the properties to be reduced. Usually we refer to these properties by concepts that classify properties at the system level, where specific patterns bring the instantiation of a property to our notice. To propose such conceptual preparations has the aim to allow conceptual transitions from the level of components to the level of systems. If, however, the conceptual 'priming procedure' fails, the corresponding reductive explanation fails, too.

Reductive explanations can be asked for in two opposite directions. Given that we already know (or assume) that some system *S* has a systemic property *P*, the task is to provide a reductive explanation for *P*. If, on the other hand, a system *S* is still being developed, as is often true with artefacts, we might at first hand only know its microstructure MS(*S*). The task, then, is to verify theoretically or to forecast whether or not *S* has some (desired or unwished) systemic property *P*.

If reductive explanations for some systemic property fail, in principle, the property targeted for reduction is *irreducible*, and thereby it is *synchronically* (i.e., in a strong sense) *emergent*.

Based on the fact that the three conditions for reductive explanations are independent of each other, there exist also three different ways in which systemic properties may be *irreducible*:

Irreducibility. A systemic property is irreducible if (1) it is not functionally construable or re-construable; if (2) the specified functional role does not result from the properties and behaviours of the system's parts and their mutual interactions; or if (3) the specific behaviour of the system's components, which micro-determines the systemic property (or behaviour), does not result from the behaviour they show in simpler systems than the one in question.

We can illustrate the situation with a graphic that takes up some ideas from one provided by Boogerd et al. (2005, see Fig. 15.4).[11]

P_R is the systemic property to be reductively explained, what is achieved only if P_R can be re-constructed or constructed via its causal role (by satisfying the conceptual condition). R(A,B,C) is the microstructure of the system that instantiates property P_R. A, B and C are the parts making up the system. S_1(A,B), S_2(A,C), and S_3(B,C) are simpler wholes than R(A,B,C) that are composed of parts that also make up R(A,B,C). The vertical arrow captures condition (2); the

as all environmental properties relevant for the system's behaviour. And this is the reason why in the philosophy of mind the third condition for emergence does not receive much attention: it cuts down the reduction base. Both Kim and Levine note, however, that in a first step we have to work the concept of the concerned property 'into shape' for reduction. Kim calls this the 'priming procedure' in which we must construe, or re-construe, the property to be reduced relationally or extrinsically (Kim 1998: 98).

[11] An in-depth analysis of these relationships is given in Boogerd et al. (2005). There we also refer to Broad's detailed considerations as presented in his article 'Mechanical Explanation and its Alternatives' (1919).

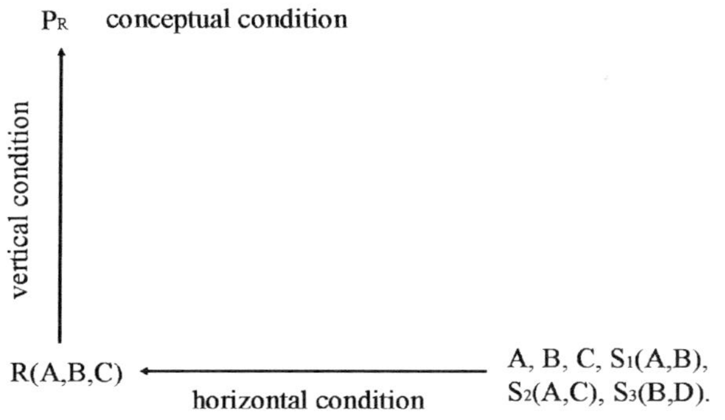

Figure 15.4. Three conditions of irreducibility.

horizontal arrow captures condition (3), which usually is not considered in the philosophy of mind.

It is very tricky to get a grasp on the problem of free will from the perspective of strong (synchronous) emergentism. In contrast to the discussion from the perspective of weak emergence, it will not suffice to consider simply the so-called 'candidates' for free decisions and actions. Here's why: The candidates are *manifest* human behaviours: decisions or actions. Suppose we succeed in reductively explaining them by finding the neurophysiological correlates that play exactly the causal role of the candidate-behaviours. Have we then also succeeded in reductively explaining that they are truly *free* decisions or actions? No. Note that there is no agreement about whether any candidate-behaviour truly has the property of being a *free* decision (or *free* action). Accordingly, there is no agreement about how we should construct or re-construct the causal role of the alleged property of being a free decision or action. Since we have no agreed starting point, we have to discuss the different positions concerning the problem of free will separately. The causal role we are looking for depends, of course, considerably on what somebody takes to be essential for being a free decision or action.

Let us start with the Libertarian position as characterized by Geert Keil. According to him, in the moment of making a free decision we have true options: as agents we have the capacity to decide to do this or that. Thus, besides the manifest action, which is presumed to be a free one, the causal role should comprise one or more alternative behavioural outputs. Now compare the presumed disposition of free agents to do this or that with dispositions of a sugar cube. The disposition of a sugar cube to dissolve in a cup of coffee, say, is synchronically determined by its microstructure and it also can be reductively explained by reference to its microstructure. Whether or not it will dissolve depends, however, on the circumstances. Analogously, it is reductively explainable that a sugar cube that in fact dissolved in a cup of tea had a structure that would have stayed stable if it had remained in a dry environment. For the libertarian, however, the situation is completely different. He does not say that under (maybe slightly) different circumstances human agents are capable of doing something different from what they in fact do; no, the libertarian claims that under *exactly the same* circumstances agents are sometimes capable of doing different things. For the sugar cube this would mean that sometimes it is able to dissolve or not dissolve under the very same circumstances, which seems quite strange for a sugar cube.

But it is also difficult to conceive of this in the case of human actions and decisions. Even in situations in which we feel as free deciders, it is not at all certain whether we really have the ability to do alternative things under the *very same* circumstances. It might be nothing but an illusion. If, on the other hand, we in fact have the ability to do this or that under the very same circumstances, it would be an ability that escapes reductive explanations: How could a (brain) *mechanism* admit two (or more) different outcomes under the very same conditions? We do not know of any indeterministic mechanism. So we get: Either there exist libertarian free decisions and actions or not. If they exist they are strongly emergent properties, and hence it would be impossible to explain them reductively.

Now, consider the 'Singer'-type hard determinist. For him, neither mental causation nor genuine free decisions do really exist. Those behaviours that are thought to be fair candidates for free actions and decisions are synchronously determined by neurophysiological processes and should be explainable reductively. In that respect, the candidates for free actions and decisions are not strongly emergent properties. What might remain a problem for the hard determinist is to explain reductively the subjective feelings agents have when they experience themselves *as if* performing free decisions. To reductively explain this kind of experience is, however, the problem of phenomenal qualities. Phenomenal experiences are also truly good candidates for emergent properties (cf. Stephan 2002 and 2004). And in that respect, the subjective feeling of being a free agent might be a strongly emergent property, too.

Then, what about the compatibilist? For him, what reductive explanations should provide us with is an account of *how* the processes that underlie our candidate actions and decisions are sensitive to arguments and rational deliberations. Thus, the issue is not to provide a mechanism for two alternative world runs, but for how we could *literally* be said to weigh arguments in a rational way in order to draw informed decisions when these behaviours are synchronically determined by brain processes. The compatibilist has no problem in accepting that the same circumstances always lead to the same results. He only wants to make sure that we as free agents are open to rational considerations and further arguments. In this case, the problem of free will transforms into the problem of mental causation.

The traditional answer is to opt for the identity theory in one way or other.

FINAL REMARKS

Neuroscientific findings show that the human brain is an extremely complex system. If we think of the billions of connections between neurons, of the multiple and parallel processing of incoming data, and the mutual connections between cortical and sub-cortical areas, we might start to grasp how intractable the *subpersonal* processes are. No single state of an organism that might correspond to a candidate for free decision will occur again; and if it could, we would not be able to notice that it had. There is no chance to know. Therefore, there is no, and there will be no, experiment that could help to decide between libertarians and hard determinists.

Thus, although both compatibilists and hard determinists might reach the conclusion that 'decision procedures' can be reductively explained in principle, this endeavour might be completely utopian: if, as it is in dynamical systems, small changes can make a huge difference—we will never be able to foretell from the brain's bottom-up perspective what decision we will eventually come up with.

The question, thus, is how much weight we are ready to give to the phenomenological perspective (or manifest image), on the one hand, and how much weight we are ready to give to the neuroscientific perspective, on the other. We experience ourselves as the authors of our decisions since we have the impression that it is us who reason, that it is us who consider pros and cons, and that it is us who decide and act on the basis of weighing reasons. Shall we go with this phenomenology as does Geert Keil, or shall we go with the neuroscientist's generalizations as suggested by Wolf Singer, where we can't be certain that his checks are all covered? I am not sure where the compatibilist ends up; he seems to try both: marry the manifest with the scientific image of ourselves. But he, too, or so it seems, has to give up essential ideas about himself . . .

What can be said is that there are more factors contributing to what we think, decide, and do than we might have thought before, than we can consciously be aware of and eventually influence. This, however, does not solve the problem of free will. It will remain unsolved. Under these circumstances, it looks a bit strange to me how far-reaching conclusions are so easily drawn by some of our leading neuroscientists . . .

REFERENCES

Beckermann, A. 2005. 'Free Will in a Natural Order of the World'. In C. Nimtz and A. Beckermann (eds), *Philosophy—Science—Scientific Philosophy. Main Lectures and Colloquia of GAP.5*. Paderborn: Mentis, 111–26.

Beckermann, A. 2006. 'Neuronale Determiniertheit und Freiheit'. In K. Köchy and D. Stederoth (eds), *Willensfreiheit als interdisziplinäres Problem*. Freiburg and Munich: Alber, 289–304.

Bedau, M. 2002. 'Downward Causation and the Autonomy of Weak Emergence'. *Principia* 6: 5–50.

Boogerd, F. C., Bruggeman, F. J., Richardson, R. C., Stephan, A., and Westerhoff, H. V. 2005. 'Emergence and its Place in Nature: a Case Study of Biochemical Networks'. *Synthese* 145: 131–64.

Broad, C. D. 1919. 'Mechanical Explanation and its Alternatives'. *Proceedings of the Aristotelian Society* 19: 86–124.

—— 1925. *The Mind and its Place in Nature*. London: Kegan Paul.

Frankfurt, H. G. 1969. 'Alternate Possibilities and Moral Responsibility'. *The Journal of Philosophy* 66: 829–39.

Kane, R. 2002. *The Oxford Handbook of Free Will*. Oxford: Oxford University Press.

—— 2002. 'Some Neglected Pathways in the Free Will Labyrinth'. In R. Kane (ed.), *The Oxford Handbook of Free Will*. Oxford: Oxford University Press, 406–37.

Keil, G. 2007. 'Mythen über die libertarische Freiheitsauffassung'. In D. Ganten, V. Gerhardt, and J. Nida-Rümelin (eds), *Die Naturgeschichte der Freiheit*. Berlin, New York: de Gruyter, 281-305.

Kim, J. 1998. *Mind in a Physical World*. Cambridge, MA: MIT Press.

McLaughlin, B. 1992. 'The Rise and Fall of British Emergentism'. In A. Beckermann, H. Flohr, and J. Kim (eds) *Emergence or Reduction? Essays on the Prospects of Nonreductive Physicalism*. Berlin, New York: de Gruyter, 49–93.

Moore, G. E. 1912. *Ethics*. London: Williams & Norgate.

O'Connor, T. 2000. *Persons & Causes: the Metaphysics of Free Will*. Oxford: Oxford University Press.

—— 2002. 'Libertarian Views: Dualist and Agent-Causal Theories'. In R. Kane (ed.), *The Oxford Handbook of Free Will*. Oxford: Oxford University Press, 337–55.

Schopenhauer, A. (2004 [1839]). 'Preisschrift über die Freiheit des Willens'. *Sämtliche Werke III. Kleinere Schriften*. Textkritisch bearbeitet und herausgegeben von W. Frhr. v. Löhneysen. Darmstadt: Wissenschaftliche Buchgesellschaft, 519–627.

Singer, W. 2005. 'Selbsterfahrung und neurobiologische Fremdbeschreibung. Zwei konfliktträchtige Erkenntnisquellen'. In H. Schmidinger and C. Sedmak (eds), *Der Mensch—ein freies Wesen?* Darmstadt: Wissenschaftliche Buchgesellschaft, 135–60.

—— [im Gespräch mit Markus C. Schulte v. Drach] 2006. 'Hirnforschung und Philosophie: "Der freie Wille ist nur ein gutes Gefühl".' sueddeutsche.de—Ressort: Wissen. <http://www.sueddeutsche.de/wissen/artikel/113/74039/>.

Stephan, A. 1998. 'Varieties of Emergence in Artificial and Natural Systems'. *Zeitschrift für Naturforschung C. A Journal of Biosciences* 53: 639–56.

—— (2007 [1999]). *Emergenz. Von der Unvorhersagbarkeit zur Selbstorganisation*. 2nd reprint: Paderborn: mentis.

—— 2002. 'Phänomenaler Pessimismus'. In M. Pauen and A. Stephan (eds), *Phänomenales Bewusstsein—Rückkehr zur Identitätstheorie?* Paderborn: mentis, 342–63.

—— 2004. 'Phänomenale Eigenschaften, Phänomenale Begriffe und die Grenzen Reduktiver Erklärung', in W. Hogrebe with J. Bromand (eds.) *Grenzen und*

Grenzüberschreitungen. XIX. Deutscher Kongress für Philosophie. Vorträge und Kolloquien. Berlin: Akademie Verlag, 404–16.

—— 2006. 'The Dual Role of "Emergence" in the Philosophy of Mind and in Cognitive Science'. *Synthese* 151: 485–98.

Walter, H. 2004. 'Neurophilosophy of Moral Responsibility: The Case for Revisionist Compatibilism'. *Philosophical Topics* 32.1–2: 477–503.

16

Strong Emergence and Freedom: Comments on Stephan

Max Kistler

In his chapter 'An Emergentist's Perspective on the Problem of Free Will', Achim Stephan asks whether new light can be shed on the free will problem by considering it from the viewpoint of different conceptions of emergence. The idea is certainly promising: part of what makes free will puzzling is the difficulty of understanding the relations between processes of deliberation and decision taking place at the psychological level and the neural processes underlying them in the brain. As Stephan reminds us, traditional answers to the free will problem can be sorted into three categories: determinism (according to which our conviction of being free is illusionary), compatibilism (according to which we can be free although all events and processes in our body are determined by earlier events and processes), and libertarianism (according to which our decisions are not determined by any earlier states of ourselves, be they psychological or physical). The hope is that the conceptual difficulties that bedevil any or all of these approaches can be partly or wholly solved in the framework of one or other conception of how psychological properties and processes *emerge from* brain properties and processes. Stephan himself takes the result of his inquiry to be negative. However, I shall at the end suggest a way in which emergence may help us make sense of freedom in a compatibilist way. Let me first make some remarks on Stephan's theory of emergence. According to Stephan (this volume), there are two independent reasons for considering a property as strongly emergent, in other words, more than weakly emergent. They correspond to two fundamental types of strong emergence, synchronic and diachronic.

The first way in which a property P can be strongly emergent is by being weakly emergent and *synchronically irreducible*: the fact that object o is P *at time t* cannot be deduced from the properties the object's parts possess *at t* together with their mutual relations *at t*.

The second way is to be weakly emergent *and (diachronically) unpredictable*: the fact that o is P *at t* cannot be deduced from the micro- and macro-properties o and its parts possess *at some earlier time t**.

1. SYNCHRONIC EMERGENCE AND IRREDUCIBILITY

Let us first look at the concept of strong emergence in terms of synchronic irreducibility. The problem for this concept is to reconcile the irreducibility of emergent properties with the hypothesis that they are synchronically *determined* by the system's parts. Synchronic determination, in the sense of nomological dependence of a systemic property on the properties of the system's parts and their interactions, is part of the concept of weak emergence: 'A system's properties and dispositions to behave depend nomologically on its micro-structure, that is to say, on its parts' properties and their arrangement' (Stephan 1999: 50–1).

Stephan takes synchronic determination to be compatible with 'synchronic irreducibility'.[1] This is indeed part of the doctrine of classical British emergentism.[2] However, today many doubt that there are any absolutely irreducible properties.[3] This change in mind is in large part due to quantum mechanics' achievement of reductively explaining chemical properties, which had been taken by emergentists such as Broad to be paradigmatic cases of properties that are irreducible although synchronically determined. It now seems that, if one takes it for granted that a given macroproperty is objectively synchronically determined by underlying (physical) microproperties, then it is a mere question of time when that determination relation will be discovered by scientific means. When we do not know how to reduce a given systemic property, this is not due to any objective feature of that property but only to our present ignorance and the imperfection of today's theories. Given any systemic property, there seems to be no reason to deny the possibility, at least in principle, that science eventually discovers its synchronic determination relation. That discovery provides scientists with the means of producing a reductive explanation of that property, in terms of the properties of the system's parts and their interactions.

One of Stephan's most interesting contributions to the analysis of emergence is his distinction between two steps of synchronic determination. This distinction might help us find out whether there can, after all, be good reasons to expect there to be absolutely (and not only provisionally) irreducible though synchronically determined properties.

Figure 16.1 (Boogerd et al. 2005) shows two ways a property can be emergent, corresponding to two steps of synchronic determination. For each step of

[1] See Stephan 1997. [2] See, e.g., Broad 2000 [1925]. [3] See McLaughlin 1992.

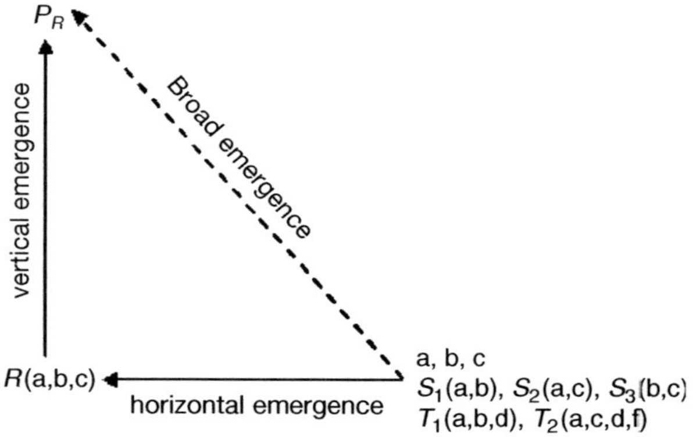

Figure 16.1. a, b, and c are the parts making up the system. S_1(a,b), S_2(a,c), and S_3(b,c) are simpler, binary, wholes including these parts. T_1(a,b,d) is a system with the same number of parts, and T_2(a,c,d,f) is a system with more parts than R(a,b,c). P_R is a systemic property. The diagonal arrow represents Broad's idea of emergence. The horizontal and vertical arrows capture the two conditions implicit in Broad.

determination it seems conceivable that it is impossible to reductively explain it. Thus it appears that there are two ways in which a systemic property P_R can be emergent, in the sense of being synchronically irreducible to properties and relations of systems' parts.

Systemic property P_R is synchronically irreducible either (1) because it is impossible to deduce the state R(a,b,c) of the interacting whole (where R(a,b,c) is taken to give rise to P_R) from the properties the parts *a*, *b*, *c* have when they are isolated, or from the properties of other systems (represented by S_1(a,b), S_2(a,b), . . .) which contain some of these parts. Boogerd et al. (2005) argue that this is the case for complex biochemical systems. In this case, P_R is a case of what they call 'horizontal emergence'. Or (2) P_R is synchronically irreducible because it is not 'behaviorally analyzable' (Stephan 1999: 52) in terms of R(a,b,c), which makes it a case of 'vertical emergence'.[4]

Let us look a little closer at the concepts of 'horizontal' and 'vertical' emergence. Take horizontal emergence first. Stephan takes the horizontal determination relation between the properties the parts have in isolation or in other circumstances (i.e. in systems S_1(a,b), S_2(a,b), . . .), and the state R(a, b, c) of the whole under consideration to be a case of 'synchronous emergence'. This raises the following difficulty. Following Humphreys (1997), we may call 'fusion' the process during which parts *a*, *b* and *c* come into interaction, and then eventually come to form a system. The problem is this: the horizontal relation cannot be synchronous because fusion takes time. Let me explain this in a little more detail.

[4] The diagonal line in Figure 16.1 represents Broad's concept of emergence, which conflates, according to Boogerd at al. (2005), horizontal and vertical emergence.

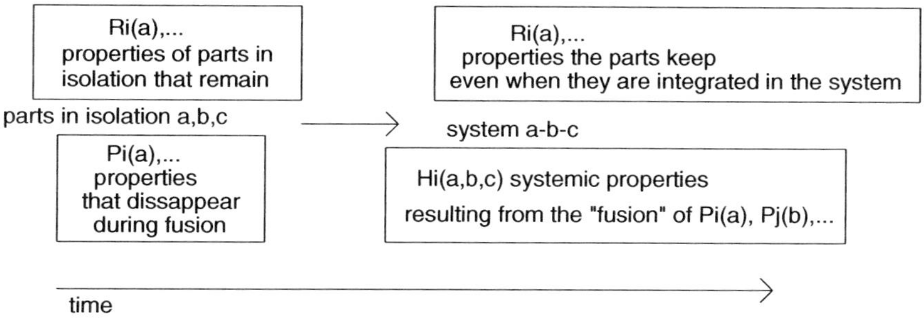

Figure 16.2. Horizontal emergence.

Under certain conditions, a stable global structure R(a,b,c) emerges ('horizontally') in such a way that, when it has emerged, *some* of the properties necessary for the interaction between the parts *a*, *b* and *c* are *lost*. As an example, take the formation of molecule electron orbits. They arise from the fusion of atomic electron orbits that *disappear* during the formation of the covalent chemical bond.

Figure 16.2 distinguishes properties P_i (a), . . . of parts that *disappear* in the process leading to an emergent property H_i (a,b,c) from properties R_i (a), . . . that remain. It is clear that those properties P_i that disappear before the emergent properties H_i comes into existence *do at no time coexist* with these emergent properties. Therefore, the relation between the properties P_i (a), . . . of the parts at the beginning of the fusion and the emergent property H_i (a,b,c) that exists at the end of the fusion process *is not synchronous*.

Let us now look at Stephan's second step of synchronic determination, and the 'vertical emergence' that arises from the impossibility to explain it reductively. He shows that there are really two ways for a systemic property P_R to be 'vertically emergent' (i.e. not to be 'behaviorally analyzable' in terms of the global state R(a,b,c)). They result from the impossibility to carry out one of the following two steps of reductive explanation.

In step (i), a systemic property is 'functionalized' in Levine's (1993) and Kim's (1998) sense, i.e. characterized by its functional role. In step (ii) it is then shown that 'the specified functional role [. . .] result[s] from the properties and behaviors of the system's parts and their mutual interactions' (Stephan, this volume: 233).

I suggest analysing step (ii) further in two substeps, so that there are more than two ways in which a property can be irreducible that make it 'vertically' emergent.[5] The first of these substeps consists in:

[5] I have argued elsewhere that neglecting this distinction creates problems for Chalmers and Jackson's (see Kistler 2005a) and Kim's (see Kistler 2005b) accounts of reductive explanation.

(iia) finding the property filling the role identified in step (i). This role-filling property is a systemic property of the interacting system, just as the role itself is played by the whole system. The second substep then consists in:

(iib) showing how properties of parts of the system and interactions between those parts bring the systemic role-filler property identified in step (iia) into being. One especially important way of doing this is by discovering a mechanism.[6]

Thus, there are really three, not just two, steps in vertical reduction.

(1) In a first step, a systemic property is functionalized by showing that the predicate expressing it does not directly denote a first-order property, but rather a role: it is equivalent to an existential quantification over some property or other that has certain causes and effects among system level properties.
(2) In a second step, the first-order property that fills the role specified in (1) is identified. This role-filler is a system level property, i.e. it belongs to the system as a whole.
(3) The role-filler property is analysed in terms of a mechanism.

Here is an example. 'Haemoglobin', though it appears to denote a substance (and 'being haemoglobin' a first-order property), really denotes the role F of being some substance or other transporting oxygen in mammal blood. What fills that role are those chemical properties M of different haemoglobin molecules that have (among others) the causal power of binding O_2 molecules. M's power of binding O_2 can then be reductively explained by interactions among the haemoglobin molecules' parts p_1, p_2, p_3. . . .[7] The mechanistic explanation of M's power of binding O_2 shows how the amino acids composing haemoglobin molecules interact ($R(p_1, p_2, \ldots)$) in such a way that the interaction gives rise to the conformation of the molecule which then explains M's binding power.

Terminology tends to obscure the difference between (1) the relation between role (F) and role-filler (M) and (2) the relation between a systemic property (M) and the elements p_1, p_2, p_3 . . . and organization $R(p_1, p_2, \ldots)$ of the mechanism: both relations are sometimes called 'realization', the discovery of both can be called 'reduction', and both are said to give rise to 'multiple realization/reduction'.

There appears to be no reason to expect any of these steps of determination to elude scientific discovery for principled reasons. There is no reason to expect any role F to be irreducible in principle, either because it is impossible in principle to find a role-filler property M or because it is impossible in principle to find a mechanistic explanation of M in terms of the system's parts and their interactions.

[6] See Machamer et al. 2000; Craver 2001; Bechtel 2006.

[7] See Rosenberg (1985: ch. 4).

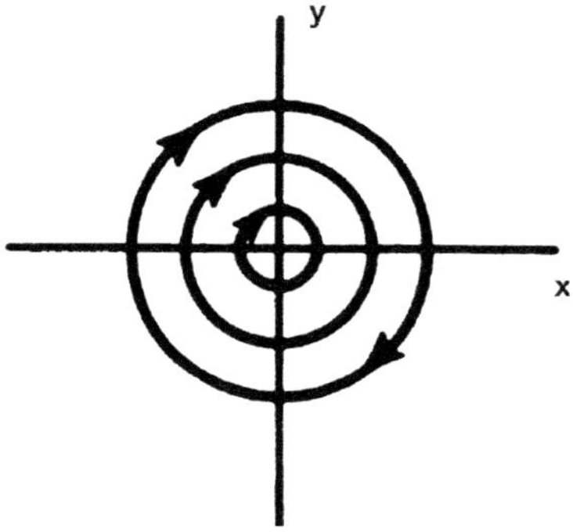

Figure 16.3. Phase space diagram of undamped harmonic oscillator (taken from Rueger 2000: 476).

It might be thought that this means that there are, after all, no strongly emergent properties. If strong emergence requires irreducibility and if there are no in principle irreducible properties, then there are no strongly emergent properties. However, this consequence is not inevitable: we can avoid it by construing strong emergence in a way compatible with reduction.

Let me mention one promising proposal of a criterion of emergence that does not require irreducibility. Rueger (2000) has suggested a *topological* criterion for diachronic emergence, which can also be used as a tool for constructing a concept of synchronic emergence compatible with reduction. A change between two dynamic states is *quantitative* if the corresponding trajectories are *topologically equivalent*. If the change is *qualitative* because the trajectories are *not* topologically equivalent, this may be taken as a ground for judging that this qualitative change is a case of emergence. Here is an example: Figure 16.3 can represent the trajectory, in phase space, of an undamped pendulum, i.e. a pendulum swinging without friction in a vacuum, if we take x to indicate angular deviation from the rest position and y to indicate angular speed. Figure 16.4 shows the trajectory of a damped oscillator. Introduction of damping causes a topological change in the form of the trajectory: it switches from circular to spiral. In terms of Rueger's criterion, this is a ground for taking the change to be qualitative. Such qualitative change can then be taken to be sufficient, together with weak emergence, for strong emergence.

2. DIACHRONIC STRUCTURE EMERGENCE

Let us now turn to the second of Stephan's concepts of strong emergence, diachronic structure emergence.[8] According to Stephan, a weakly emergent

[8] Stephan's concept of diachronic structure emergence is similar to Bedau's (1997) notion of 'weak emergence'.

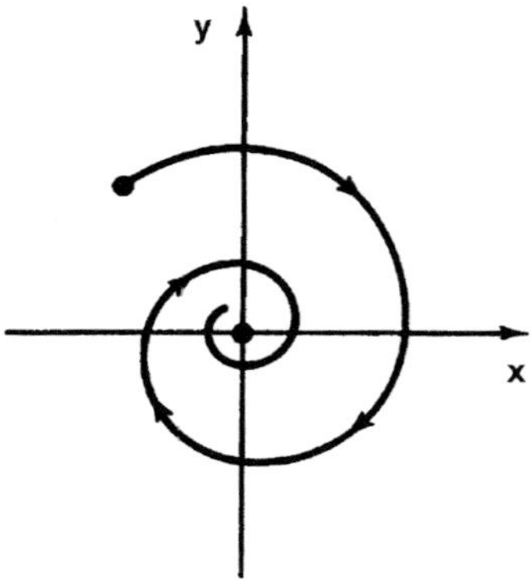

Figure 16.4. Phase space diagram of damped harmonic oscillator (taken from Rueger 2000: 475).

systemic state is diachronically structure emergent if and only if its formation obeys laws of deterministic chaos and is unpredictable unless by simulation. It is open to the charge that unpredictability unless by simulation seems to be neither necessary nor sufficient for emergence.

It is not necessary because the emergence of many properties is predictable without simulation: take a crystal that appears in the process of cooling a liquid. The coming into being of the crystal's observable macroproperties such as its colour, form, and hardness is a paradigmatic case of diachronic emergence. Although the movement of the molecules obeys deterministic chaos, the presence of attractors in such chaotic systems makes their evolution predictable.

Unpredictability is not sufficient either for emergence: if there is no point attractor, the evolution of a system may be unpredictable though nothing emerges. Take the system of air molecules in the atmosphere. The trajectories of the air molecules are not in the basin of any point attractor so that it is impossible to predict them in the long run. However, no qualitatively new property emerges from the evolution of this chaotic system.

3. FREEDOM AND CONSCIOUSNESS

Let me now turn to the main issue raised by Stephan's paper. All versions of the three replies to the free will problem Stephan mentions acknowledge the supervenience of mental processes on physical processes in the brain. Figure 16.5 sketches the compatibilist position advocated by Beckermann (2005). To take a free decision is a mental process, represented by a sequence of mental events m_0, m_1, . . . , which supervenes on a parallel series of physical events p_0, p_1, ...

Keil's (2007) 'libertarian' conception, sketched in Figure 16.6, does not differ from Beckermann's in respect of the relation between mental and underlying

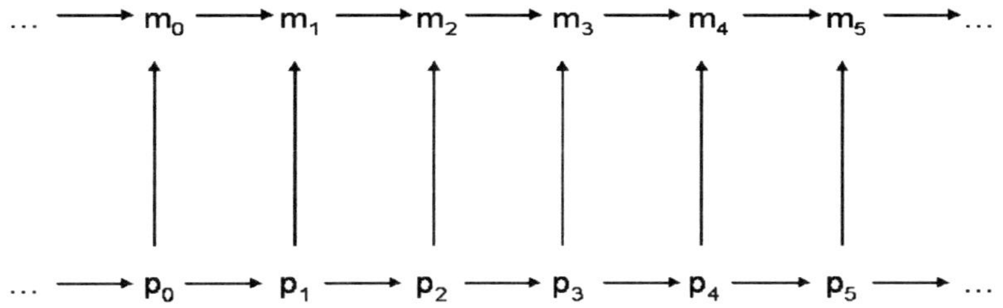

Figure 16.5. Schema of Beckermann's compatibilism (taken from Stephan, this volume).

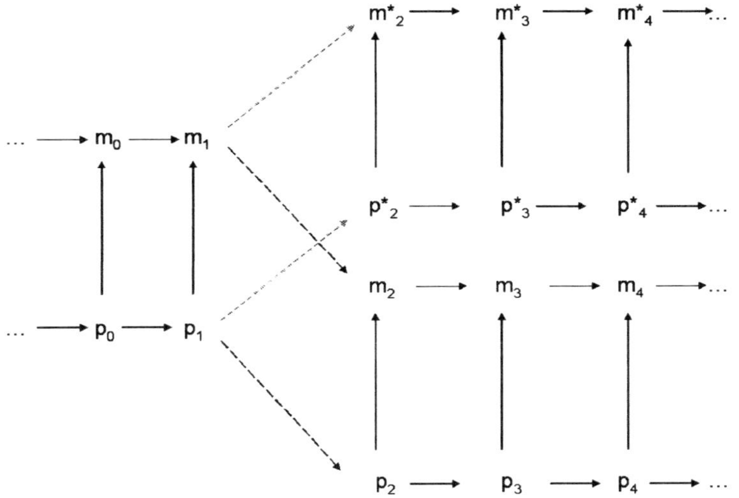

Figure 16.6. Schema of Keil's libertarianism (taken from Stephan, this volume).

physical properties: a series of mental events supervenes on a parallel series of physical events. The difference lies in Keil's thesis that both series contain bifurcation points. Free decisions are supposed to take place at these points.

Stephan's as well as Beckermann's and Keil's formulations leave open two interpretations of the relation between the physical events p_i and the mental events m_i in these schemas. According to the first, the mental events are distinct from the physical events though the former 'rest on' and supervene on the latter. The second interpretation has it that there is just one process that is both mental and physical, its constitutive events having both mental and physical properties. These interpretations correspond to two ways of conceiving of events: on Davidson's (1989 [1970]) conception, mental and physical events are token identical, in the sense that there is really just one event that can be described alternatively in mental or physical vocabulary. On Kim's conception, the relation between a mental event *m* and the physical event *p* it supervenes on is that

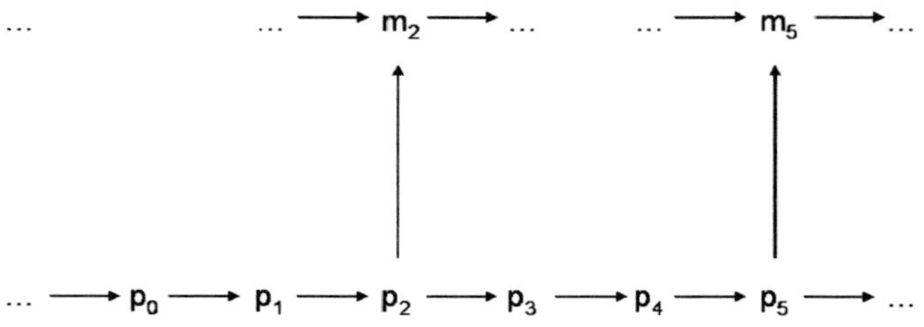

Figure 16.7. Stephan's (this volume) sketch of Singer's hard determinism.

between a role and what fills the role: *m* is the functional description of a mental role and *p* the physical role-filler. However, both interpretations have in common that there could be no causal interaction between the mental events m_i in the process leading to an action and the physical events p_i on which they supervene.

Let us now turn to Singer's position, according to which our belief in free will is an illusion. Singer holds that what distinguishes apparently free actions from other actions is that part of the process leading to the action is *conscious*. Stephan's interpretation, sketched in Figure 16.7, misrepresents Singer's position by assimilating it to Beckermann's and Keil's with respect to the relation between mental events and their underlying physical events.

It is incompatible with Singer's position to represent, as does Stephan in Figure 16.7, the relation between event p_2 (that is part of the chain causing some muscle movement constitutive of a given action) and event m_2, which is the event of the agent becoming conscious of p_2, by the same symbol (vertical arrow) as the relation between p_i and m_i in the schemas (Figures 16.5 and 16.6) representing Beckermann's and Keil's positions.

The incompatibility stems from the fact that the relation between a mental event m_i and the underlying physical event p_i is, as we have seen, not causal, whereas the process making a state conscious *is* causal.

According to the main psychological model for (access-)[9] consciousness, a mental state *m* is conscious if its content is *accessible* for information processing. In functional terms, *m* is conscious if and only if it is situated in a 'global work space'.[10] If a state is situated in the global work space, it has the capacity to interact with many functional subsystems of the mind/brain—both on the input (vision, language understanding) and output (motor system) side. This capacity rests on the configuration and strength of neural connections. Therefore, an event m_i of becoming (access-) conscious of an event p_i is not only mentally but also physically different from p_i, and the relation $p_i \rightarrow m_i$ is causal.

[9] Block (1995) has introduced a famous distinction between access and phenomenal consciousness. Psychological research focuses on access consciousness.

[10] See Baars 1997; Dehaene et al. 2003.

4. FREEDOM DESPITE DIACHRONIC AND SYNCHRONIC DETERMINATION

Let me end with sketching a way in which the concept of emergence might indeed be used in constructing a compatibilist solution to the free will problem. Emergence can help us understand how a complex system such as a human being could be free, although it is composed exclusively from physical parts subject to deterministic laws.

The first hypothesis I will use is that the human body, and in particular its brain, is a complex system that obeys the laws of deterministic chaos. In that case, for a given precision of description, it is impossible to predict, from a description of the conjunction of the states of the parts of the system (the neurons and synapses) at time t, a description of the conjunction of the states of the parts at $t + \Delta t$ if the time span Δt is long enough. This impossibility to predict is common for complex systems exhibiting deterministic chaos. The evolution of the conjunction of microstates is 'undetermined' in this sense. For any given set of global states of the body $\{S_i(t)\}$ that belong, at time t, to a given type T, these states have at much later times evolved into states $\{S_i\ (t+\Delta t)\}$ that do not any longer belong to any common type T*.

The second hypothesis is that mental states emerge from brain states, either by 'horizontal' emergence, i.e. through fusion, or by 'vertical' emergence, i.e. by the systematic interaction of the parts of the brain in a mechanism.

The third hypothesis is that these mental states obey to 'system laws',[11] in this case psychological laws. Those laws impose constraints on the evolution of the system and thus contribute to determining the evolution of (1) the system properties and (2) the state of the system's parts (neurons and synapses).

In this framework, the conviction that our actions are free, i.e. determined at a psychological level by our preferences and beliefs, can be reconciled with the conviction that all parts of our bodies and brains obey deterministic laws. The state of the body of a person at time t_3 is determined jointly by two constraints: (1) by physical laws in virtue of the physical properties of the parts of the system at t_2 (this is short-term diachronic determination because t_2 must immediately precede t_3), and (2) by the psychological laws applying to the person by virtue of systemic properties of the system at t_1 (preceding t_3 by a longer time span). A description of all parts of the person's body and their interactions does not suffice to predict a description of all parts at a much later time. In this sense, the state of our body does not on its own determine the state of our

[11] The term has been coined by Schurz (2002). Such laws are valid for specific types of system, such as the organisms of a certain biological species or ruby lasers. If the evolution of a system is regular enough that it obeys such a law, it is what Cartwright (1999) has called a 'nomological machine'.

body over long time spans. In particular, it does not on its own determine our actions. The determination of our actions is mediated by emergent psychological properties of our body and by psychological processes such as deliberation and decision.[12]

REFERENCES

Baars, B. 1997. *In the Theater of Consciousness*. New York: Oxford University Press.

Bechtel, William 2006. *Discovering Cell Mechanisms*, Cambridge: Cambridge University Press.

Beckermann, A. 2005. 'Free Will in a Natural Order of the World'. In C. Nimtz and A. Beckermann (eds), *Philosophy—Science—Scientific Philosophy. Main Lectures and Colloquia of GAP 5*. Paderborn: Mentis, 111–26.

Bedau, M. A. 1997. 'Weak Emergence'. *Philosophical Perspectives: Mind, Causation, and World* 11: 375–99.

Block, N. 1995. 'On a Confusion about a Function of Consciousness'. Repr. in N. Block, O. Flanagan, and G.Güzeldere (eds), *The Nature of Consciousness*. Cambridge (MA): MIT Press, 1997, ch. 20.

Boogerd, F.C., Bruggemann, F.J., Richardson, R.C., Stephan, A., and Westerhoff, H. 2005. 'Emergence and its Place in Nature'. *Synthese* 145: 131–64.

Broad, C. D. 2000 [1925]. *The Mind and its Place in Nature*. London: Harcourt, Brace and Co.; repr. London: Routledge.

Cartwright, N. 1999. *The Dappled World: A Study of the Boundaries of Science*. Cambridge: Cambridge University Press.

Craver, C. F. 2001. 'Role Functions, Mechanisms and Hierarchy'. *Philosophy of Science* 68: 31–55.

Davidson, D. 1989 [1970]. 'Mental Events'. In idem, *Essays on Actions and Events*. Oxford: Oxford University Press, 207–27.

Dehaene, S., Sergent, C. and Changeux, J.-P. 2003. 'A Neuronal Network Model Linking Subjective Reports and Objective Physiological Data during Conscious Perception'. *Proceedings of the National Academy of Science* (USA), 100.14: 8520–8525.

Humphreys, P. 1996. 'Aspects of Emergence'. *Philosophical Topics* 24: 53–70.

—— 1997. 'How Properties Emerge'. *Philosophy of Science* 64: 1–17.

Keil, G. 2007. 'Mythen über die libertarische Freiheitsauffassung'. In D. Ganten, V. Gerhardt, and J. Nida-Rümelin (eds), *Die Naturgeschichte der Freiheit*. Berlin and New York: de Gruyter.

Kim, J. 1998. *Mind in a Physical World*. Cambridge, MA: MIT Press.

Kistler, M. 2005a. 'Réduction "rôle-occupant", réduction "micro-macro", et explication réductrice a priori'. *Dialogue : Canadian Philosophical Review* 44: 225–48.

—— 2005b. 'Is Functional Reduction Logical Reduction?'. *Croatian Journal of Philosophy* 5: 219–34.

—— 2006. 'The Mental, the Macroscopic, and their Effects'. *Epistemologia* (Genoa, Italy), 29: 79–102.

[12] I have developed this idea in Kistler (2006) and (2007).

—— 2007. 'La réduction, l'émergence, l'unité de la science et les niveaux de réalité'. *Matière Première* (Paris, France) 2: 67–97.

Kronz, F. M. and Tiehen, J. T. 2002. 'Emergence and Quantum Mechanics'. *Philosophy of Science* 69: 324–47.

Levine, J. 1993. 'On Leaving out What it's Like'. In M. Davies and G. Humphreys (eds), *Consciousness: Psychological and Philosophical Essays*. London: Blackwell, 121–36.

Machamer, P., Darden, L., and Craver, C. F. 2000. 'Thinking about Mechanisms'. *Philosophy of Science* 67: 1–25.

McLaughlin, B. 1992. 'The Rise and Fall of British Emergentism'. In A. Beckermann, H. Flohr, and J. Kim (eds), *Emergence or Reduction?—Essays on the Prospects of Nonreductive Physicalism*. Berlin and New York: de Gruyter, 49–93.

Rosenberg, A. 1985. *The Structure of Biological Science*. Cambridge: Cambridge University Press.

Rueger, A. 2000. 'Robust Supervenience and Emergence'. *Philosophy of Science* 67: 466–89.

Schurz, G. 2002. '*Ceteris paribus* Laws: Classification and Deconstruction'. *Erkenntnis* 57: 351–72.

Stephan, A. 1997. 'Armchair Arguments against Emergentism'. *Erkenntnis* 46: 305–14.

—— 1999. 'Varieties of Emergentism'. *Evolution and Cognition* 5: 49–59.

17

Rationality, Reasoning, and Group Agency*

Philip Pettit

INTRODUCTION

Agents have to display a modicum of rationality in the formation and enactment of their attitudes, else they will not pass as agents at all; their performance will be too random or erratic to count as action. This is true of group agents as well as of individual agents and so it raises the question that this chapter addresses. How do group agents achieve the rationality that their status as agents requires? Can we expect rationality to emerge spontaneously when groups are organized properly? Or does the maintenance of a rational group configuration require a continuing surveillance on the part of members? Does it require this as a matter of feasible practice, if not as a matter of logical necessity?

The rationality of individual agents is secured for the most part by their make-up or design. Some agents, however—in particular, human beings—rely on the intentional exercise of thinking or reasoning in order to ensure the maintenance of rationality, and to further its improvement. This is the activity that is classically exemplified in Rodin's sculpture of *Le Penseur*; we all recognize what the bent-over, almost clichaic figure is doing: he is lost in thought. This use of reasoning sharpens the question addressed in the paper. Do group agents have to rely on reasoning in order to maintain a rational configuration of attitudes? Or can such a configuration be maintained without an analogue of thinking?

The first section of this chapter sets up the distinction between rationality and reasoning, as it applies with individual subjects, and then the second and

* This chapter was presented as the Dialectica Lecture 2006 at the *Gesellschaft fuer Analytische Philosophie* meeting in Berlin, September 2006, and was published in *Dialectica* 2007; it is reprinted here with permission. It also had an airing at the Inter-University Center, Dubrovnik, the University of Sydney, the Massachusetts Institute for Technology, Brown University, Queen's University, Belfast, and Stanford University. My thanks for the many useful comments I received in discussion at those events, in particular to respondents and co-symposiasts: Katie Steele, Jenann Ismael, and John Sutton in Sydney, Avi Tucker in Belfast. The paper is related to ongoing work with Christian List and, apart from the debt that I owe to our incorporated, shared self, I am greatly indebted to him for detailed criticisms of an earlier draft.

third sections ask after its application to the case of group agents. A principal requirement of reasoning, according to the first section, is access on the part of reasoners to the content of their existing commitments; they must have feedback on the things that they already hold and seek, believe and plan. The second section argues, by contrast with the individual case, that under plausible conditions the absence of feedback to members of a group on their existing group commitments is going to make it hard, perhaps even effectively impossible, for them to achieve group rationality. And the third section then shows that the presence of such system-level feedback, and specifically its use in group reasoning, would suffice to make group rationality accessible.

The argument exploits a lesson of recent results on the aggregation of judgements and provides a novel perspective on the emergence of agency among groups of individuals (Pettit 2003c; List and Pettit 2005, 2006). A brief conclusion construes the upshot in terms of a distinction that is familiar from other areas of discussion, between self-organizing and self-governing systems. This serves to highlight the core message: group agency among human beings does not emerge without effort; group agents are made, not born.

1. RATIONALITY AND REASONING

Simple agents

To begin with, some basics.[1] If a creature is to count as an intentional agent, then it must have desires or goals for which it is disposed to act and it must form beliefs about its environment to guide its action, identifying suitable opportunities and strategies. Such desires and beliefs can be characterized as attitudes towards propositions, with the desire consisting in the targeting of a proposition, the belief in the acceptance of a proposition, and with the distinction between targeting and acceptance being given by a difference in direction of fit. An agent will act to make the world fit a targeted proposition—a would-be goal—and will adjust to make its mind fit a proposition it accepts: a would-be fact (Anscombe 1957; Searle 1983; Smith 1987). It will act for the realization of its desires, seeking to bring the world in line with them; and it will act in this way according to its beliefs, where its beliefs are brought into line by the world. This will be so, at any rate, within what we think of as feasible limits and favourable conditions.

If a system is to fulfil world-changing desires according to world-tracking beliefs, then it must satisfy three sorts of standards. Attitude-to-evidence standards will require, among other things, that the system's beliefs be responsive to

[1] These are rehearsed more fully in List and Pettit (forthcoming).

evidence. Attitude-to-attitude standards will require that, even as they adjust under evidential inputs, its beliefs and desires do not assume such an incoherent form that they support inconsistent options. And attitude-to-action standards will require that the system tend to act, and to form intentions to act, on the lines that its beliefs and desires support.

These are standards of rationality, as I understand the term. Rational standards are nothing more nor less than desiderata of agency: standards such that agents will generally do better as agents by robustly satisfying them, at least within relevant limits and conditions.[2] Let an agent form beliefs that run counter to the evidence and it will tend to adopt strategies that do not satisfy its desires. Let it form attitudes that support inconsistent lines of action and it will fail even more dramatically in the pursuit of desire-satisfaction. And let it not be disposed to act or intend to act as its beliefs and desires require, and it will fail more theatrically still. Further intuitive desiderata of rationality may be less important but, other things being equal, their satisfaction will also enhance agency. One example is the attitude-to-evidence desideratum that an agent form those general beliefs for which a run of evidence provides inductive support. And another is the attitude-to-attitude desideratum that the agent form beliefs and desires with respect to propositions that are entailed by propositions already believed or desired, or reject the belief or desire held for the entailing propositions.

Quite a simple system can merit the ascription of propositional attitudes, and characterization as an agent. Consider the little robot that navigates a table top on wheels, scanning various cylinders on the table with bug-like eyes, and moving to set upright any cylinder that falls or lies on its side. Even a system as rudimentary as this can be said to accept propositions to the effect that this or that or another cylinder is upright or on its side and to be disposed, with any cylinder on its side, to target or realize a proposition to the effect that it is upright once again.

Any creature, including one as simple as this robot, will have to display a minimal level of rational competence, if it is to deserve the name of agent. The movement of the robot's eyes will have to pick up relevant evidence about the orientation and location of cylinders on their sides. Its cognitive processing will have to ensure that it forms a set of consistent representations as to where the cylinders are. And the representations will have to interact with its overall goal to generate attempts to set those cylinders back in an upright position. In other words it will have to display a minimal level of rationality in attitude-to-evidence, attitude-to-attitude, and attitude-to-action relations.

Or at least it will have to do this within what we take to be intuitively feasible limits and intuitively favourable conditions. Suppose that the robot tends to

[2] Some of these standards might be taken as relevant, of course, to systems that do not qualify fully as agents. Thus we might invoke the standards of belief-formation that would be appropriate with an agent in assessing a system that can form beliefs but not act upon them.

knock cylinders at the edge of the table onto the floor. Do we say that besides the desire on which it generally acts, it has a desire to knock such cylinders off the table? That will depend on our knowledge of its design specifications. It will be plausible that there is no specification for such a goal, if the actions it uses with cylinders at the edge of the table are just the same as those it uses with other cylinders. In that case we will treat the condition where a cylinder is at the edge of the table as less than favourable and will stick with the original characterization of its intentional attitudes.

Sophisticated agents

Non-human creatures, certainly non-human animals, get to be much more complex agents than the robot imagined. There are a number of ways in which the robot might be designed to more complex specifications. It might be built to search out and pick up other behavioural strategies, on a trial-and-error basis, if its existing efforts at raising cylinders run into problems—say, if they knock cylinders onto the floor. It might form beliefs about other objects besides the cylinders or about other properties besides the location and orientation of the cylinders. And it might embrace a number of purposes, not just the single goal of setting certain cylinders upright.

But even if such complexities are introduced, robotic agents in this mould will remain simple in one salient respect. They will form propositional attitudes of belief and desire with respect to concrete objects like the cylinders and the salient properties of such objects; they will come to form beliefs in, and desires for, various propositions involving such items and such features. But nothing in the story told means that they will conceptualize and attend to those very propositions, forming attitudes about their properties and relations in turn. Nothing entails that they will practice what we might describe as propositional ascent.

These agents will form ordinary propositional attitudes but not attitudes of a meta-propositional character. An ordinary propositional attitude is an attitude towards a proposition in which only concrete objects and their properties figure, whether that proposition be singular or existential or universal in form. A meta-propositional attitude is an attitude towards a proposition—if you like, a meta-proposition—in which propositions themselves may figure as objects of which properties and relations are predicated. Some meta-propositions will ascribe properties like truth and evidential support, and relations like consistency and entailment, to propositions, as in the claim that 'p' is true or that it is inconsistent with 'q'. Others will identify propositions as scenarios that are believed or desired by agents, or as scenarios that are credible or desirable, as in the assertion that someone believes the proposition 'p', or that 'p' is credible. Others again will identify them as scenarios that the agent is disposed to realize, or as scenarios the agent invites others to rely on his or her realizing, as in the claim that someone is going to make 'p' true or that 'p' is something someone intends or promises to make true.[3]

The absence of meta-propositional attitudes in the robot and its more flexible counterparts means that they are subject to a salient restriction. A robot might ask itself a question, as we can put it, when it registers a movement in its peripheral visual field and then focuses its eyes on the relevant location out of a desire, derivative from its cylinder-raising desire, to know whether or not a cylinder has fallen; this desire will be satisfied so far as the focusing of the eyes has the effect of letting a belief form on the matter. But because the robot and its counterparts don't have meta-propositional attitudes, they cannot ask themselves similar questions about connections between propositions, say about whether they are consistent or inconsistent, and then do something—pay attention to the inter-propositional relations—out of a desire to have a belief form one way or the other.

This restriction means that the robotic creatures cannot reason. To be rational, as we saw, is to satisfy certain desiderata in the attitude-to-evidence, attitude-to-attitude, and attitude-to-action categories. To be able to reason, under the model I shall adopt here, is to be able to conduct an intentional activity that is designed—and perhaps explicitly intended—to raise the chance of satisfying such desiderata (Pettit 1993). Specifically, it is to be able to ask oneself questions about certain propositional properties and relations; to be able thereby to let beliefs form on such matters; and to be disposed to adjust rationally to whatever beliefs one forms. This is a sort of activity that the robot clearly cannot conduct.

Suppose I currently believe that p and that q. Perhaps because of worrying about the conditions in which I formed those beliefs, I may ask myself whether 'p' and 'q' are consistent propositions, setting in train a process of forming a belief in answer to the question and adjusting to the result. And I may do so with benefit. For if I come to form the belief that the propositions are inconsistent, I will have brought to the surface the fact that I believe that p, that q and that 'p' and 'q' are inconsistent. And, if things go right, I will then be prompted to eliminate the inconsistency among those beliefs and achieve a higher degree of rationality; in the ordinary course of events, I will be prompted to give

[3] We often ascribe meta-propositional attitudes, not by identifying a proposition and the property that is predicated of it, but by using propositional operators that make the attitudes seem to be of a regular sort. Instead of saying that people believe that 'p' is true, signalling their capacity to think about the proposition 'p', we say they believe that it is true that p; instead of ascribing a belief that someone believes 'p' we say that they believe that the person believes that p; instead of taking them to see desirability in the realization of a proposition 'q', we say that they believe that it is desirable that q. But I take it that such ascriptions are plausible only with creatures that are capable of meta-propositional thinking. It would be misleading to ascribe beliefs that it is true that p and that it is false that not-p, for example, to a creature that is incapable of recognizing any common element in those beliefs (Evans 1982). And in order to be able to recognize a common element, the creature would have to be able to have beliefs about the proposition 'p'.

up on my belief that p or that q, depending on which is the more weakly supported.

The process I will have gone through in such a case constitutes what counts, on the model adopted here, as reasoning.[4] In the example given the reasoning is theoretical in character insofar as the meta-propositional interrogation involved is deployed in the service of checking on the formation of beliefs. An otherwise similar process would constitute practical reasoning if the interrogation were deployed in the service of checking on the formation of intentions. I reason theoretically when the fact of forming the belief that 'p' and 'q' are inconsistent as a result of meta-propositional interrogation leads me to reject at least one of those propositions. I reason practically when the fact of forming the belief that making 'r' true is the only way of making 's' true as a result of meta-propositional interrogation leads me from an intention to realize 's' to an intention to realize 'r'. And so on.[5]

How can subjects like you and me form meta-propositional beliefs and engage in reasoning? The answer surely has to do with the fact that we have language and can use sentences as ways of making propositions into objects of attention. The normal assertoric use of sentences like 'p' or 'q' will be to report the fact that p or that q, expressing a belief in that fact. But the sentences can also be used to exemplify what is said in normal assertoric usage, allowing the speaker to think about those propositions and to let beliefs form as to whether they are consistent or not (Davidson 1984). I say to myself 'p', and I say to myself 'if p, then q', and, registering their relationship, conclude 'so, q'. The robotic agent may be more or less perfectly designed to form the belief that q whenever it forms the belief that p and that if p, q, but it will not register the relationship between those

[4] The model should prove relatively uncontroversial, on two counts. First, it is a model of reasoning, not of good reasoning; and second, it gives an account of reasoning as such, not of the special exercise in which someone makes a judgement as to what good reasons require overall. I reason whenever I set out to form meta-propositional beliefs and let them play out as checks on the process whereby my attitudes form. The exercise will be good reasoning when it takes me where I ought to go or where there is good reason to go, but the theory of reasoning as such need not offer an account of where precisely that is. Thus I do not engage with the issues dividing Niko Kolodny and John Broome, for example (see Kolodny 2005; Broome 2007). Moreover, to go to the second point, when I reason in the sense at issue, I do not need to make any judgement about where there is good reason to go. Why should we say that the exercise constitutes reasoning only in the presence of the rider? And why should we think that we have given an account of reasoning only when we have given an account of what can be put forward as purportedly good reasons? On these matters, see Burge (1998).

[5] The idea is not that the meta-propositional belief will have to be correct and will naturally be authoritative in determining the adjustment for which it argues. The presence of that belief is only meant to put a further check in place, however fallible, by providing another locus at which any irrationality ought to show up. Finding myself with the beliefs that p, that q, and that 'p' and 'q' are inconsistent, I might be led to question the belief in the inconsistency, not the belief that p or that q. The possibility is unlikely to materialize to the extent that the inconsistency is a priori demonstrable, or the belief in the inconsistency is the product of more reflective consideration than that which led to the other beliefs; but it is certainly not closed. I am grateful to Rory Pettit for pressing me on this point.

propositions; it will not go through any process that would enable it to use the word 'so' or some cognate.

Without a linguistic or symbolic means of objectifying the propositions, it is hard to imagine how any meta-propositional thinking could emerge (Pettit 1993: ch. 2; Dennett 1994; Clark 1999). Thus it is difficult to see how non-human animals can reason in the active sense described. None ever appears to conduct the activity we naturally ascribe to *Le Penseur*. A dog may perk up its ears on hearing a sound and attend to what is happening, asking itself whether the family is home or dinner is being served. But no dog does anything that we might interpret as asking itself whether certain propositions are really supported by the evidence, whether realizing one will help realize another, whether they are consistent, and so on. Or, at least, not outside of Gary Larson cartoons.

Reasoning in the sense characterized may be pursued in a content-based or a form-based manner. Standard reasoning involves an activity in which people think about the scenarios that answer semantically to various sentences and reflect on how far they are supported by the evidence to hand, how far the scenarios are consistent with one another, whether one scenario entails another, and so on. But given that many of those sorts of relationships are guaranteed to hold in virtue of the form of the corresponding sentences, reasoning can also be conducted in abstraction from the content, as in seeing that since certain scenarios relate in the form of 'p and q' and 'p', then regardless of how these schematic letters are interpreted, the first must entail the second. Numerical calculation offers an everyday example of form-based reasoning. Using the multiplication rule to work out that 11 times 11 is 121 can be seen as an exemplar of reasoning that abstracts from content.[6]

I said earlier that a simple agent like the robot, or indeed the dog, will have to display a minimal level of rationality in order to earn the name of agent. It must be clear that there are goals that it pursues, that it pursues them according to certain beliefs, and that those beliefs are not randomly or rigidly formed, but display some sensitivity to evidence. Does the same lesson apply to symbol-using creatures like you and me?

Yes and no. Yes, we must display some sensitivity to the demands of rationality. But no, this need not be displayed in the spontaneous, behaviourally vindicated achievement of rational performance. We may display the required sensitivity through showing ourselves to be ratiocinatively if not behaviourally responsive—that is, responsive at the meta-propositional level—to rational requirements. We may not behave as if 'p' and 'if p, q' entail 'q', for example, but that will raise no doubts about our status as agents so far as we can be challenged, forced to recognize the entailment, and led to criticize our own

[6] Such intentional exercises in calculation should be distinguished, of course, from the computations in an uninterpreted language of thought that are sometimes postulated as the means whereby intentionality is subpersonally realized (Fodor 1975).

behaviour. Being symbolic creatures, we can have our attitudes identified on the basis of our avowals as well as our actions, and we can vindicate our claim to be agents, not just by acting appropriately, but also by being disposed to recognize the implications of our avowals, to criticize our own performance on that basis and, at least ideally, to defer to those implications, bringing ourselves into line with what they require (Levi 1991; Bilgrami 1998; McGeer and Pettit 2002). We shall notice an analogue of this observation towards the end of the discussion of groups.

Simple and sophisticated compared

The rationality of the simple creature is realized subpersonally so far as there is nothing the creature can do in order to improve its rational performance (Dennett 1969). The robot and its counterparts have to be more or less rational in order to count as agents but they cannot exercise their agency with a view to ensuring or enhancing that rationality. They can act on the cylinders and other concrete objects, having beliefs and desires that bear on the properties of those objects. And in order to do this, they must have beliefs and desires that minimally satisfy attitude-to-evidence, attitude-to-attitude and attitude-to-action desiderata. But, not having attitudes that bear on the propositional objects of those beliefs and desires, they cannot do anything to check or channel the process in which the beliefs and desires form, connect and occasion action. It is by grace of their design, artificial or natural, that they generally satisfy the desiderata of rationality, not by virtue of anything they themselves can do.

We reasoning creatures transcend this limitation. We do not have to rely entirely on the processing for which our nature programmes in order to be rational. We can do something about it, as we check out the meta-propositional constraints and connections that are relevant to what we should believe, desire, or do. We can monitor and hope to improve our own performance, putting extra checks in place in order to guard against rational failure.

The transcendence of the non-intentional that we thereby achieve is only partial, of course (Carroll 1895). Suppose that paying attention to the relations between the propositions 'p', 'if p, q' and 'q' is to have a rationality-enhancing effect on me, triggering me to move from beliefs in the premises to a belief in the conclusion or to inhibit one of the beliefs in the premises. It can have that effect only if two conditions hold, both of which involve my non-intentional, rational processing. First, I must be able to rely on the attentional activity to generate a correct meta-propositional belief and, second, I must be able to rely on that meta-propositional belief having the required effect. Reasoning does not work in parallel to non-intentional rational process, seeking the same end by different means; it exploits rational process in a novel way, searching out extra inputs to impose as checks on how the process is operating.

Even if it is partial, however, the transcendence that reasoning achieves gives us a degree of personal control over our own rational performance. The control deserves to be described as personal on two counts. First, it is a form of control in which I intentionally pursue the satisfaction of rational desiderata, rather than merely relying on my non-intentional processing. I act as an intentional system with a view to achieving rationality rather than leaving the task just to subsystems within me. I am a systemic agent in this domain, not just a site of subsystemic activity.

The second count on which this form of control deserves to be described as personal rather than subpersonal is that we each exercise it in sensitivity to what we as an intentional system already believe and desire; implicitly or explicitly it requires that we process feedback on where we are already committed, and keep track on those commitments. In principle there might be a creature that asks itself meta-propositional questions without any awareness of what it already believes or desires, and without any awareness of whether the questions asked are relevant to its own attitudes. The creature might even do this with the effect—intended or unintended—of putting extra rational checks on its attitude-formation. But clearly we ordinary reasoners are not like that. If we ask meta-propositional questions about various propositions, at least outside the classroom, then we will do so because those propositions already figure as the objects of our attitudes or are propositions about which we are trying to form attitudes. We will ask meta-propositional questions, and look for the benefits of reasoning, under the guidance of feedback on where we already stand. Thus, if we ask whether 'p' and 'q' together entail 'p and q', that will typically be because of feedback awareness that by our existing lights it is the case that p and it is the case that q.

Not only do I pursue rationality as a systemic agent, then, rather than just leaving it to subsystems within me, I do it out of a sense of myself as a systemic agent with a record of attitudinal commitment. The first aspect of personal control means that it is I as an intentional agent who seeks to exercise control. And the second means that it is over me, conceptualized as a centre of enduring, available-in-feedback attitudes that I seek to exercise that control.[7]

Our interest in the sections following is in how far the divergence between subpersonally rational and personally ratiocinative agents is replicated with group agents. The most straightforward way of addressing this question will be by looking at the possibility of group agents in which members do not have feedback on the existing commitments of the group and then at group agents where members do enjoy such feedback. The negative thesis of the chapter, defended in section 2, is that under a range of plausible conditions, groups that lack system-level feedback—and so cannot reason in the ordinary sense—are

[7] These considerations do not exhaust all aspects of what is required for personal control: a form of control in which I can properly be said to assume responsibility. The discussion of that topic would take me too far afield. For more, see Pettit (2001: ch. 5).

not going to be able to perform satisfactorily as agents. The positive thesis, defended in section 3, is that groups that have access to such feedback and are able to reason are likely to be capable of a satisfactory performance under those conditions.

2. GROUP AGENTS WITHOUT SYSTEM-LEVEL FEEDBACK

A group of individuals will succeed in becoming a single agent or agency to the extent that the members can coordinate with one another and replicate the performance of an individual agent. The question I address in this section is whether the members could form such an agent without any system-level feedback at any stage on where it stands in the space of commitments—on what it already believes and desires. Could they form an agent in a manner that mimics the performance of the simple robot or animal?

The issue

In order to replicate the performance of a single agent, the members of a group will have to subscribe, directly or indirectly, to a common set of goals, plus a method for revising those goals, and to a common body of judgements, plus a method for updating those judgements. And in addition they will have to endorse a method of ensuring that one or more of them—or an appointed deputy—is selected to form and enact any intention, or perform any action, that those group attitudes may require. Or at least they will have to take steps that provide for these results, within feasible limits and under intuitively favourable conditions. Within such constraints, the group as a whole will have to be robustly disposed to achieve the minimal rationality that is required of any unreasoning agent; I put aside for the moment the question of whether the ability to reason weakens this requirement, as it does with individual human beings.

This analysis of group agency highlights the fact that if we take any grouping or organization of people to constitute an agent—in particular, an unreasoning agent—then not only must we hold that, as a matter of fact, its collective behaviour is representable as the pursuit of plausible goals in accordance with plausible beliefs. We must also expect it to be robustly representable in such a manner. As we imagine various changes in its circumstances, for example, we must expect it to adjust so as to continue to act sensibly in pursuit of the goals attributed, or perhaps to alter the goals it pursues. This means that in looking at any group of people, the current evidence may leave it underdetermined whether or not it is a group agent in the sense defined. We may lack data on how it would adjust in counterfactual circumstances and may not be in a position, therefore, to say whether or not it is truly an agent.

How might a group be organized to meet agency requirements without any members having feedback on where it is already committed? Assume, plausibly, that the group agent does not emerge behind the backs of its members, so to speak, whether on the basis of some selectional pressure or the devious plan of some organizing genius; assume, in other words, that the members are each aware of being part of a group. This assumption makes for a point of contrast with the individual, subpersonally rational agent, but the analogy with such an agent will still remain alive insofar as the following scenario obtains. The members act on the shared intention that they together should establish and enact a plan or constitution under which their efforts are coordinated suitably.[8] And, crucially, that plan or constitution does not require any of them to monitor the attitudes or actions of the group as a whole, gathering information on how things are going at the system level. The group operates without any system-level feedback on the attitudinal configuration generated under the constitution.

Were such an arrangement in place, then each member would play his or her local part and the global consequence would be the appearance of a pattern of rational attitude formation and enactment at the group level. The group's attitudes would form and unform in a rational pattern, and would rationally prompt and direct action. And this would happen without any members having to monitor or regulate how things transpire at the group level. The members would each look after their own local business and the global business of the group would take care of itself.

Under the constitution or arrangement envisaged, some individuals would have to play a special role in triggering the group procedure needed to revisit a goal or judgement, in enacting a decision of the group on one or another issue, or on any of a number of fronts. But much of the business of the group would consist in the formation of the attitudes required for agency. How might those attitudes be formed, then, without anyone having system-level feedback? I shall consider this question with particular reference to judgements. How might the members ensure that on every issue that comes before it, the group will form a suitable judgement? How, in particular, might they ensure this without having

[8] I like an account of acting on a shared intention that is broadly in line with Bratman (1999). Set out in Pettit and Schweikard (2006), it stipulates that for individuals to act on an unforced, shared intention to X they must each intend that they together X; intend to do their individual part in a presumptively salient plan; believe that others each intend to do their part too; intend to do their part because of believing this; and enjoy a common form of awareness that those conditions are fulfilled. Those who act on such an intention may do so reluctantly or without relish, so that the account in the text does not suppose an equal commitment on that part of all members. It may even be that some members do not even share properly in the intention but acquiesce in the shared intention of others; the acquiescence will mean that they play their allotted parts and will make them indistinguishable from those who endorse the intention.

any feedback, and so without having any record, of the judgements that the group has already formed?

The voting proposal

The salient method or strategy whereby a group might seek to achieve this result is by recourse to voting. The members of the group are its eyes and ears and voting will enable them each to register their evidence on any issue confronted.[9] Suppose the group has to determine whether p or q or r, then, establishing the answers on which members or their deputies are to act in pursuing the group's goals. The members will determine the group's judgement on such an issue by holding a vote on whether p or q or r, and then aggregating those votes according to some acceptable procedure. They may employ any of a variety of voting procedures, some centralized, others decentralized, for this purpose. Everyone may be invited to vote on every issue, whether in a process of majoritarian or non-majoritarian voting. Or issues may be segregated so that different subunits—at the limit, singletons—are given different questions to resolve, and one or another process of voting is adopted in each. Propositions will be presented to one or more members of the group and they will be treated as matters of belief or desire, depending on whether they command appropriate support.

The recourse to voting will rule out one salient possibility for establishing the attitudes of the group. This is that members might put degrees of belief or probability together, thereby generating a system of probability for the group as a whole.[10] But this is not a serious loss. Voting can take place over probabilistic issues, such as whether it is more likely than not that p, whether it is nearly certain that p, and the like. And in any case the possibility ruled out is quite fanciful. Even if people each have fine-grained degrees of belief on propositions considered by the group, it is not clear how they could know what they are, and so not clear how they could communicate them to one another in the fashion required. Such degrees of belief may show up in behaviour, particularly in dispositions to accept various gambles, but they need not be available to introspection (Harman 1986).[11]

[9] The members might deliberate with one another before voting, thereby pooling information and comparing their responses. But whether or not deliberation occurs, there is likely to be disagreement on any issue and so a need to determine the non-unanimously supported group position (see Pettit 2003b).

[10] This possibility is significant in the following respect. If the representations that individuals aggregate into group representations come in degrees, and the aggregation reflects those degrees, then there need not be any problem like the discursive dilemma discussed below; this presupposes attitudes of an on-off kind. But the aggregation may be subject to other difficulties. For examples of some difficulties, see Raiffa (1968) and Hylland and Zeckhauser (1979: 220–37); I am grateful to Arthur Applebaum for drawing my attention to this work.

[11] Frank Ramsey describes a procedure whereby it is, in principle, possible for an interpreter to construct both a utility and probability function for an individual subject, on the basis of the

The question before us, then, is whether the members of a group might be able to use a voting procedure—a procedure of voting without feedback—so as to determine the judgements of the group after a rational pattern. Could it organize itself on the basis of no-feedback voting? Could it relate in this manner to its own members and satisfy the conditions for rational agency at the group level?

There would certainly be problems under an uncoordinated arrangement that allowed one subunit to decide whether p, another whether q, and a third whether p&q; these subunits might commit the group to an inconsistent set of judgements. But might a group guard against such inconsistency without any voting members having feedback on the commitments of the group as a whole? Might the subunits in a networked agency be coordinated so as to avoid the problem? Or might the members assemble so as to decide by centralized voting on each of the issues it has to resolve?

We may concentrate, without a serious loss of generality, on the assembly case. If a group cannot operate satisfactorily without feedback in this case, then it is unlikely to be able to operate without feedback in any other mode. The smaller subunits in any networked group will almost certainly have to face the same problem that arises for the assembly, if they are each charged with making judgements on logically connected propositions like 'p', 'q', and 'p & q'. And in any event, there will be the extra problem of how to ensure that the different propositions they support do not form an inconsistent set.

The theory of judgement-aggregation

Recent results on the aggregation of judgements are relevant to the issue we have raised and the possibility of a satisfactory voting solution. Those results reveal that there are severe constraints on how far a group can attain rationality in its judgements over logically connected issues and remain responsive in intuitively important ways to its members: specifically, responsive in ways that voting procedures automatically tend to implement. The results argue that the space for simultaneously ensuring both group rationality and individual responsiveness is very restricted.

There are broadly three respects in which we might expect that a group agent should be responsive to its membership. First, it should be robustly

subject's expression of binary preference as between different items and different gambles over items; for a summary description of the procedure see Pettit (2002: part 2, essay 2). Might it be applied with a group? No. Extracting a set of coherent binary preferences from a group will be subject to the same problem as that which arises, as we shall see later, with extracting a coherent set of judgements. The discursive dilemma that is used below to illustrate the problem with judgements can be extended readily to binary preferences; the group may prefer that p, that q, that r, and that not-p & q & r.

responsive, not just contingently so; the group judgements should be determined by the judgements of members, more or less independently of the form those judgements take. Second, the group should be inclusively responsive, not just responsive to a particular member—a dictator—and not just responsive to named individuals; otherwise it would fail to use its members as its eyes and ears, as epistemic considerations suggest it should do, as well as failing on a democratic count. Third, the group should be issue-by-issue responsive—if you like, proposition-wise responsive (List and Pettit 2006)—with its judgement on any question being determined by the judgements of its members on that very question.

The recent results on the aggregation of judgements show that under a wide range of specifications of these responsiveness conditions, some quite weak, it is impossible to have a procedure for determining group judgements that both satisfies those conditions and ensures, over issues that are connected in one or another degree, that the judgements will be complete and consistent. One example of such a result is proved in List and Pettit (2002), and others have followed.[12] It demonstrates the relevant sort of impossibility under the following precisifications of the three responsiveness conditions.

- Robust responsiveness: the procedure works for every profile of votes among individuals (universal domain);
- Inclusive responsiveness: the procedure treats individuals as equal and permutable (anonymity); and
- Issue-by-issue responsiveness: the group judgement on each issue is fixed in the same way by member judgements on that very issue (systematicity).

The best way to communicate the lesson of these results is by way of an example. Take the connected set of issues: whether p, whether q, whether r, and whether p & q & r. Suppose that a group of three individuals, A, B, and C, wishes to form a rational set of judgements over those issues, say because the question of how best to promote some group goals depends on the answers. And imagine now that the group follows a procedure of majority voting. Such a procedure will be robustly responsive in the sense that it will work under any profile of consistent member votes; it will be inclusively responsive in the sense that it gives everyone an equal vote; and it will be issue-by-issue responsive in the sense that it lets the judgement on every issue be determined by the votes of members on that very issue. And because of ensuring such full-blown responsiveness it may generate an inconsistent and therefore irrational set of judgements on the issues considered. The members may vote as follows.

[12] See in particular Pauly and Van Hees (2006) and Dietrich and List (2007). Notice that the three dimensions of responsiveness are not always reflected in a one-to-one fashion by three exactly corresponding conditions.

	p?	q?	r?	p&q&r?
A	No	Yes	Yes	No
B	Yes	No	Yes	No
C	Yes	Yes	No	No
A-B-C	Yes	Yes	Yes	No

In a situation like this the group will face a 'discursive dilemma'.[13] Either it secures majoritarian responsiveness to the views that are registered in the votes of members, in which case it will have to endorse the inconsistent set of judgements in the last row and fail to be rational. Or it ensures its own coherence, revising one of the judgements in the last row, in which case it will have to offend at least against majoritarian, issue-by-issue responsiveness; it will have to break with the majority view on 'p' or on 'q' or on 'r' or with the unanimously supported verdict on 'p & q & r'.

Back to the voting proposal

Back now to the question of whether a group can attain rationality on the basis of a voting procedure, and do so without feedback to its members on where it is already committed as a group. Such a group will confront a growing number of issues over time: now whether p, as it might be; now whether q; and so on. I assume that two conditions will generally hold with such a group. First, the issues it addresses will tend, sooner or later, to form connected sets and to give rise to the sorts of problem addressed in the theory of judgement aggregation (List 2006). And second, it will be important for the group to form complete and consistent judgements over those issues, since it will generally form judgements only on a need-for-action basis; thus, important decisions are liable to be jeopardized by any failures of completeness or consistency.

Under these assumptions, any responsive voting procedure, whether it has a majoritarian or non-majoritarian or mixed character, is liable to lead the group into forming inconsistent sets of judgements. That is the lesson of the theorems on judgement aggregation. Suppose that the group faces a new issue that is logically connected with issues that have been resolved under prior votes. Lacking feedback on the prior resolutions, members won't know what the group already judges and will have to vote blindly on the issue before them. Hence they are quite likely to vote in such a way that the group ends up with an inconsistent set of judgements. To return to the schematic example given, the members may

[13] See Pettit 2001a, 2003c. The idea of the discursive dilemma is a generalization of the legal idea of a doctrinal paradox (see Kornhauser and Sager 1993). For an overview of the topic, and of other issues, see List (2006).

vote that not-p & q & r, even when prior voting has committed the group to judgements that p, that q, and that r.

An unsatisfactory solution

Is there any way out of this problem under the no-feedback stricture with which we have been working? There is one class of solutions available but I shall argue that they are unsatisfactory in a distinct manner.

The most salient of these solutions is the sequential priority rule (List 2004).[14] This would organize issues so that whenever a group faces an issue on which its prior judgements dictate a resolution, then voting is suspended or ignored and the judgement recorded on that issue is the one that fits with existing judgements. There are a number of technologies whereby such an organization of issues might be realized without anyone in the group having to get feedback on the judgements already made by the group. And so the rule may seem to illustrate a procedure whereby the group might ensure its rationality without introducing feedback and so without activating any sort of group reasoning.

Assume that the group registers its views on 'p' and 'q' and 'r' before it confronts the issue of whether p & q & r, so that its existing commitments dictate that it should make a positive judgement. Under the organization postulated, the judgement recorded will be that p & q & r, independently of whether, or how, the members vote.[15] In following the rule, the group is bound to display a suitable sensitivity to meta-propositional constraints, in particular the constraint of formal consistency, but members need not have any feedback on its existing commitments. They may play their local parts blindly, without anyone keeping track of the group as a whole, yet the global upshot will be the formation of reliably consistent sets of judgements.

The reason why the sequential priority rule enables a group to be consistent, evading the difficulties identified in the impossibility results on judgement aggregation, is that while it forces the group to be robustly and inclusively responsive to its members, on intuitive interpretations of those conditions, it allows failures of issue-by-issue responsiveness. On any question where prior

[14] A variant on this procedure would divide issues into basic, mutually independent premise-issues and derived issues—this will be possible with some sets of issues, though not with all—and treat those judgements as prior, letting them determine the group's judgements on derived issues (see Pettit 2001b; List 2004).

[15] The group would reach the same judgement, if it worked with the (rather implausible) rule that would cast issues involving logically simple propositions as basic and that let other issues be fixed by its commitments on such premises; see footnote 14. But those rules might come apart in other cases. Suppose that the group registers positive views on 'p' and 'if p, q' before it confronts the issue of whether q. The regular sequential priority rule would deem it to judge that q, even if members are disposed to vote against 'q', but the variant, premise-rule would have the group vote on q and, given commitments for 'p' and against 'q', would deem the group to reject 'if p, q'.

judgements dictate a certain line, the group may adopt a position that goes against the views of the members on that particular issue. The position taken will be driven by the positions that members take on other issues but not by their positions on that issue itself (List and Pettit 2006).

It should be clear that a group might avoid inconsistency by having all of its attitudes formed under the sequential priority rule, or suitable variants. But would it be a rationally satisfactory agent? I argue not.

The problem is that while such a group would reliably achieve consistency in the judgements it forms, it would be entirely inflexible in its responses and potentially insensitive to the overall requirements of evidence. When I realize that some propositions that I believe entail a further proposition, the rational response may well be to reject one of the previously accepted propositions rather than to endorse the proposition entailed. Those are the undisputed lessons of any coherence-based methodology, and the group that operates under a sequential priority rule, or under any variant, will be unable to abide by them; it will not be robustly sensitive to the requirement of attitude-to-evidence rationality.

The evidential insensitivity of the sequential priority rule is apparent from the path-dependence it would induce.[16] One and the same agent, with access to one and the same body of evidence, may be led to form quite different views, depending on the order in which issues present themselves for adjudication. The group agent that follows the rule will be required to respond to essentially conflicting bodies of testimony—conflicting majority judgements among its members—without any consideration as to which judgement it seems best to reject. It will be forced by the order in which issues are presented not to give any credence to the judgement its members may be disposed to support on the most recent issue before it. And this is so, regardless of the fact that often it will be best to reject instead a judgement that was endorsed at an earlier stage.

The path-dependence imposed under the sequential priority rule can be illustrated with the now familiar, schematic example. Let the group confront the three atomic issues before it faces the issue of whether p & q & r and it will judge that p & q & r. Let it confront that compound issue earlier, say before the issue of whether r, and it will judge that not-p & q & r. When it comes to the issue of whether r, the rule will force it in consistency with prior commitments to judge that not-r. Thus the judgements with which it ends up will vary with the order in which the judgemental issues are presented. And yet, on any intuitive

[16] For ways of mitigating the effects of path-dependence see List (2004). The evidential insensitivity of the sequential priority rule appears in other ways, too. Suppose, for example, that the A-B-C group judges in favour of 'p', 'q', and 'r' and is then tested on a conjunction of those propositions with an incontestable, empirical truth: say, 'p & q & r & $e=mc^2$'. A majority will vote against the compound proposition, since it is disposed to vote against 'p & q & r' alone; and that judgement can stand since it is not consistent with the earlier judgements. But what now will the rule dictate about the group's judgement on '$e=mc^2$', if it is next faced with that issue on its own? The group will have to be deemed, preposterously, to judge that it is not the case that $e=mc^2$. I am grateful to Caspar Hare for discussion on this point.

conception of evidential rationality, the order of presentation ought not to be relevant in this manner.

3. GROUP AGENTS WITH SYSTEM-LEVEL FEEDBACK

These observations suggest that the search for a no-feedback constitution that would have groups perform on the model of rational but unreasoning agents is likely to be a futile enterprise. More specifically, the search is likely to be futile in any scenario where the groups face logically connected issues, and where they aspire to achieve a significant degree of responsiveness in the dimensions given. No doubt there are other possible approaches, on a par with the sequential priority rule, which would have a group seek to operate without feedback. But it is hard to see how any could deal satisfactorily with the evidential problem raised for that rule. For how could a group display evidential sensitivity across a number of issues without keeping track of those issues, and of the responses that they each elicited from the group?

The straw-vote procedure

This is not bad news for group-formation as such, however. For it turns out that once we allow members to have feedback on where a group is committed, and once we make arrangements for that feedback to have an effect, then group rationality ceases to be so elusive. The possibility can be illustrated with what we may call a straw-vote procedure. This is a procedure that a group might implement in a centralized assembly where every member votes on every issue; but it can also stand in for analogous procedures with groups of a networked, non-assembling kind.

The idea in the straw-vote procedure is to have members take a straw vote on every new issue; consider whether the judgement supported is consistent with existing judgements; and, if it is not, revise one of its conflicting judgements so as to ensure consistency. The judgement revised is not necessarily the judgement just supported in voting, as it would be under the sequential priority rule. If that is what seems evidentially more appropriate, one of the earlier judgements may be revised instead.

The straw-vote procedure might be detailed in this set of instructions to members of the group:

(1) With every issue that comes up for judgement, take a majority vote on that issue and, as issues get progressively settled in this way, keep a record of the accumulating body of judgements.
(2) With every new issue that is voted on, check to see if the judgement supported is consistent with the existing commitments of the group.

(3) If majority voting generates an inconsistency, treat the judgement supported and the set or sets of judgements with which it is inconsistent in the record as candidates for reversal.
(4) Identify the problematic judgements—say, the judgements that p, that q, that r, and that not-p & q & r—and address the question of how to resolve the inconsistency.
(5) Take a vote on where it would be best to revise the judgements: whether, in the simple example considered, it would be best to revise the judgement that p, that q, that r, or that not-p & q & r.
(6) Take the proposition identified in this way, and hold another vote on how the group should judge that proposition.
(7) If the group reverses its previous judgement, treat the new verdict on that proposition as the one to be endorsed by the group.
(8) If the previous judgement is not reversed in that vote, go back to stage 3 and try again.
(9) If it appears that there is no prospect of success in this process, try to quarantine the inconsistency, and the area of decision it would affect, so that it does not generate problems elsewhere.
(10) If this quarantining is not possible, perhaps because the area of action affected is important to the group's aims, there is no alternative but to declare defeat on the issues under consideration, even perhaps to disband.

The procedure outlined in these instructions is not a particularly surprising proposal and we can well imagine a group adopting it. The procedure requires a group to treat the appearance of every new, connected issue as introducing a problem of how best to judge, not just over that particular issue, but over that issue together with the previously considered, logically connected issues. The group will recognize the inconsistency of the judgements that were elicited from it in separate votes and will resolve that problem of inconsistency insofar as members can converge on one presumptively best set of judgements overall. The set of judgements it adopts at the end of the exercise will be fixed by the pattern of member judgements over those issues but in a way that violates issue-by-issue responsiveness in the minimal measure required for group rationality.

The straw-vote procedure shows how feedback can make group rationality accessible in circumstances where a no-feedback rule would lead to problems. But the strategy it illustrates not only involves the use of feedback; it also displays the exercise of a sort of group reasoning.

The requirement that the group consider every proposition that is supported by a straw vote for whether it is consistent with propositions already endorsed amounts to a requirement that it practise semantic ascent and look to relations between propositions. And the requirement that it adjust to a judgement that there is an inconsistency amongst the propositions endorsed amounts to a requirement that it respond appropriately to any observed irrationality, removing the inconsistency while also respecting other constraints of rationality like that of evidential sensitivity. If it follows the straw-vote procedure, then the group will count as a reasoning subject in the image of the reasoning subjects that we individuals constitute. It will exercise a sort of control over its own processes of judgement-formation that resembles the personal control associated with individual reasoning. The members will act together in implementing an intentional exercise of group control. And they will do this in respect of themselves considered as a unified centre of attitude formation and enactment.

There is no saying exactly how the members of a group will adjust so as to rectify perceived inconsistencies, of course, or other forms of irrationality; that is a matter of variable psychology. But the fact that they can be relied on to adjust in such a manner means that the group can go through the exercise of reasoning with confidence that it may prove beneficial. This should not be surprising, since the situation is similar on the personal front. There is no saying how I as an individual may adjust, by grace of my subpersonal nature, to one or another perceived inconsistency. But the fact that I can be relied upon to adjust in a suitable manner means that I can go through the exercise of reasoning with a similar confidence that it will bear fruit. As it is on the group front, so it is on the personal.

Although I have used the straw-vote procedure to illustrate how group reasoning may help to resolve issues of group rationality, it should be made clear that there are many variations possible on that particular approach. These variations will apply, not just with the assembled group, but also with groups that are networked out of more or less independent subunits. The simplest analogue would be a networked group in which one subunit, with the authorization of the group as a whole, plays the role that is played by the assembly under the straw-vote procedure. This unit would review the commitments that would materialize on the basis of voting in one or more other subunits and revise the emerging commitments in the minimal measure required to secure consistency in the commitments of the group as a whole. It would serve in a role analogous to that served by the highest court in a system of judicial review.[17]

Under this variant system, as under the straw-vote procedure, the group can be said to reason about what judgements to support, rather than relying on rational judgements to bubble up without any need of monitoring. It intentionally

[17] Stearns (2000) draws attention to a way in which a group might maintain its rational configuration, thanks to the unauthorized intervention of some members. Take the case of a collegial court that has to make judgements on related issues; say, the case where it has to judge in matters of tort on whether there was harm done, whether there was a duty of care, whether there was negligence, and so, whether the defendant was liable; these issues relate under legal doctrine like 'p', 'q', 'r', and 'p & q & r'. It is possible for a court as a group to vote in a case like this for each of the atomic propositions but against the compound. And in some such cases, there is evidence that at the last minute one or another judge votes inconsistently with his or her commitments on the atomic issues in order to preserve the rational configuration of the court.

conducts an exercise—in this case, via an authorized subunit—that meets the specifications for reasoning. It asks meta-propositional questions out of a desire to identify the answers. And it does this in confidence that the answers will have an appropriate impact on the processes whereby group attitudes form and have an effect.

The struggle for group rationality

But though group rationality may be achieved through recourse to reasoning, it is sometimes achieved only with difficulty. Groups may see in rationality what is required of them and yet, like individual agents, fail or falter on this front. They may display group *akrasia*—collective weakness of will (Pettit 2003a).

Imagine a non-commercial academic journal with an editorial committee of three members that resolves all the issues it faces by majority vote. Suppose that the committee votes in January for promising subscribers that there will be no price rise within five years. Suppose that it votes in mid-year that it will send papers to external reviewers and be bound by their decision as to whether or not to publish any individual piece. And suppose that in December it faces the issue as to whether it should be just as prepared to publish technical papers that involve costly typesetting as it is to publish other papers: whether it should treat them as equal. The earlier votes will argue against its being prepared to do this, since a rise in the number of technical papers submitted and endorsed by reviewers—endorsed, without an eye to overall production costs—might force it to renege on one or other of those commitments. But, nonetheless, a majority may support the even-handed treatment of technical papers, without any individual being in any way irrational. The members of the committee may vote as follows.

	Price freeze?	External review?	Technical papers equal?
A.	Yes	No	Yes
B.	No	Yes	Yes
C.	Yes	Yes	No

The group now faces a hard choice of broadly the kind we have been discussing. Suppose that the members operate with the straw-vote procedure and that they agree that the issue on which the group should revise its view is that of whether to treat technical papers on a par with other papers; they may vote unanimously that it is impossible to revise its position on either of the other issues, perhaps because the editorial position on those questions has already been made public. How, then, may we expect the consequent vote to go?

If members are individually dedicated to the group and are in no way tempted to defect from what it requires of them, then of course they will each vote for offering less than equal treatment to technical papers; they will reverse the previous group position. A group whose members were dedicated in this way would operate like a perfectly virtuous agent, always spontaneously supporting what the balance of available reasons requires of the group. But not all members need be so devoted to the group in which they figure; and when something less than full collective devotion is on offer, then it may prove very difficult for members to get their act together and ensure that the group lives up to the considerations that it endorses.

Take the majority, A and B, who originally supported an open policy on technical papers. That majority may remain individually and stubbornly inclined to support the equal treatment of technical papers. We can imagine them turning their eyes from the group as a whole, and sticking to their votes when the issue is raised again. We can imagine them refusing to hear the call of the group and acting like encapsulated centres of voting who are responsive only to their own modular prompts. As we imagine this, we envisage the group failing to reverse its judgement on an issue where every member of the group thinks it is desirable to reverse judgement. The recalcitrant majority in this sort of case might be moved by a more or less selfish inclination or identification, being technically minded themselves, or they might be moved by a sense of fairness towards those who would be disadvantaged; personal virtue is as likely as personal vice to source recalcitrance towards the collectivity.

Could it really be rational, however, for the recalcitrant members to stick to a deviant pattern of voting, whether out of individual bias or virtue? I don't see why not. They would satisfy their private motives, partial or impartial, by doing so. And they might individually expect to get away with such voting, being outvoted by the others; they might each expect to be able to free-ride. Or they might hope that even if a majority remains recalcitrant, this will not cause problems: there will not be a deluge in the number of technical papers submitted and accepted, and the committee can get away with holding by all of the three commitments involved.

The possibility of people remaining encapsulated in their personal identities in this way, and the danger that that holds out for the survival of the group, shows that it is essential in general that the members should break out of their capsules. If the group is to evolve as a centre of agency, with a capacity to be held responsive to the demands of consistency, then it must be able to discipline members into supporting only certain patterns of judgement. And if a group is to have that capacity, then its members must be willing to put their own views aside and identify with the group as a whole, whether spontaneously or under the impact of institutional incentives. They must be ready to reason and act from the perspective of that common centre.

What should we say about the editorial group, however, if it just fails to get its act together and lives on the wild side, exposed to a constant danger of bankruptcy? Should we think that it is not a group agent after all but only a collection of individuals, like those who live in the same zip code, who should not be held to expectations of consistency and the like? Or should we think of it as a group agent that is failing on this front and that can be held up to criticism for the failure?

Clearly we would think the latter; and, equally clearly, we should do so. The reason we would hold it to expectations of consistency is that the tasks entrusted to the group, and embraced by it, mark it off from the zip-code population, giving it the cast of an agent. And the reason we should hold it to these expectations is that even while it breaches a constraint of rationality, it can acknowledge the relevant standards, like any symbol-using, reasoning creature, and can display its agency by doing so. We saw earlier that the reasoning human being can vindicate his or her agency, unlike the simple animal, even while failing to conform to certain standards of rationality. The same is true with the reasoning groups that human beings form.

CONCLUSION

Jenann Ismael (n.d.) makes a useful distinction between self-organizing and self-governing systems and it may be useful, in conclusion, to summarize the upshot of the argument in these terms.

Consider the sort of entity that generates a global, agential profile as a result of the locally stimulated responses of its members or parts or subsystems. Take, for example, the insect colony in which individual insects combine to produce a field of chemical stimuli such that when individuals respond to that field—for each insect, to the local stimuli provided by that field—the result is a coherent, self-organizing pattern of action. At any moment in the life of the colony the insects act together in different roles to advance a coherent set of goals in a coherent way. And the inputs they make in doing so generate a field of stimuli that elicits suitable responses at the next stage and continues into the future to support the agential profile. A system of this kind is self-organizing. Its behaviour as a whole, which displays a striking coherence, is a function of the behaviour of the subunits. Yet those units produce that behaviour as a result of each of them marching to its own drum, without any of them processing system-level feedback.

The self-governing agent, in contrast with this merely self-organizing system, gives some subunits a special, regulatory role. They keep track of the performance of the system—this is at least in the spirit of Ismael's account—and intervene at certain junctures in order to ensure that it displays the character of a rational agent.[18] They monitor the system as a whole, gathering feedback on

[18] The systems considered here are agents, but not all self-organizing or self-governing systems have to be agents. The free market in which individually self-seeking agents sustain equilibrium prices is not an agent, yet ordinarily it counts as a self-organizing system. Nor is the command economy an agent, though it should probably count as a self-governing system.

its dispositions to behaviour, whether by observation or by using a predictive or simulative device (Grush 2004). And they manage the system by making interventions that correct for ways in which it may be disposed to act out of rational type.

The lesson of this discussion is that in more or less standard conditions a collectively rational and individually responsive group agent will have to be self-governing rather than self-organizing. This lesson holds, more specifically, for conditions where logically connected issues present themselves for resolution, rational agency requires a complete, consistent set of answers, and it is important that those answers are rationally sensitive to the overall evidence available. Whether it has an assembly or a network character, the group agent will have to organize itself for such conditions so that some or all of its members can keep track of its accumulating judgements and take steps to guard against the onset of inconsistency. The lesson may not have the full-dress credentials of an a priori necessity but it is as safe a bet as we are likely to be able to identify in this area.

REFERENCES

Anscombe, G. E. M. 1957. *Intention*. Oxford: Blackwell.

Bilgrami, A. 1998. 'Self-knowledge and Resentment'. In B. Smith, C. Wright, and C. Macdonald (eds), *Knowing Our Own Minds*. Oxford: Oxford University Press, 207–41.

Bratman, M. 1999. *Faces of Intention: Selected Essays on Intention and Agency*. Cambridge: Cambridge University Press.

Broome, J. 2007. 'Wide or Narrow Scope?'. *Mind and Language* 116: 359–70.

Burge, T. 1998. 'Reason and the First Person'. In B. Smith, C. Wright, and C. Macdonald (eds), *Knowing Our Own Minds*. Oxford: Oxford University Press.

Carroll, L. 1895. 'What the Tortoise said to Achilles'. *Mind* 4 : 278–80.

Clark, A. 1999. 'Leadership and Influence: The Manager as Coach, Nanny and Artificial DNA'. In J. Clippinger (ed.), *The Biology of Business: De-coding the Natural Laws of Enterprise*. San Francisco: Jossey-Bass, 46–66.

Davidson, D. 1984. *Inquiries into Truth & Interpretation*. Oxford: Oxford University Press.

Dennett, D. 1994. 'Learning and Labelling: A Commentaary on A. Clark and A. Karmiloff-Smith' *Mind and Language* 8: 540–8.

Dennett, D. C. 1969. *Content and Consciousness*. London: Routledge.

Dietrich, F. and List, C. 2007. 'Arrow's Theorem in Judgment Aggregation'. *Social Choice and Welfare* 29.1: 19–33.

Evans, G. 1982. *The Varieties of Reference*. Oxford: Oxford University Press.

Fodor, J. 1975. *The Language of Thought*. Cambridge: Cambridge University Press.

Grush, R. 2004. 'The Emulation Theory of Representation: Motor Control, Imagery, and Perception'. *Behavioral and Brain Sciences* 27: 377–96.

Harman, G. 1986. *Change in View*. Cambridge, MA: MIT Press.

Hylland, A. and Zeckhauser, R. 1979. 'The Impossibility of Bayesian Group Decision Making with Separate Aggregation of Beliefs and Values'. *Econometrica* 47: 1321–36.
Ismael, J. (n.d.). 'Selves and the Limits of Self-organization'. Unpublished manuscript.
Kolodny, N. 2005. 'Why be Rational?'. *Mind* 114: 509–63.
Kornhauser, L. A. and Sager, L. G. 1993. 'The One and the Many: Adjudication in Collegial Courts'. *California Law Review* 81: 1–59.
Levi, I. 1991. *The Fixation of Belief and its Undoing: Changing Beliefs through Inquiry*. Cambridge: Cambridge University Press.
List, C. 2004. 'A Model of Path-Dependence in Decisions over Multiple Propositions'. *American Political Science Review* 98: 495–513.
—— 2006. 'The Discursive Dilemma and Public Reason'. *Ethics* 116: 362–402.
—— and Pettit, P. 2002. 'Aggregating Sets of Judgments: An Impossibility Result'. *Economics and Philosophy* 18: 89–110.
—— 2005. 'On the Many as One'. *Philosophy and Public Affairs* 33.
—— 2006. 'Group Agency and Supervenience'. *Southern Journal of Philosophy* 45 (Spindel supplement).
—— (forthcoming). 'Group Agency'. Oxford: Oxford University Press.
McGeer, V. and Pettit, P. 2002. 'The Self-regulating Mind'. *Language and Communication* 22: 281–99.
Pauly, M. and Van Hees, M. 2006. 'Logical Constraints on Judgment Aggregation'. *Journal of Philosophical Logic* 35: 569–85.
Pettit, P. 1993. *The Common Mind: An Essay on Psychology, Society and Politics*. New York: Oxford University Press.
—— 2001a. *A Theory of Freedom: From the Psychology to the Politics of Agency*. Cambridge and New York: Polity and Oxford University Press.
—— 2001b. 'Deliberative Democracy and the Discursive Dilemma'. *Philosophical Issues* supplement, *Noûs* 11: 268–99.
—— 2002. *Rules, Reasons, and Norms: Selected Essays*. Oxford: Oxford University Press.
—— 2003a. 'Akrasia, Collective and Individual'. In S. Stroud and C. Tappolet (eds), *Weakness of Will and Practical Irrationality*. Oxford: Oxford University Press.
—— 2003b. 'Deliberative Democracy, the Discursive Dilemma, and Republican Theory'. In J. Fishkin and P. Laslett (eds), *Philosophy, Politics and Society*, Vol 7: *Debating Deliberative Democracy*. Cambridge: Cambridge University Press, 138–62.
—— 2003c. 'Groups with Minds of their Own'. In F. Schmitt (ed.), *Socializing Metaphysics*. New York: Rowan and Littlefield.
—— and Schweikard, D. 2006. 'Joint Action and Group Agency'. *Philosophy of the Social Sciences* 36 (2006).
Raiffa, H. 1968. *Decision Analysis*. New York: Addison-Wesley.
Searle, J. R. 1983. *Intentionality*. Cambridge: Cambridge University Press.
Smith, M. 1987. 'The Humean Theory of Motivation'. *Mind* 96: 36–61.
Stearns, M. L. 2000. *Constitutional Process: A Social Choice Analysis of Supreme Court Decision Making*. Ann Arbor: Michigan University Press.

Index